Progress in Optical Science and Photonics

Volume 7

Series Editors

Javid Atai, Sydney, NSW, Australia

Rongguang Liang, College of Optical Sciences, University of Arizona, Tucson, AZ, USA

U. S. Dinish, Singapore Bioimaging Consortium (SBIC), Biomedical Sciences Institutes, A*STAR, Singapore, Singapore

The purpose of the series Progress in Optical Science and Photonics is to provide a forum to disseminate the latest research findings in various areas of Optics and its applications. The intended audience are physicists, electrical and electronic engineers, applied mathematicians, biomedical engineers, and advanced graduate students.

More information about this series at http://www.springer.com/series/10091

Mithun Kuniyil Ajith Singh
Editor

LED-Based Photoacoustic Imaging

From Bench to Bedside

 Springer

Editor
Mithun Kuniyil Ajith Singh
Cyberdyne Inc.
Rotterdam, The Netherlands

ISSN 2363-5096   ISSN 2363-510X   (electronic)
Progress in Optical Science and Photonics
ISBN 978-981-15-3986-2   ISBN 978-981-15-3984-8   (eBook)
https://doi.org/10.1007/978-981-15-3984-8

This Springer imprint is published by the registered company Springer Nature Singapore Pte Ltd.
The registered company address is: 152 Beach Road, #21-01/04 Gateway East, Singapore 189721,
Singapore

Foreword

Medical imaging provides critical information about diseases and consequently assists in selecting the right course of action and appropriate therapy. Sensitive imaging techniques can help in early and accurate detection of deadly diseases, resulting in timely intervention and reducing the burden of medication and surgery at later stages. All imaging modalities aim to achieve these goals at an affordable cost. Most commonly used clinical imaging modalities include ultrasound imaging, X-ray computed tomography, magnetic resonance imaging, and positron emission tomography. All these modalities are different in terms of the physics, contrast offered, expense of installation and operation, and sensitivity for different applications. An ideal medical imaging system should be able to diagnose diseases with high sensitivity and specificity at a cost that is affordable for all clinics around the world.

Photoacoustic imaging is an emerging hybrid technology that offers rich optical contrast and scalable ultrasonic resolution and imaging depth. This modality beats the optical diffusion limit, which is the main barrier that hinders deep-tissue imaging of any high-resolution optical techniques. It is also straightforward to implement dual-mode photoacoustic and ultrasonic imaging in one single point-of-care platform due to the shared ultrasonic detection. Image contrast of photoacoustic imaging is based on either endogenous chromophores such as hemoglobin or exogenous contrast agents with possibilities of molecular targeting. Photoacoustic imaging with excellent functional and molecular imaging capabilities is quite mature in research settings and has already shown great potential in myriads of preclinical applications and early clinical trials. However, compared to scientific developments, clinical translation of this technology is still in a premature stage. While approximately 20 companies are active pursuing commercialization of photoacoustic imaging, no product has been approved by the US FDA yet. One of the limiting factors in photoacoustic imaging is the necessity of expensive and bulky solid-state lasers for tissue illumination. It would be ideal to develop affordable and miniaturized light sources that can enable point-of-care imaging capabilities in a resource-limited setting. Tremendous advancements in solid-state device technology have recently resulted in the development of high-power pulsed light emitting diodes (LED) for photoacoustic imaging. These diodes are affordable, portable, and potentially eye/skin safe and hence hold great potential for clinical translation.

This book is timely as its publication occurs in a period when photoacoustic imaging is facing an exciting transition from the benchtop to bedside. It has been assembled and edited by a young leading scientist in the field, Dr. Mithun Kuniyil Ajith Singh, who is working in the intersection of research and business and is immensely focused on clinical translation of LED-based photoacoustics. With multiple chapters from the key leaders in the field, this book highlights the use of LEDs in biomedical photoacoustic imaging. The 16 chapters cover the entire span of fundamentals, principles, instrumentation, image reconstruction and data/image processing methods, preclinical, and clinical applications of LED-based photoacoustic imaging. Apart from academic contributions, the chapter provides an industry perspective on opportunities and challenges in clinical translation. This opus will be undoubtedly of interest to experts in academia, industry, and medicine interested in clinical translation of photoacoustic imaging.

Photoacoustics offers a rare combination of highly specific optical contrast and spatial resolution and imaging depth of ultrasound (in centimeter scales), making it one of the fastest growing medical imaging modalities of the decade. We all are fortunate to work on this technology as it has an enormous potential for touching and saving lives. Let us unite our efforts and work toward eradicating the most challenging diseases and build a healthy and happy world together.

Lihong V. Wang, Ph.D.
Bren Professor of Medical Engineering
and Electrical Engineering California
Institute of Technology
Pasadena, CA, USA

Contents

About the Editor

Dr. Mithun Kuniyil Ajith Singh is an engineering scientist with extensive experience in preclinical and clinical photoacoustic and ultrasound imaging. He is presently working as a Research and Business Development Manager at CYBERDYNE, INC, the Netherlands. In his current role, he initiates and coordinates various scientific projects in collaboration with globally renowned research groups, especially focusing on the clinical translation of LED-based photoacoustic imaging technology. Mithun received his Technical Diploma in Medical Electronics from Model Polytechnic College, Calicut, India, in 2004; his Bachelor's degree in Biomedical Instrumentation and Engineering from SRM University, Chennai, India, in 2008; and his Master's degree in Bio-photonics from the École normale supérieure de Cachan, France, in 2012. From January 2013 to December 2016, he pursued research for his Ph.D. thesis at the Biomedical Photonic Imaging Group, University of Twente, the Netherlands, under the supervision of Professor Wiendelt Steenbergen. While working on his Ph.D., he invented a new method called PAFUSion (photoacoustic-guided focused ultrasound) for reducing reflection artifacts and improving contrast and imaging depth in photoacoustic imaging, and demonstrated its potential in clinical pilot studies. Mithun has published over 40 international journal articles and conference proceedings papers. He won the Seno Medical Best Paper Award at the annual conference 'Photons plus Ultrasound: Imaging and Sensing' during the SPIE Photonics West 2016 event in San Francisco. As one of many academic achievements and scholarships, SRM University, India, presented him with a Commendation Award for his outstanding contributions to the field of biomedical engineering in 2017. Mithun is currently serving as a reviewer for several leading journals in the field of optics and won Elsevier's Outstanding Reviewer Award in 2018.

Fundamentals and Theory

Fundamentals of Photoacoustic Imaging: A Theoretical Tutorial

Mayanglambam Suheshkumar Singh, Souradip Paul, and Anjali Thomas

Abstract We report a study on theoretical aspects of the generation of initial photoacoustic (PA) pressure waves and its propagation in a mechanical medium, with detailed derivations of associated equations, which is the basis of all PA imaging modalities (both microscopy and tomography). We consider the tissue sample for imaging as a hydrostatic (*PVT*) thermodynamic system. The phenomenon of the generation of initial pressure wave, due to transient ($\sim$ns) illumination by electromagnetic (EM) waves and subsequent rapid heating, is assumed as a thermodynamic process. For the propagation of PA wave, tissue sample is considered as a biomechanical system that supports the propagation of mechanical disturbances from one point to another. The derived equations are in agreement with standard equations that are commonly employed in PA imaging systems, therefore our assumptions of considering the system as a hydrostatic thermodynamic system and PA effect as a thermodynamic process are validated. This chapter will be of great value to the PA imaging research community for an in-depth theoretical understanding of the subject.

1 Introduction

PA imaging modality stands as a promising imaging technology to address the long-standing challenge of achieving microscopic ($\sim$µm) resolution at depths beyond optical transport mean free-path ($\sim$1 mm), in real time [1, 2]. Moreover, this imaging technology—as a single imaging unit— can provide multiple structural, functional,

M. Suheshkumar Singh (✉) · S. Paul · A. Thomas
Biomedical Imaging and Instrumentation Laboratory (BIIL), School of Physics (SoP), Indian Institute of Science Education and Research Thiruvananthapuram (IISER-TVM), Thiruvananthapuram 695551, India
e-mail: suhesh.kumar@iisertvm.ac.in

S. Paul
e-mail: souradip.rkm16@iisertvm.ac.in

A. Thomas
e-mail: anjalithomas16@iisertvm.ac.in

© Springer Nature Singapore Pte Ltd. 2020
M. Kuniyil Ajith Singh (ed.), *LED-Based Photoacoustic Imaging*,
Progress in Optical Science and Photonics 7,
https://doi.org/10.1007/978-981-15-3984-8_1

and molecular information about tissues, non-invasively and non-destructively. It is of great clinical interest and value [3–5] in diagnosis, staging, monitoring and therapeutic treatments of various diseases in early stages [4]. Recently, this imaging modality has been extended from clinical applications to biological applications (more specifically, in molecular and cellular imaging at the sub-microscopic resolution but achievable penetration depth <1 mm) [5–10].

PA imaging technology is rapidly growing since the last decade in laboratory research studies (in late 1990s [5, 8, 11]). Meanwhile more than 100 research laboratories around the globe are dedicated to the study of its biomedical and clinical applications. This surge in interest over this short span of time can be attributed to its potential and promising features of biomedical and clinical interest [5, 8]: (1) high contrast and high spatial resolution obtainable from this imaging modality at high penetration depths, not achievable with other conventional imaging modalities [confocal, two-photon, and optical coherence tomography (OCT)], (2) high scalability of imaging, ranging from the individual cell to the entire body, (3) imaging—with multiple resolution levels of structural anatomy and tissue patho-physiology, (4) obtainable patho-physiological information, i.e., pathological stages of tissues through measurement of functional parameters (Hb, HbO_2, SO, and Total Hb), which control physiological activities (metabolism, molecular and genetic activities), (5) it is non-invasive, non-destructive and non-hazardous in nature. PA imaging modality has been exploited for various biological, pre-clinical, and clinical (oncology, ophthalmology, dermatology, gastroenterology, cardiology and osteoarthritis) studies. Applications of this imaging modality to diagnostic and therapeutic treatments–employing target specific (light absorbing) contrast agents–were also reported [12, 13]. PA imaging technology has also been employed to study recovery of acoustic property, temperature, blood flow velocity and elastic property of soft biological tissues [14–30]. On the other hand, with the advancement of computational reconstruction techniques, real time imaging (frame rate $\sim$100 Hz equivalent to $\sim$10 µs) at microscopic spatial resolution ($\sim$40 µm) and reasonably high penetration depths (cm) has been achieved [1, 31]. So, it is evident that advances in the PA-imaging modality have mostly found their way into its experimental and technological aspects. But, the theoretical understanding has mostly been limited to the establishment of mathematical models (which are all 2nd order partial differential equations in some way or the other) for computational reconstruction algorithms and their implementations (delay-and-sum, iterative finite element method (FEM) [32], Green's function or Born-approximation [33, 34] and beam forming [31]). To the best of our knowledge, theoretical studies of the PA effect and wave propagation with detailed accounts of derivations and associated physical phenomena—are very few [30, 32, 35]. Even in these studies, the second order wave equation has mostly been derived from a set of first order partial differential equations (PDEs) [33] without proper understanding of the associated physical phenomena and assumptions. This is mandated for a thorough understanding of the subject and its applicability to a greater extent. In this chapter, we report the detailed derivation of the second order PDE that governs the generation and propagation of the initial PA pressure wave in mechanical medium from the set of first order PDEs, with sequential mathematical steps and the associated physical interpretations and assumptions.

In our study, we consider the tissue sample as a hydrostatic *(PVT)* thermodynamic system in thermodynamic equilibrium, which is completely characterized by macroscopic variables namely, temperature (T), pressure (P), and volume (V). In the meantime, the phenomenon of the generation of initial pressure wave—due to transient (~ns) illumination by electromagnetic (EM) waves, and subsequent absorption and heating for a very short duration—is considered as a thermodynamic process which is like a perturbation on the equilibrium state of the system. For the propagation of PA-wave, tissue sample is considered as a bio-mechanical system that supports the propagation of mechanical disturbances from one point to another though mechanical interactions among constituent particles or molecules.

Rest of this chapter is organised as: Sect. 2 gives a detailed theory of the generation of PA wave and its propagation through a mechanical tissue medium. More specifically, derivation of generation of initial PA pressure (P_0)—due to transient illumination by electromagnetic (EM) waves—is given in Sect. 2.1. Section 2.2 gives an elaborate account of derivation of the second order wave equation that governs the propagation of PA wave in a bio-mechanical medium.

2 Theory of Photoacoustic Wave Generation and Propagation

Photoacoustic imaging, which is broadly classified into photoacoustic microscopy (PAM) and photoacoustic tomography (PAT), is fundamentally based on the PA effect, which was propounded by Alexander Graham Bell in 1880 [36, 37]. The discovery and its exploitation for biomedical and clinical applications have a deep history. It was only in the late 1990s that PA effect was first exploited for studies in biomedical imaging and its clinical applications.

PA effect is the generation of acoustic wave in a sample material due to the rapid heating of the sample through transient illumination and absorption of short pulse (ns) electromagnetic (EM) radiation. This light stimulated ultrasound wave is, generally, known as PA signal (denoted by P_0). A mechanical medium (like tissue sample) supports the propagation of PA wave which is dependent on the tissue physical properties (including thermodynamic, acoustic, and mechanical). From this sequence of boundary measurements of time-resolved PA signal, the distributions of the initial pressure (P_0) and its derivatives (including, optical absorption coefficient (μ_a), acoustic velocity (v_{ac}), elastic coefficient (E), flow velocity, and Grueneisen parameter (Γ) are obtained using various computational techniques. In this way, from the physics point of view, the entire process of PA imaging can be grouped into four distinctive stages: (i) Transient illumination of a specific region in the tissue sample with short pulse of laser light or LED (pulse width of few ns). This stage is governed by propagation of electromagnetic waves in a medium (more specifically, by Maxwell's equation of electromagnetic theory) and, hence characterized by the optical properties of the propagating medium, namely, optical absorption (μ_a), scattering (μ_s)

coefficients, and refractive index (n). Detailed study of this stage is not studied in our present chapter. It may be referred to somewhere in other research domain of diffuse optical tomography (DOT) [38]. (ii) Generation of high-frequency acoustic signal (of the order of MHz–GHz). Due to intrinsic optical absorption coefficient $(\mu_a(\vec{r}))$ of the laser irradiating medium, deposited optical energy gets absorbed, which is then converted into heat energy or thermal energy through the oscillational relaxation of the constituent particles/molecules of the sample, thereby, inducing a localized rise of temperature over the irradiated region. This temperature rise $(\Delta T(\vec{r}))$ is dependent on the optical absorption coefficient distribution $\mu_a(\vec{r})$. Associated with rapid heating, under the given physical constraint of short pulsed laser illumination with duration being less than time scales of thermal and stress confinements, the irradiated tissue undergoes thermoelastic expansion [39] and, subsequently, it induces a transient rise in pressure. A thorough study, both from physical (thermodynamics) and mathematical aspects, is given in Sect. 2.1. (iii) Isotropic propagation of the PA wave in mechanical tissue medium. In the tissue sample, which is a viscoelastic medium, optically stimulated mechanical PA wave propagates through back-and-forth contraction and rarefaction of mechanically coupled constituent particles about their respective thermal (mean) positions [23]. This propagation of mechanical PA wave is characterized by acoustic property distribution in the medium, i.e., acoustic velocity, impedance, absorption, and scattering. Mathematically, propagation of the acoustic wave is governed by the second order wave equation, derivation of which is addressed in Sect. 2.2, as a primary objective of our present article. (iv) Detection of the PA-wave from the tissue boundary and image reconstruction, the PA signal is picked-up by keeping a single transducer unit or an array of ultrasound transducer elements around specimen boundary. Distribution of initial PA-pressure (P_0) and its derivative (including physical properties and patho-physiological information) are reconstructed computationally. This stage is not studied here in this chapter and it may be referred to [33, 40] for further study.

2.1 Generation of Photoacoustic Wave (Initial PA-Pressure)

One may explain the generation of initial PA pressure (P_0), due to the transient light illumination generation as a thermodynamic process. As mentioned above, in PA imaging—both microscopy and tomography—a material sample of interest (over a pre-specified region) is irradiated with an optical beam (laser source or LED) for a very short duration (pulse width $\sim$few ns). Due to transient absorption of light energy, a rapid change in kinetic energy $(K.E.)$ and subsequently the internal energy (U) of constituent particles/molecules in the tissue materials over irradiated region of interest takes place. Meanwhile, internal energy is directly related to the absolute temperature (T) (example, for ideal gas, $U = P.E. + K.E. \approx \frac{3}{2} N \kappa_B T$, where κ_B and N are Boltzmann constant and number of molecules respectively. Here interaction among constituent particles, i.e., potential energy $(P.E.)$ is neglected, in comparison to kinetic energy $(K.E.)$). In other words, optical energy is converted to heat energy through change in kinetic energy of constituent particles/molecules of the sample material [41] which results in thermal expansion. In this way, due to transient optical illumination and subsequent transient optical absorption, the sample undergoes a rapid heating and subsequent cooling, thereby, inducing rapid expansion and contraction, which is thermoelastic expansion [41]. Thermodynamically, this can be characterised by the change in volume (V), pressure (P) and temperature (T) provided by transient absorption of optical energy (say, optical fluence (φ)) [41]. In other words, the system undergoes a thermodynamic process that is governed and characterized by a set of thermodynamic equations (Maxwell's equations of thermodynamics) giving the relationship among the various macroscopic variables, such as pressure (P), volume (V), temperature (T), entropy (S), number of particle (N), chemical potential (μ), and mass density (ρ) [42].

To a good approximation, for PA imaging in biomedical/clinical and biological applications, one may neglect changes in entropy (S), number of particles (N) and hence, mass (m) in the tissue sample of interest. Under such conditions, the tissue system can be considered as a hydrostatic (PVT) thermodynamic system that is completely characterized (in all of the aspects of chemical, mechanical and thermal processes) by pressure, (P), volume (V) and temperature (T) [42]. In this process, a system of tissue sample that is enclosed by an imaginary boundary of light irradiation can be considered as the thermodynamic system while the remaining tissue material as the surroundings, in the thermodynamic sense.

Let us consider a thermodynamic equation of state for hydrostatic system in thermodynamic equilibrium (in our case, arbitrarily chosen elementary volume element (V) in material sample) and it is expressed as:

$$V \equiv V(T, P). \tag{1}$$

Note that over these individual elemental volumes (V) in of thermodynamic equilibrium, macroscopic variables (say, T and P) are average measures over the individual volume elements, i.e., distribution of state variables are uniform over volume (V) and hence, are considered independent of time (t) and space $(\vec{r})$. So, in Eq. 1, T and P are expressed without dependence on space and time.

Upon perturbation of thermodynamic equilibrium of the hydrostatic thermodynamic system—that is induced by irradiation with electromagnetic waves (or optical energy) for an infinitesimal period of time ($\sim$ns)—differential change in volume (ΔV) from its equilibrium state (V) can be derived, using Taylor expansion, as:

$$\Delta V \approx \left.\frac{\partial V}{\partial T}\right|_P \Delta T + \left.\frac{\partial V}{\partial P}\right|_T \Delta P,$$

$$\Rightarrow \frac{\Delta V}{V} = \left.\frac{1}{V}\frac{\partial V}{\partial T}\right|_P \Delta T + \left.\frac{1}{V}\frac{\partial V}{\partial P}\right|_T \Delta P, \ V \neq 0,$$

$$= \beta \Delta T - \kappa_T \Delta P, \tag{2}$$

where ΔP and ΔT are differential changes in pressure (P) and temperature (T) from their corresponding equilibrium values; $\frac{\Delta V}{V}$ is the fractional change in volume; β ($=\frac{1}{V}\frac{\partial V}{\partial T}\Big|_P$) is the thermal coefficient of expansion (also called as volume expansibility) and κ_T ($=-\frac{1}{V}\frac{\partial V}{\partial P}\Big|_T$) is the isothermal compressibility.

For acquiring PA-signal, data acquisition (DAQ) system of sampling frequency ($\sim$MHz) that corresponds to data acquisition period $\sim\mu$s is typically employed. Shortly, time scale for acquisition of PA-data ($\sim\mu$s) is of the order of magnitudes higher than that of volume expansion-contraction ($\sim$ns). In other words, for a particular (time-resolved) measurement, acquired signals are averaged over several oscillatory volume expansion-contraction and thus, relative volume change ($\frac{\Delta V}{V}$) is negligible. Under these physical conditions, in above Eq. 2, one can neglect the fractional volume change ($\frac{\Delta V}{V}$) in comparison to other two terms. Therefore, Eq. 2 is reduced:

$$\Delta P \approx \frac{\beta}{\kappa_T}\Delta T, \tag{3}$$

which shows that ΔP is practically large ($\sim10^6$ Pa) for an infinitesimal temperature change (ΔT) that can be estimated from practical value of β ($\sim10^{-4}$ K^{-1}) and κ_T ($\sim10^{-10}$ Pa^{-1}) for soft tissues [41].

From thermodynamic heat transfer relation ($\Delta Q_{dens} = \rho c_V \Delta T$), where derivation is provided in Appendix 1 (Eq. 33), that relates heat density (ΔQ_{dens}) absorbed in a thermodynamic system to (absolute) temperature rise (ΔT) through specific heat capacity at constant volume (c_V), we obtain:

$$\Delta T = \frac{1}{\rho c_V}\Delta Q_{dens}, \tag{4}$$

where c_V ($=C_V/m$) is the heat capacity per unit mass (called as specific heat capacity) at constant volume and $\Delta Q_{dens} = \Delta Q/V$ is change in heat density at constant volume [42].

Now, considering optical energy absorbed by thermodynamic system (which can be expressed as ($\mu_a\varphi$)) is completely converted into heat energy—neglecting all the non-thermal effects (including florescence, photo-luminescence, and chemical reaction)—we can express the conservation of energy as:

$$\Delta Q_{dens} = \mu_a\varphi = A_e, \tag{5}$$

where φ is the optical fluence that gives the measure of optical energy per unit time of transient optical beam being incident on thermodynamic tissue sample; A_e ($=\mu_a\varphi$) is the specific volumetric optical absorption which measures effective optical energy absorbed by the system upon illumination.

Using Eqs. 4 and 5 in Eq. 3, we obtain:

$$\Delta P = \frac{\beta}{\kappa_T} \frac{1}{\rho c_V} \mu_a \varphi = \frac{\beta}{\kappa_T} \frac{1}{\rho c_V} A_e. \tag{6}$$

Considering residual pressure prior to transient optical illumination as reference (whereby, ΔP is represented by P_0), Eq. 6 can be rewritten as:

$$P_0 = \frac{\beta}{\kappa_T} \frac{1}{\rho c_V} \mu_a \varphi, \tag{7}$$

$$= \frac{\beta}{\kappa_T} \frac{1}{\rho c_V} A_e, \tag{8}$$

$$= \Gamma \mu_a \varphi, \tag{9}$$

which is what we call the initial pressure of PA-signal or the generation of PA-signal that is induced by short laser pulse or LED illumination. Γ $(= \frac{\beta}{\kappa_T} \frac{1}{\rho c_V})$ is the constant of proportionality which is known as Grueneisen parameter. All of the above three equations (Eqs. 7–9) are commonly adopted in literature and imply that the strength of initial PA pressure (P_0) is dependent on the intensity of incident pulse optical beam and it is characterized by tissue's optical, thermal, and mechanical properties. In PA imaging, it is a primary task to map distribution of P_0 and subsequently, tissue physical and bio-physical properties as obtainable from P_0.

Again, from relationship of c_V and c_P (i.e., $\frac{c_V}{c_P} = \frac{1}{\rho \kappa_T v_{ac}^2}$, where derivation is provided in Appendix 2 (Eq. 37)):

$$\kappa_T \rho c_V = \frac{c_P}{v_{ac}^2}. \tag{10}$$

Now, using Eq. 10, we can re-write initial PA pressure (given in Eq. 9) as:

$$P_0 = \frac{\beta v_{ac}^2}{c_P} \mu_a \varphi. \tag{11}$$

where, c_P is the specific heat capacity at constant pressure while Gruneisen parameter (Γ) can also be re-written as $\Gamma = \frac{\beta}{\kappa_T \rho c_V} = \frac{\beta v_{ac}^2}{c_P}$.

We know that acoustic, thermal, and mechanical properties are dependent on (absolute) temperature (T) [40, 43]. For spatial and temporal distribution of incident optical energy, one can express initial pressure rise (P_0) in functional forms either:

$$P_0(\vec{r}, t, T) = \frac{\beta(\vec{r}, T)}{\kappa_T(\vec{r}, T)} \frac{1}{\rho(\vec{r}, T) c_V(\vec{r}, T)} \mu_a(\vec{r}) \varphi(\vec{r}, t) = \Gamma(\vec{r}, T) A_e(\vec{r}, t), \tag{12}$$

or,

$$P_0(\vec{r}, t, T) = \frac{\beta(\vec{r}, T)v_{ac}^2(\vec{r}, T)}{c_P(\vec{r}, T)}\mu_a(\vec{r})\varphi(\vec{r}, t) = \Gamma(\vec{r}, T)A_e(\vec{r}, t). \tag{13}$$

The above two equations show that, for a given intensity of incident optical beam, strength of initial pressure of PA signal is characterized spatial distribution of not only physical properties of imaging (tissue) sample but also its temperature (T).

2.2 Propagation of Photoacoustic Wave

Upon illumination of light absorbing tissue sample by an intense (coherent or incoherent) light beam ($\varphi(\vec{r})$) for a short duration ($\sim$ns), as it is discussed above (Sect. 2.1), an initial pressure rise (known as initial photoacoustic signal ($P_0(\vec{r}, t, T)$)) is induced over the light illuminated region. In other words, there exists an initial pressure gradient or non-uniform distribution of pressure—which measures force per unit volume—given by $\vec{\nabla}P$. This force, which is resulted from the variation in spatial distribution of pressure, ($\vec{\nabla}P(\vec{r}, t, T)$) is the pressure-gradient force acting on constituent particles/molecules and directed from region of higher pressure to region of lower pressure and it can be expressed as:

$$\vec{F}_{pg} = -\vec{\nabla}P, \tag{14}$$

which is force density or force per unit volume. For the sake of mathematical simplicity, we drop out argument dependence on space ($\vec{r}$), time (t), and temperature (T) (say, Eq. 14).

Considering only the pressure-gradient force, while neglecting other forces (including body forces (such as, gravitation and electromagnetic) and externally applied forces), the net force density in the hydrostatic system under consideration can be written as $\vec{F}_{eff} = \vec{F}_{pg}$, i.e., $\vec{F}_{eff} = -\vec{\nabla}P$. From Newton's 2nd law of motion (i.e., $\frac{d\vec{p}}{dt} = \vec{F}_{eff}$), equation of motion of constituent particles with mass density (ρ) can be expressed as:

$$\frac{d(\rho\vec{v})}{dt} = -\vec{\nabla}P,$$

$$\text{i.e., } \vec{v}\frac{d\rho}{dt} + \rho\frac{d\vec{v}}{dt} = -\vec{\nabla}P. \tag{15}$$

where $\vec{p}$ ($=\rho\vec{v}$) is momentum density of thermodynamic hydrostatic system while $\vec{v}$ is velocity of constituent particles [40, 43]. In left hand side (Eq. 15), 1st term and 2nd term govern flow of mass and acceleration of constituent particles, both of which terms are induced by pressure gradient ($\vec{\nabla}P$) following stimulation of thermodynamic system with short-pulse optical beam.

Again, from the conservation of mass, equation of continuity can be written as [43]:

$$\frac{\partial \rho}{\partial t} + \vec{\nabla}.\vec{J} = 0,$$

$$\text{i.e., } \frac{\partial \rho}{\partial t} + \vec{\nabla}.(\rho \vec{v}) = 0, \tag{16}$$

where $\vec{J} = \rho \vec{v}$ is the mass current density which can also be considered as momentum density.

From continuum hypothesis, in association with assumption of local equilibrium [42], we can deduce a general expression for time-derivative of mass density ($\rho \equiv \rho(\vec{r}, t)$) (detail derivation is accomplished in Appendix "Continuum Hypothesis and Assumption of Local Equilibrium" (Eq. 41)):

$$\frac{\partial \rho}{\partial t} = \rho \left[\kappa_T \frac{\partial P}{\partial t} - \beta \frac{\partial T}{\partial t} \right], \tag{17}$$

$$\text{i.e., } -\vec{\nabla}.(\rho \vec{v}) = \rho \left[\kappa_T \frac{\partial P}{\partial t} - \beta \frac{\partial T}{\partial t} \right], \text{ using Eq. 16,} \tag{18}$$

$$\Rightarrow \left(\vec{\nabla} \rho \right).\vec{v} + \rho \vec{\nabla}.\vec{v} = -\rho \left[\kappa_T \frac{\partial P}{\partial t} - \beta \frac{\partial T}{\partial t} \right]. \tag{19}$$

Taking partial derivative on both side of above equation (Eq. 19) with respect to time (t), we get:

$$\left[\left(\vec{\nabla} \frac{\partial \rho}{\partial t} \right).\vec{v} + (\vec{\nabla} \rho).\frac{\partial \vec{v}}{\partial t} \right] + \left[\frac{\partial \rho}{\partial t} (\vec{\nabla}.\vec{v}) + \left(\vec{\nabla}.\frac{\partial \vec{v}}{\partial t} \right) \rho \right] = -\rho \left(\kappa_T \frac{\partial^2 P}{\partial t^2} - \beta \frac{\partial^2 T}{\partial t^2} \right)$$
$$- \frac{\partial \rho}{\partial t} \left(\kappa_T \frac{\partial P}{\partial t} - \beta \frac{\partial T}{\partial t} \right),$$

$$\left(\vec{\nabla} \frac{\partial \rho}{\partial t} \right).\vec{v} + (\vec{\nabla} \rho).\frac{\partial \vec{v}}{\partial t} + \frac{\partial \rho}{\partial t} (\vec{\nabla}.\vec{v}) + \left(\vec{\nabla}.\frac{\partial \vec{v}}{\partial t} \right) \rho = \rho \left(\beta \frac{\partial^2 T}{\partial t^2} - \kappa_T \frac{\partial^2 P}{\partial t^2} \right)$$
$$- \rho \left(\kappa_T \frac{\partial P}{\partial t} - \beta \frac{\partial T}{\partial t} \right)^2, \text{ using Eq. 17}$$

$$\text{say, } LHS \approx \rho \left(\beta \frac{\partial^2 T}{\partial t^2} - \kappa_T \frac{\partial^2 P}{\partial t^2} \right), \tag{20}$$

where higher degree term, $\rho \left(\kappa_T \frac{\partial P}{\partial t} - \beta \frac{\partial T}{\partial t} \right)^2$, is neglected and *LHS* is written as:

$$
\begin{aligned}
LHS &= \left(\vec{\nabla} \frac{\partial \rho}{\partial t} \right).\vec{v} + (\vec{\nabla}\rho).\frac{\partial \vec{v}}{\partial t} + \frac{\partial \rho}{\partial t}(\vec{\nabla}.\vec{v}) + \left(\vec{\nabla}.\frac{\partial \vec{v}}{\partial t} \right)\rho, \\
&= -\left[\vec{\nabla}(\vec{\nabla}\rho.\vec{v}) \right].\vec{v} - (\vec{\nabla}\rho.\vec{v})(\vec{\nabla}.\vec{v}) - \left[\vec{\nabla}(\rho\vec{\nabla}.\vec{v}) \right].\vec{v} - \rho(\vec{\nabla}.\vec{v})(\vec{\nabla}.\vec{v}) \\
&\quad + (\vec{\nabla}\rho).\frac{\partial \vec{v}}{\partial t} - \rho\vec{\nabla}.\left(\frac{\vec{\nabla}P}{\rho} \right) + \rho\left[\vec{\nabla}(\vec{\nabla}.\vec{v}) \right].\vec{v} + \rho(\vec{\nabla}.\vec{v})(\vec{\nabla}.\vec{v}) \\
&\quad - \rho\vec{\nabla}.\left[(\vec{v}.\nabla)\vec{v} \right].
\end{aligned}
\tag{21}
$$

where simplification of *LHS* is given in Appendix 4 (Eq. 46). Now, using Eqs. 21 and 20 can be expressed as:

$$
\begin{aligned}
\rho \left(\beta\frac{\partial^2 T}{\partial t^2} - \kappa_T\frac{\partial^2 P}{\partial t^2} \right) &= -\left[\vec{\nabla}((\vec{\nabla}\rho).\vec{v}) \right].\vec{v} - ((\vec{\nabla}\rho).\vec{v})(\vec{\nabla}.\vec{v}) - \left[\vec{\nabla}(\rho\vec{\nabla}.\vec{v}) \right].\vec{v} \\
&\quad + (\vec{\nabla}\rho).\frac{\partial \vec{v}}{\partial t} - \rho\vec{\nabla}.\left(\frac{\vec{\nabla}P}{\rho} \right) + \rho\left[\vec{\nabla}(\vec{\nabla}.\vec{v}) \right].\vec{v} \\
&\quad - \rho\vec{\nabla}.\left[(\vec{v}.\vec{\nabla})\vec{v} \right],
\end{aligned}
$$

$$
\begin{aligned}
\text{i.e.,} \quad \frac{1}{v_{ac}^2}\frac{\partial^2 P}{\partial t^2} - \rho\vec{\nabla}.\left(\frac{\vec{\nabla}P}{\rho} \right) &= \rho\beta\frac{\partial^2 T}{\partial t^2} + \left[\vec{\nabla}((\vec{\nabla}\rho).\vec{v}) \right].\vec{v} + ((\vec{\nabla}\rho).\vec{v})(\vec{\nabla}.\vec{v}) \\
&\quad + \left[\vec{\nabla}(\rho\vec{\nabla}.\vec{v}) \right].\vec{v} - (\vec{\nabla}\rho).\frac{\partial \vec{v}}{\partial t} \\
&\quad - \rho\left[\vec{\nabla}(\vec{\nabla}.\vec{v}) \right].\vec{v} + \rho\vec{\nabla}.\left[(\vec{v}.\vec{\nabla})\vec{v} \right].
\end{aligned}
\tag{22}
$$

Speed of sound in a (fluid) medium is given by $v_{ac} = \sqrt{\frac{1}{\kappa_S \rho}} \approx \sqrt{\frac{1}{\kappa_T \rho}}$ [41] (see Appendix 2 (Eqs. 35 and 36)). We consider two physical assumptions: (1) Mass density fluctuation in tissue medium under external perturbation of short pule laser stimulation is very small relative to ambient or equilibrium (mass) density ($\sim$1000 kg/m^3) [44]. (2) Velocity of constituent particles ($\sim$1 m/s [44]) under external perturbation is very small when compared to the speed of sound ($\sim$1500 m/s for soft tissue) [44]. Under these physical assumptions, in Eq. 22, terms involving particle velocity ($\vec{v}$) and density variation along direction of particle velocity can be neglected and subsequently, we can obtain wave equation that governs the propagation of mechanical PA wave:

$$
\frac{1}{v_{ac}^2}\frac{\partial^2 P}{\partial t^2} - \rho\vec{\nabla}.\left(\frac{\vec{\nabla}P}{\rho} \right) = \rho\beta\frac{\partial^2 T}{\partial t^2}.
\tag{23}
$$

which is a general wave equation that holds true for medium with variation in spatial distribution of mass density (ρ), i.e., $\vec{\nabla}\frac{1}{\rho} \neq 0$.

Again, from the heat transfer equation ($\Delta Q_{dens} = \rho c_P \Delta T$ that relates heat transfer (ΔQ) in a thermodynamic system to temperature change (ΔT)), we obtain heating function (H) [42]:

$$H = \lim_{\Delta t \to 0} \frac{\Delta Q_{dens}}{\Delta t} = \lim_{\Delta t \to 0} \rho c_P \frac{\Delta T}{\Delta t} = \rho c_P \frac{\partial T}{\partial t}, \tag{24}$$

which is defined as heat energy (Q) per unit time per unit volume, i.e., heat density (Q_{dens}) per unit time [40].

Combining Eqs. 23 and 24, wave equation can further be deduced:

$$\frac{1}{v_{ac}^2} \frac{\partial^2 P}{\partial t^2} - \rho \vec{\nabla} \cdot \left(\frac{\vec{\nabla} P}{\rho} \right) = \rho \beta \frac{\partial}{\partial t} \left(\frac{H}{\rho c_P} \right) = \frac{\beta}{c_P} \frac{\partial H}{\partial t}, \tag{25}$$

Equation 25 can also be written—expressing P, H, and ρ in functional forms, i.e., $P \equiv P(\vec{r}, t)$, $H \equiv H(\vec{r}, t)$, and $\rho \equiv \rho(\vec{r})$—in functional form as:

$$\frac{1}{v_{ac}(\vec{r})^2} \frac{\partial^2 P(\vec{r}, t)}{\partial t^2} - \rho(\vec{r}) \vec{\nabla} \cdot \left(\frac{\vec{\nabla} P(\vec{r}, t)}{\rho(\vec{r})} \right) = \rho(\vec{r}) \beta(\vec{r}) \frac{\partial}{\partial t} \left(\frac{H(\vec{r}, t)}{\rho(\vec{r}) c_P} \right),$$
$$= \frac{\beta(\vec{r})}{c_P(\vec{r})} \frac{\partial H(\vec{r}, t)}{\partial t}. \tag{26}$$

Left hand side in Eq. 25 or Eq. 26 gives wave equation while right hand side represents source term, i.e., differential change in heat density per unit time (or heating function) with respect to time serves as source of PA-wave which is stimulated due to transient illumination of optical beam. The equations are in consistent with the PA wave equation given in Refs. [32, 35]

Now, for homogeneous thermodynamic system, where spatial distribution of mass density (ρ) is uniform ($\vec{\nabla}\rho$ or $\vec{\nabla}\frac{1}{\rho} \approx 0$), Eq. 25 becomes:

$$\frac{1}{v_{ac}^2} \frac{\partial^2 P}{\partial t^2} - \nabla^2 P = \frac{\beta}{c_P} \frac{\partial H}{\partial t}, \tag{27}$$

where ρ is spatial independent, i.e., $\rho \neq \rho(\vec{r})$. In functional form, one may write:

$$\frac{1}{v_{ac}^2(\vec{r})} \frac{\partial^2 P(\vec{r}, t)}{\partial t^2} - \nabla^2 P(\vec{r}, t) = \frac{\beta(\vec{r})}{c_P(\vec{r})} \frac{\partial H(\vec{r}, t)}{\partial t}. \tag{28}$$

Equations 25 and 27 are in agreement with wave equations [32, 35] which are commonly employed (as standard PA wave equations) in PA imaging (in general) and reconstruction algorithms (in particular), i.e., generally, distribution of mass density (ρ) is assumed to be uniform or space independent over the entire region of interest for imaging. Therefore, our hypothesis of considering PA imaging system as a hydrostatic thermodynamic system and PA effect as a hydrostatic thermodynamic process is validated.

3 Conclusion

In this chapter, we provide a proper mathematical formulation and a clear understanding of the underlying physical processes that are the basis of PA imaging modalities (both microscopy and tomography). Tissue sample is considered as a hydrostatic (PVT) thermodynamic system and the phenomenon of the generation of initial pressure wave, due to transient optical ($\sim$ns) illumination and subsequent rapid heating, as a thermodynamic process. We also describe the PA wave propagation, which is nothing but the propagation of mechanical disturbances from one point to another through the vibration of constituent particles or molecules, in a bio-mechanical system (soft tissue). Mathematical equations, derived under the physical assumptions made, are in agreement with standard PA equations (initial PA-pressure and its subsequent propagation) which are commonly employed in photoacoustic imaging (in general) and conventional reconstruction algorithms (in particular). Therefore, our physical assumptions and formulation of PA wave generation and its propagation stands validated. This article will be beneficial to the PA-imaging community, for a thorough understanding of the mechanism of PA wave generation and its propagation in the tissue medium.

Acknowledgements Authors acknowledge Dr. Joy Mitra and Tathagata Sarkar, School of Physics (SoP), Indian Institute of Science Education and Research Thiruvananthapuram (IISER-TVM), Thiruvananthapuram, Kerala, India for the technical discussion extended to us.

Appendix 1: Thermodynamic Heat Transfer Equation

From the 1st law of thermodynamics [42], heat transfer to thermodynamic system is given by:

$$\Delta Q = \Delta U + \Delta W,$$
$$= \Delta U + P\Delta V, \tag{29}$$

where $\Delta W \, (= P\Delta V)$ is (thermodynamic) work done by the thermodynamic system; ΔU is change in internal energy; and ΔV is change in volume. On the other hand, heat capacity at constant volume (i.e., $\Delta V = 0$) is defined as:

$$C_V = \left.\frac{\Delta Q}{\Delta T}\right|_V = \left.\frac{\Delta U}{\Delta T}\right|_V, \quad \text{for } \Delta V = 0,$$
$$i.e., \ \ \Delta U = C_V \Delta T. \tag{30}$$

Now, using Eq. 30, we re-write Eq. 29:

$$\frac{\Delta Q}{V} = \frac{\Delta U}{V} + P\frac{\Delta V}{V}, \quad V \neq 0. \tag{31}$$

Neglecting fractional change in volume ($\frac{\Delta V}{V}$), as it is done above Sect. 2.1 (Eq. 3), we can re-write Eq. 31 as:

$$\frac{\Delta U}{V} \approx \frac{\Delta Q}{V}, \tag{32}$$

$$\Rightarrow \frac{C_V \Delta T}{V} = \frac{\Delta Q}{V}, \quad \text{using Eq. 30,}$$

$$\text{i.e., } \Delta Q_{dens} = \rho c_V \Delta T, \tag{33}$$

where c_V is the heat capacity per unit mass (called as specific heat capacity) at constant volume and $Q_{dens} = Q/V$ is the heat density [42].

Appendix 2: Relationship of Specific Heat Capacities at Constant Volume and Pressure

Speed of sound (v_{ac}) with which particular mechanical waves is propagating in a medium is characterized by mass density (ρ) and elastic coefficient (E) of the medium, and it can be expressed as [42]:

$$v_{ac} = \sqrt{\frac{E_S}{\rho}}, \tag{34}$$

where E_S is isentropic bulk modulus (sometimes, it is also called as adiabatic bulk modulus) and is related to isothermal compressibility (κ_T) (also known as isothermal bulk modulus) [42]:

$$\frac{E_S}{E_T} = \frac{\kappa_T}{\kappa_S} = \frac{C_P}{C_V} = \gamma. \tag{35}$$

For water, in room temperature (25–30 °C), isentropic compressibility (κ_S) and isothermal compressibility (κ_T) are comparable, i.e., $\gamma \approx 1$ [42]. In the human body, water is the main fluid content [44], so that we assume $\kappa_T \approx \kappa_S$. Using Eq. 34 and Eq. 35, we get:

$$v_{ac}^2 \approx \frac{1}{\kappa_S \rho}, \tag{36}$$

and hence,

$$\kappa_T c_V \rho = \frac{c_P}{v_{ac}^2}. \tag{37}$$

Appendix 3: Local Thermodynamic Properties

Equilibrium Thermodynamics

Considering mass density ($\rho = \frac{m}{V}$)—which is mass (m) averaged over volume (V) of thermodynamics system in equilibrium—to be hydrostatic thermodynamic parameter being characterized by temperature (T) and pressure (P), one can write equation of state (under equilibrium) in functional form as:

$$\rho \equiv \rho(P, T). \tag{38}$$

Under quasi-static condition [42], differential change in ρ-while undergoing the system (in general) and elemental volume (in particular) from one thermodynamic equilibrium state to another equilibrium state of the system (in our case, due to perturbation induced by transient illumination of short pulsed optical beam) can be deduced as:

$$
\begin{aligned}
\Delta\rho &= \left.\frac{\partial\rho}{\partial T}\right|_P \Delta T + \left.\frac{\partial\rho}{\partial P}\right|_T \Delta P, \\
&= \left.\frac{\partial}{\partial T}\left(\frac{m}{V}\right)\right|_P \Delta T + \left.\frac{\partial}{\partial P}\left(\frac{m}{V}\right)\right|_T \Delta P, \\
&= -\left.\frac{m}{V^2}\frac{\partial V}{\partial T}\right|_P \Delta T - \left.\frac{m}{V^2}\frac{\partial V}{\partial P}\right|_T \Delta P, \\
&= \frac{m}{V}[\kappa_T \Delta P - \beta \Delta T],
\end{aligned}
\tag{39}
$$

Here, mass (m) of the hydrostatic thermodynamic system is assumed to be constant. β and κ_T are volume expansibility and isothermal compressibility respectively and can be expressed as:

$$\beta = \left.\frac{1}{V}\frac{\partial V}{\partial T}\right|_P,$$

$$\text{and, } \kappa_T = -\left.\frac{1}{V}\frac{\partial V}{\partial P}\right|_T.$$

Continuum Hypothesis and Assumption of Local Equilibrium

Although a physical thermodynamic system (like, fluid, semi-fluid, and gas, etc.) consists of an infinite number of constituent atoms and/or molecules, statistical mechanics proves that (at macroscopic scale) a thermodynamic system in equilibrium can be explicitly described by a set of thermodynamic variables (such as pressure (P), volume (V), temperature (T), entropy (S), number of particle (N), and chemical potential (μ) which are generally called as macrostate variables). Shortly, in statistical sense, any physical properties of interest of a thermodynamic system (in equilibrium) are obtainable from measurements of the macroscopic variables. On the other hand, thermodynamic equilibrium explicitly means that physical thermodynamic system under consideration is equilibrium from chemical, mechanical, and thermal aspects. In this way, measurements of thermodynamic macroscopic variables—averaged over the system—give measurement of any (physical or chemical) properties of interest of the system, i.e., any physical properties of interest is characterised by the macrostate variables and thus can be expressed in terms of the variables. To extend this concept of global equilibrium to any other thermodynamic systems, that are not in thermo-dynamic equilibrium, one can adopt principle of local thermodynamic equilibrium [45]. Specifically, in this local thermodynamic equilibrium, one can consider that entire thermodynamic system is constituted by an infinite number of imaginary sub-systems—that occupy an infinitesimally small volume relative to that of entire system but contain a sufficiently large number of constituent particles/molecules for a valid statistical formulation—such that one can assume thermodynamic equilibrium over such individual sub-systems and thus, one can define [at any given time (t)] local thermodynamic properties or macrostate variables that can be formulated with equi-librium statistical physics. This means to say that, from aspects of thermodynamics, local thermodynamic state variables can be characterized as functions of space $(\vec{r})$ and time (t). Again, assumption of local equilibrium says that local thermodynamic properties, defined for infinitesimal sub-systems, and their derivatives satisfy classical thermodynamic relations (more specifically, Maxwell's equations of thermodynam-ics) which governs any physical process in thermodynamic equilibrium [45]. In this way, continuum hypothesis [45] permits us to replace thermodynamic parameters by corresponding thermodynamic fields as continuous functions of space and time (say, $P \equiv P(\vec{r}, t), V \equiv V(\vec{r}, t), T \equiv T(\vec{r}, t), S \equiv S(\vec{r}, t), N \equiv N(\vec{r}, t),$ and $\mu \equiv \mu(\vec{r}, t)$) and follow thermodynamic equations of equilibrium.

Now, for an arbitrarily chosen (locally equilibrium) thermodynamic sub-system, we can re-write Eq. 39:

$$\Delta\rho(\vec{r}, t) = \rho(\vec{r}, t)[\kappa_T \Delta P(\vec{r}, t) - \beta \Delta T(\vec{r}, t)]. \tag{40}$$

Here, we consider thermodynamic sub-systems as hydrostatic (as it is discussed in Sect. 2.1). Taking temporal change relative to Δt in Eq. 40, we obtain:

$$\frac{\Delta\rho(\vec{r}, t)}{\Delta t} = \rho(\vec{r}, t)\left[\kappa_T \frac{\Delta P(\vec{r}, t)}{\Delta t} - \beta \frac{\Delta T(\vec{r}, t)}{\Delta t}\right],$$

$$\Rightarrow \lim_{\Delta t \to 0} \frac{\Delta\rho(\vec{r}, t)}{\Delta t} = \lim_{\Delta t \to 0} \rho(\vec{r}, t)\left[\kappa_T \frac{\Delta P(\vec{r}, t)}{\Delta t} - \beta \frac{\Delta T(\vec{r}, t)}{\Delta t}\right],$$

$$\text{i.e., } \frac{\partial\rho(\vec{r}, t)}{\partial t} = \rho(\vec{r}, t)\left[\kappa_T \frac{\partial P(\vec{r}, t)}{\partial t} - \beta \frac{\partial T(\vec{r}, t)}{\partial t}\right]. \tag{41}$$

Note that thermodynamic coefficients κ_T and β are, in principle, spatial and temporal fields. However, in practical applications, dependence of these thermodynamic coefficients on space $(\vec{r})$ and time (t) are often neglected (as it is assumed in our case (Eq. 41)). Equation 41 implies that a thermodynamic system, which is perturbed from its thermodynamic equilibrium state by an external agency or disturbance (transient optical irradiation, in our case of PA imaging), has a tendency to bring back to its thermodynamic equilibrium state (which is stable) through transfer of thermodynamic physical parameters (say, $T(\vec{r}, t), P(\vec{r}, t)$, and $V(\vec{r}, t)$ eventually resulted from transport of mass density $(\rho(\vec{r}, t)))$ from one point $(\vec{r}, t)$ to another (in general) and one macroscopic element (defined by $(\vec{r}, t)$) of local thermodynamic equilibrium to another (in particular).

Appendix 4: Identities of Vector and Scalar Quantities

For a given quantity (scalar or vector)—say $\phi(x, y, z, t)$ or $\vec{A}(x, y, z, t)$—one can obtain identities (from Taylor's expansion) as [46],

$$\Delta\phi = \frac{\partial\phi}{\partial t}\Delta t + \frac{\partial\phi}{\partial x}\Delta x + \frac{\partial\phi}{\partial y}\Delta y + \frac{\partial\phi}{\partial z}\Delta z,$$

$$\Rightarrow \lim_{\Delta t\to 0}\frac{\Delta\phi}{\Delta t} = \frac{\partial\phi}{\partial t} + \frac{\partial\phi}{\partial x}\frac{\Delta x}{\Delta t} + \frac{\partial\phi}{\partial y}\frac{\Delta y}{\Delta t} + \frac{\partial\phi}{\partial z}\frac{\Delta z}{\Delta t},$$

$$\Rightarrow \frac{d\phi}{dt} = \frac{\partial\phi}{\partial t} + \frac{\partial\phi}{\partial x}\frac{dx}{dt} + \frac{\partial\phi}{\partial y}\frac{dy}{dt} + \frac{\partial\phi}{\partial z}\frac{dz}{dt},$$

$$\Rightarrow \frac{d\phi}{dt} = \frac{\partial\phi}{\partial t} + \frac{\partial\phi}{\partial x}v_x + \frac{\partial\phi}{\partial y}v_y + \frac{\partial\phi}{\partial z}v_z,$$

$$\Rightarrow \frac{d\phi}{dt} = \frac{\partial\phi}{\partial t} + \vec{v}.\vec{\nabla}\phi, \tag{42}$$

where $\vec{v} = v_x\hat{i} + v_y\hat{j} + v_z\hat{k}$ is velocity of constituent particle with individual (velocity) components $v_x = \frac{dx}{dt}; v_y = \frac{dy}{dt}$; and $v_z = \frac{dz}{dt}$.

Similarly, for any vector quantity (say, $\vec{A}$), we get:

$$\Delta\vec{A} = \frac{\partial\vec{A}}{\partial t}\Delta t + \frac{\partial\vec{A}}{\partial x}\Delta x + \frac{\partial\vec{A}}{\partial y}\Delta y + \frac{\partial\vec{A}}{\partial z}\Delta z,$$

$$\Rightarrow \lim_{\Delta t\to 0}\frac{\Delta\vec{A}}{\Delta t} = \frac{\partial\vec{A}}{\partial t} + \frac{\partial\vec{A}}{\partial x}\frac{\Delta x}{\Delta t} + \frac{\partial\vec{A}}{\partial y}\frac{\Delta y}{\Delta t} + \frac{\partial\vec{A}}{\partial z}\frac{\Delta z}{\Delta t},$$

$$\Rightarrow \frac{d\vec{A}}{dt} = \frac{\partial\vec{A}}{\partial t} + \frac{\partial\vec{A}}{\partial x}\frac{dx}{dt} + \frac{\partial\vec{A}}{\partial y}\frac{dy}{dt} + \frac{\partial\vec{A}}{\partial z}\frac{dz}{dt},$$

$$\Rightarrow \frac{d\vec{A}}{dt} = \frac{\partial\vec{A}}{\partial t} + \left(v_x\frac{\partial}{\partial x} + v_y\frac{\partial}{\partial y} + v_z\frac{\partial}{\partial z}\right)\vec{A},$$

$$\Rightarrow \frac{d\vec{A}}{dt} = \frac{\partial\vec{A}}{\partial t} + (\vec{v}.\vec{\nabla})\vec{A}. \tag{43}$$

Appendix 5: Simplification of LHS

Using above two identities (Eqs. 42 and 43) in Eq. 15, we obtain:

$$\vec{v}\left(\frac{\partial \rho}{\partial t} + \vec{v}.\vec{\nabla}\rho\right) + \rho\left(\frac{\partial \vec{v}}{\partial t} + (\vec{v}.\vec{\nabla})\vec{v}\right) = -\vec{\nabla}P, \tag{44}$$

$$\Rightarrow \vec{v}\left(-\vec{\nabla}.(\rho\vec{v}) + \vec{v}.\vec{\nabla}\rho\right) + \rho\left(\frac{\partial \vec{v}}{\partial t} + (\vec{v}.\vec{\nabla})\vec{v}\right) = -\vec{\nabla}P, \quad \text{using Eq. 16}$$

$$\Rightarrow \qquad \vec{v}\left(-\rho(\nabla.\vec{v})\right) + \rho\left(\frac{\partial \vec{v}}{\partial t} + (\vec{v}.\vec{\nabla})\vec{v}\right) = -\vec{\nabla}P,$$

$$\Rightarrow \qquad -\rho\vec{v}(\vec{\nabla}.\vec{v}) + \rho\frac{\partial \vec{v}}{\partial t} + \rho(\vec{v}.\vec{\nabla}\vec{v}) = -\vec{\nabla}P,$$

$$\Rightarrow \qquad \left[-\frac{\vec{\nabla}P}{\rho} + \vec{v}(\vec{\nabla}.\vec{v}) - (\vec{v}.\vec{\nabla})\vec{v}\right] = \frac{\partial \vec{v}}{\partial t}. \tag{45}$$

Rewriting *LHS* in Eq. 21, we obtain:

$$LHS = \left(\vec{\nabla}\frac{\partial \rho}{\partial t}\right).\vec{v} + (\vec{\nabla}\rho).\frac{\partial \vec{v}}{\partial t} + \frac{\partial \rho}{\partial t}(\vec{\nabla}.\vec{v}) + \left(\vec{\nabla}.\frac{\partial \vec{v}}{\partial t}\right)\rho,$$

$$= \left(\vec{\nabla}\frac{\partial \rho}{\partial t}\right).\vec{v} + (\vec{\nabla}\rho).\frac{\partial \vec{v}}{\partial t} + \frac{\partial \rho}{\partial t}(\vec{\nabla}.\vec{v}) + \left(\vec{\nabla}.\frac{\partial \vec{v}}{\partial t}\right)\rho,$$

$$= \left[\left(\vec{\nabla}\frac{\partial \rho}{\partial t}\right).\vec{v} + \frac{\partial \rho}{\partial t}(\vec{\nabla}.\vec{v})\right] + (\vec{\nabla}\rho).\frac{\partial \vec{v}}{\partial t} + \rho\left(\vec{\nabla}.\frac{\partial \vec{v}}{\partial t}\right), \text{ using Eq. 45}$$

$$= \vec{\nabla}.\left[\left(\frac{\partial \rho}{\partial t}\right)\vec{v}\right] + (\vec{\nabla}\rho).\frac{\partial \vec{v}}{\partial t} + \rho\vec{\nabla}.\left[-\frac{\vec{\nabla}P}{\rho} + \vec{v}(\vec{\nabla}.\vec{v}) - (\vec{v}.\vec{\nabla})\vec{v}\right], \text{ using Eq. 16,}$$

$$= \vec{\nabla}.\left[-\left(\vec{\nabla}.(\rho\vec{v})\right)\vec{v}\right] + (\vec{\nabla}\rho).\frac{\partial \vec{v}}{\partial t} + \rho\vec{\nabla}.\left[-\frac{\vec{\nabla}P}{\rho} + \vec{v}(\vec{\nabla}.\vec{v}) - (\vec{v}.\vec{\nabla})\vec{v}\right],$$

$$= -\vec{\nabla}.\left[\left((\vec{\nabla}\rho).\vec{v} + \rho(\vec{\nabla}.\vec{v})\right)\vec{v}\right] + (\vec{\nabla}\rho).\frac{\partial \vec{v}}{\partial t} + \rho\vec{\nabla}.\left[-\frac{\vec{\nabla}P}{\rho} + \vec{v}(\vec{\nabla}.\vec{v}) - (\vec{v}.\vec{\nabla})\vec{v}\right],$$

$$= -\vec{\nabla}.\left[((\vec{\nabla}\rho).\vec{v})\vec{v}\right] - \vec{\nabla}.\left[\rho(\vec{\nabla}.\vec{v})\vec{v}\right] + (\vec{\nabla}\rho).\frac{\partial \vec{v}}{\partial t} + \rho\vec{\nabla}.\left[-\frac{\vec{\nabla}P}{\rho} + \vec{v}(\vec{\nabla}.\vec{v}) - (\vec{v}.\vec{\nabla})\vec{v}\right],$$

$$= -\left[\vec{\nabla}((\vec{\nabla}\rho).\vec{v})\right].\vec{v} - ((\vec{\nabla}\rho).\vec{v})(\vec{\nabla}.\vec{v}) - \left[\vec{\nabla}(\rho\vec{\nabla}.\vec{v})\right].\vec{v} - \rho(\vec{\nabla}.\vec{v})(\vec{\nabla}.\vec{v}) + (\vec{\nabla}\rho).\frac{\partial \vec{v}}{\partial t}$$

$$-\rho\vec{\nabla}.\left(\frac{\vec{\nabla}P}{\rho}\right) + \rho\vec{\nabla}.\left[\vec{v}(\vec{\nabla}.\vec{v})\right] - \rho\vec{\nabla}.\left[(\vec{v}.\vec{\nabla})\vec{v}\right],$$

$$= -\left[\vec{\nabla}((\vec{\nabla}\rho).\vec{v})\right].\vec{v} - ((\vec{\nabla}\rho).\vec{v})(\vec{\nabla}.\vec{v}) - \left[\vec{\nabla}(\rho\vec{\nabla}.\vec{v})\right].\vec{v} - \rho(\vec{\nabla}.\vec{v})(\vec{\nabla}.\vec{v})$$

$$+(\vec{\nabla}\rho).\frac{\partial \vec{v}}{\partial t} - \rho\vec{\nabla}.\left(\frac{\vec{\nabla}P}{\rho}\right) + \rho\left[\vec{\nabla}(\vec{\nabla}.\vec{v})\right].\vec{v} + \rho(\vec{\nabla}.\vec{v})(\vec{\nabla}.\vec{v}) - \rho\vec{\nabla}.\left[(\vec{v}.\nabla)\vec{v}\right]. \tag{46}$$

References

1. X.L. Dean-Ben, T.F. Fehm, S.J. Ford, S. Gottschalk, D. Razansky, Spiral volumetric optoacoustic tomography visualizes multi-scale dynamics in mice. Light. Sci. Appl. **6**, 1–8 (2017)
2. X.L. Dean-Ben, E. Bay, D. Razansky, Functional optoacoustic imaging of moving objects using microsecond-delay acquisition of multispectral three-dimensional tomographic data. Sci. Rep. **4**, 1–6 (2014)
3. J.M. Yang, C. Favazza, R. Chen, J. Yao, X. Cai, K. Maslov, Q. Zhou, K.K. Shung, L.V. Wang, Simultaneous functional photoacoustic and ultrasonic endoscopy of internal organs. Nat. Med. **18**, 1297–1302 (2012)
4. M.S. Singh, A. Thomas, Photoacoustic elastography imaging: a review. J. Biomed. Opt. **24**, 040902 (2019)
5. E.M. Strohm, M.J. Moore, M.C. Kolios, Single cell photoacoustic microscopy: a review. IEEE J. Sel. Top. Quant. Electron. **22**, 6801215 (2015)
6. A. Danielli, K. Maslov, A. Garcia-Uribe, A.M. Winkler, C. Li, L. Wang, Y. Chen, G.W. Dorn II, L.V. Wang, Label-free photoacoustic nanoscopy. J. Biomed. Opt. **19**, 086006 (2014)
7. C. Zhang, K. Maslov, S. Hu, R. Chen, Q. Zhou, K.K. Shung, L.V. Wang, Reflection-mode submicron-resolution in vivo photoacoustic microscopy. J. Biomed. Opt. Lett. **19**, 020501 (2012)
8. L.V. Wang, S. Hu, In vivo imaging from organelles to organs. Science **335**, 1458–1462 (2012)
9. C. Zhang, K. Maslov, L.V. Wang, Sub-wavelength-resolution label-free photoacoustic microscopy of optical absorption in vivo. Opt. Lett. **35**, 3195–3197 (2010)
10. B. Rao, F. Soto, D. Kerschensteiner, L.V. Wang, Integrated photoacoustic, confocal, and two-photon microscope. J. Biomed. Opt. **19**, 36002 (2014)
11. https://www.pubfacts.com/search/photoacoustic+imaging
12. J. Song, J. Kim, S. Hwang, M. Jeon, S. Jeong, C. Kim, S. Kim, Smart gold nanoparticles for photoacoustic imaging: an imaging contrast agent responsive to the cancer microenvironment and signal amplification via pH-induced aggregation. Chem. Comm. **52**, 8287–8290 (2016)
13. E.E. Connor, J. Mwamuka, A. Gole, C.J. Murphy, M.D. Wyatt, Gold nano-particles are taken up by human cells but do not cause acute cytotoxicity. Chem. Comm. **1**, 325 (2005)
14. H. Jiang, Z. Yuan, X. Gu, Spatially varying optical and acoustic property reconstruction using finite-element-based photoacoustic tomography. J. Opt. Soc. Am. **23**, 878 (2006)
15. J. Shah, S. Park, S. Aglyamov, T. Larson, L. Ma, K. Sokolov, K. Johnston, T. Miller, S.Y. Emelianov, Photoacoustic imaging and temperature measurement for photothermal cancer therapy. J. Biomed. Opt. **13**, 034024 (2008)
16. S. Sethuraman, S.R. Aglyamov, J.H. Amirian, R.W. Smalling, S.Y. Emelianov, Intravascular photoacoustic imaging using an ivus imaging catheter. IEEE Trans. Ultras. Ferroel. Freq. Cont. **54**, 978–986 (2007)
17. W. Xia, M.K.A. Singh, E. Maneas, N. Sato, Y. Shigeta, T. Agano, S. Ourselin, S.J. West, A.E. Desjardins, Handheld real-time led-based photoacoustic and ultrasound imaging system for accurate visualization of clinical metal needles and superficial vasculature to guide minimally invasive procedures. Sensors (Basel) **18**, 1394 (2018)
18. A. Buehler, M. Kacprowicz, A. Taruttis, V. Ntziachristos, Real-time handheld multispectral optoacoustic imaging. Opt. Lett. **38**, 1404–1406 (2013)
19. Z. Zhang, Y. Shi, S. Yang, D. Xing, Sub-diffraction-limited second harmonic photoacoustic microscopy based on nonlinear thermal diffusion. Opt. Lett. **43**, 2336–2339 (2018)
20. L. Xi, L. Zhou, H. Jiang, C-scan photoacoustic microscopy for in vivo imaging of *Drosophila pupae*. Appl. Phys. Lett. **101**, 013702 (2012)
21. M.S. Singh, H. Jiang, Estimating both direction and magnitude of flow velocity using photoacoustic microscopy. Appl. Phys. Lett. **104**, 253701 (2014)
22. G. Gao, S. Yang, D. Xing, Viscoelasticity imaging of biological tissues with phase-resolved photoacoustic measurement. Opt. Lett. **36**, 3341–3343 (2011)
23. M.S. Singh, H. Jiang, Elastic property attributes to photoacoustic signals: an experimental phantom study. Opt. Lett. **39**, 3973 (2014)

24. N. Wadamori, Non-restrained measurement of young's modulus for soft tissue using a photoacoustic technique. Appl. Phys. Lett. **20**, 103707 (2014)
25. Y. Zhao, S. Yang, C. Chen, D. Xing, Simultaneous optical absorption and viscoelasticity imaging based on photoacoustic lock-in measurement. Opt. Lett. **39**, 2565–8 (2014)
26. Y. Liu, Z. Yuan, Multi-spectral photoacoustic elasticity tomography. Biomed. Opt. Exp. **7**, 3323–3334 (2016)
27. M.S. Singh, H. Jiang, Ultrasound (US) transducer of higher operating frequency detects photoacoustic (PA) signals due to the contrast in elastic property. AIP Adv. **6**, 025210 (2016)
28. Q. Wang, Y. Shi, F. Yang, S. Yang, Quantitative photoacoustic elasticity and viscosity imaging for cirrhosis detection. Appl. Phys. Lett. **112**, 211902 (2018)
29. P. Hai, Y. Zhou, L. Gong, L.V. Wang, Quantitative photoacoustic elastography of young's modulus in humans. J. Biomed. Opt. **112**, 10064 (2017)
30. P. Hai, Y. Zhou, J. Liang, C. Li, L.V. Wang, Photoacoustic tomography of vascular compliance in humans. J. Biomed. Opt. **20**, 126008 (2015)
31. H. Zafar, A. Breathnach, H.M. Subhash, M.J. Leahy, Linear-array-based photoacoustic imaging of human microcirculation with a range of high frequency transducer probes. J. Biomed. Opt. **20**(5), 051021 (2015)
32. H. Jiang, *Photoacoustic Tomography* (CRC Press, Taylor and Francis Group ISBN-13, 2015)
33. B.E. Treeby, B.T. Cox, k-wave: Matlab toolbox for the simulation and reconstruction of photoacoustic wave fields. J. Biomed. Opt. **15**(2), 021314 (2010)
34. B.T. Cox, P.C. Beard, Fast calculation of pulsed photoacoustic fields in fluids using k-space methods. J. Acoust. Soc. Am. **117**(6), 3616–3627 (2005)
35. L.V. Wang, *Photoacoustic Imaging and Spectroscopy*, vol. 20 (CRC Press, Taylor and Francis Group, 2009), ISBN-13 978-1-4200-5991-5
36. A.G. Bell, On the production and reproduction of sound by light. Am. Ass. Adv. Sc. **XX**, 305–324 (1880)
37. A.G. Bell, The production of sound by radiant energy. Science **49**, 242–253 (1881)
38. S.R. Arridge, Optical tomography in medical imaging. Inverse Prob. **15**, R41–R93 (1999)
39. P. Beard, Biomedical photoacoustic imaging. Science **1**, 602 (2011)
40. L.V. Wang, H. Wu, *Biomedical Optics: Principles and Imaging* (Wiley, Hoboken, 2007), p. 2006030754
41. J. Xia, J. Yao, L.V. Wang, Photoacoustic tomography: principles and advances. Electrom. Wales (Cambridge) **147**, 1–22 (2014)
42. M.W. Zemansky, R.H. Dittman, *Heat and Thermodynamics* (1997)
43. M. Pramanik, L.V. Wang, Thermoacoustic and photoacoustic sensing of temperature. J. Biomed. Opt. **14**, 054024 (2009)
44. A. Brahme, D. Panetta, M. Demi, *Comprehensive Biomedical Physics* (Elsevier Science, 2014), p. 320
45. J.M.O. de Zárate, J.V. Sengers, *Hydrodynamic Fluctuations in Fluids and Fluid Mixtures* (Elsevier Science, 2006), p. 320
46. M.R. Spiegel, *Vector Analysis* (Schaum's Outlines, 1959), p. 320

High-Power Light Emitting Diodes; An Alternative Excitation Source for Photoacoustic Tomography

Thomas J. Allen

Abstract Photoacoustic tomography is a non-invasive imaging modality that is based on laser-generated ultrasound and which can provide high quality, 3D images of soft biological tissue. Photoacoustic signals are typically generated using Q-switched lasers, which are relatively bulky and expensive; so far, this has hindered the translation of the technique from the laboratory into a clinical environment. An alternative is to use light emitting diodes (LEDs) as excitation sources; these devices have the advantage of being compact, inexpensive, and available in a wide range of wavelengths (visible and NIR), all of which makes them well suited to clinical applications. The main drawback of LEDs is their low pulse energy (a few μJ), which is significantly below the tens of mJ provided by Q-switched lasers. However, a range of studies have demonstrated the possibility of using LEDs to generate and detect photoacoustic signals with a sufficient SNR for in-vivo imaging of the superficial vasculature. This chapter reviews key developments in LED-based photoacoustic imaging that have occurred over the past decade.

1 Introduction

Photoacoustic tomography is a relatively new biomedical imaging modality [1], which is based on laser generated ultrasound. It is a hybrid modality which combines the high contrast of optical imaging techniques with the high spatial resolution (<100 μm) of ultrasound imaging. It offers the possibility of acquiring high-quality 3D images of the internal structure of soft biological tissues such as blood vessels. The technique provides not only structural but also functional information through methods such as spectroscopy [2], flow measurements [3, 4], or thermometry [5]. Photoacoustic tomography has the potential to be used in a wide range of clinical applications, such as imaging skin pathologies [6], cardiovascular disease

T. J. Allen (✉)
University College London, London, UK
e-mail: t.allen@ucl.ac.uk

© Springer Nature Singapore Pte Ltd. 2020
M. Kuniyil Ajith Singh (ed.), *LED-Based Photoacoustic Imaging*,
Progress in Optical Science and Photonics 7,
https://doi.org/10.1007/978-981-15-3984-8_2

23

[7, 8], oncology [9], abnormalities of the microcirculation (e.g. diabetes [10]), arthritis [11] and other conditions. Photoacoustic signals are typically generated using Q-switched Nd:YAG pumped OPO, Ti:Sapphire or dye laser systems as they provide the necessary high pulse energies (mJ) and short pulse durations (ns) required for photoacoustic tomography. However, these excitation sources suffer from a range of limitations; they tend to be bulky, expensive, require water cooling and regular maintenance (e.g. alignment), and have a low pulse repetition frequency (PRF) (<200 Hz) which limits the achievable imaging speed. These practical limitations have so far inhibited the translation of photoacoustic tomography from a laboratory technique to one which can be easily implemented in a clinical environment. There is therefore a need for novel excitation sources.

An alternative is to use semiconductor devices such as laser diodes [12–14] and light emitting diodes (LEDs) [15–24], as they provide the means to overcome these limitations; they are compact (on the sub-mm scale), relatively cheap (a single element device costs a couple hundreds of USD or less), robust, have a high wall-plug efficiency (>10%) and do not require regular maintenance. In addition, they are available over a wide range of wavelengths (see Fig. 1), making them well suited for spectroscopic applications. The main drawback of semiconductor devices is their relatively low peak powers, compared to the mega Watts provided by Q-switched lasers sources. For example, laser diodes can provide peak powers ranging from tens of Watts for a single element device to tens of kWs for a stack or bar of laser diodes, resulting in pulse energies ranging from tens of μJ to a couple of mJs when operating with pulse durations of tens of ns. The peak powers of LEDs are even lower, a few Watts for a single device when driven in continuous wave (CW) mode. However, it is possible to overdrive them with current pulses tens of times their nominal rating, allowing for pulse energies on the order of a few μJ to be achieved when driven with

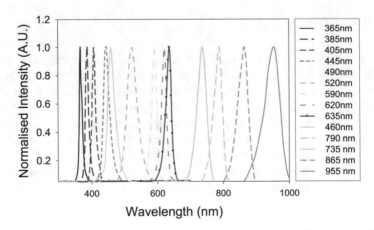

Fig. 1 Optical spectra for a range of commercially available LEDs [20]. (Figure adapted with permission from [20])

a short pulse duration (a couple of hundreds of ns or less). Also, their low pulse energies can to some extent be mitigated through strategies such as combining the output of several devices, exploiting their high PRF to acquire and signal average many signals over a relatively short period of time, and optimizing their pulse duration [13, 25]. A significant advantage of LEDs over laser diodes is that they are cheaper, with devices available for a few tens of USD.

The purpose of this book chapter is to provide a brief historical perspective of the application of LEDs as excitation sources for photoacoustic imaging. Four key areas of development will be described (Sect. 4); (1) the very possibility, through single point measurements, of generating detectable photoacoustic signals with LEDs; (2) the possibility of obtaining tomographic images from those measurements; (3) some major achievements in spectroscopic photoacoustic imaging based on LEDs, which open up applications such as measuring SO_2 levels in a vessel; (4) the ability of implementing novel excitation schemes for LED-based photoacoustic imaging. These developments have been based on the widefield illumination of the sample, which is required for photoacoustic tomography.

Before presenting the above, however, a brief introduction to photoacoustic tomography will be given (Sect. 2) so as to equip the reader with the necessary scientific background. This will cover the basic image formation principles of photoacoustic imaging, including the source of contrast and insights into the factors and mechanisms that limit imaging depth and resolution. Not only will this set the scene for this chapter, it will also be a useful source of information for the remainder of this book. In addition to the photoacoustic imaging technique itself, a brief overview of high-power LEDs will be provided (Sect. 3). This section will cover, in a concise manner, their operational principle and basic characteristics. It will also describe the possibility of overdriving LEDs when operated in pulsed mode. The chapter concludes with a summary and an outlook of potential developments that apply to high-power LEDs in the context of photoacoustic imaging (Sect. 5).

2 Photoacoustic Tomography

In biomedical photoacoustic tomography, a tissue sample is typically illuminated by a short nanosecond pulse of light, such that the whole sample is flooded with light (see Fig. 2a). The light then undergoes multiple scattering within the tissue before being absorbed by a chromophore, such as blood, melanin, water or lipids, which are the major chromophores present in biological tissue. As the absorbed light is converted to heat via vibrational and collisional relaxation, a small, localized temperature increase (<0.1 K) is induced, resulting in a pressure increase (P_0). This pressure then relaxes into broadband (tens of MHz) acoustic waves, which propagate to the tissue surface where they are recorded as A-lines by a single mechanically scanned ultrasound detector or an array of ultrasound detectors. By measuring the time of arrival of these waves and making the assumption that the speed of sound in biological tissue

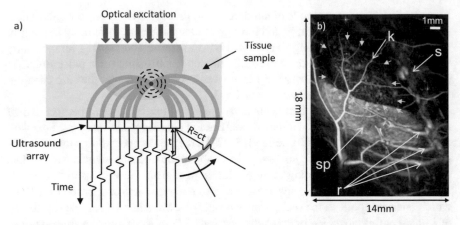

Fig. 2 Photoacoustic tomography; **a** Schematic representation; **b** photoacoustic image of the vasculature around the abdomen of a mouse [26] using a photoacoustic scanner based on a planar geometry and a Q-switched Nd:YAG pumped OPO laser system with an incident fluence below the safe maximum permissible limit for skin (<6 mJ/cm^2). As well as the superficial vasculature, several deeper lying organs can be visualised such as the spine (s), the kidneys (k, outlined with yellow arrows), the spleen (sp) and the ribs (r). (Figure adapted with permission from [26])

is homogenous (e.g. 1485 m/s), it is then possible, using a reconstruction algorithm, to form a photoacoustic image representing the absorbed optical energy distribution.

The reconstruction algorithm can be conceptually understood as follows. The amplitude of a recorded signal at time t is given by the integral of the initial pressure distribution (P_0) that lies on the surface of a sphere centred on a detector with a radius ct, where c is the speed of sound. If the pressure is measured at several points over a line or surface, an image of the optical absorbed energy distribution can then be obtained by back-projecting the detected pressure signals. This is illustrated in Fig. 2a, where an array of detectors is used to measure the photoacoustic pressure over a line. Each element of the array records a time series of the generated photoacoustic signal. The photoacoustic signal is incident at the detectors with varying time delays due to their different distances from the acoustic source. If the time series are then back projected over a spherical surface, the photoacoustic signal present in each individual time series will coherently interfere at the location of the acoustic source, creating an image of the absorbed optical energy distribution. This type of back-projection approach is equivalent to the delay-and-sum [27] algorithm used in phased array US imaging and has been widely used for photoacoustic tomography as it is relatively simple to implement and intuitive. However, it is non-optimal in terms of accuracy and computational expense and therefore more advanced reconstruction algorithms have been developed such as time-reversal [28], and Fast Fourier Transform-based methods [29].

Figure 2b shows an example of a photoacoustic image of the microvasculature of a mouse acquired with a Fabry–Perot sensor based photoacoustic scanner [30]. The simultaneous high contrast and resolution of the blood vessels are clearly visible.

The photoacoustic image represents a map of the initial pressure distribution (P_0), which is directly related to the heating produced by the absorption of light. Under the assumption of instantaneous heating, which in practice requires that the duration of the excitation pulse is significantly shorter than the time it takes for an acoustic wave to travel through the heated region (a condition known as stress confinement), the initial pressure distribution (P_0) can be related to the absorbed optical energy $H(r)$ via the following equation:

$$P_0 = \Gamma H(r) = \Gamma \mu_a(r) \Phi(r; \mu_a, \mu_s, g). \tag{1}$$

Here, Γ is the Grüneisen coefficient, a measure of the conversion efficiency of heat energy to pressure, and $H(r)$ is the absorbed optical energy distribution, which in turn is a product of the local absorption coefficient $\mu_a(r)$ and the optical fluence $\Phi(r; \mu_a, \mu_s, g)$, where $\mu_a(r)$ and $\mu_s(r)$ are the absorption and scattering coefficients of the illuminated tissue volume, and g is the anistropy factor. Equation 1 implies that P_0 is dependent on the product of three terms, the Grüneisen parameter (Γ), the absorption coefficient ($\mu_a(r)$) and the fluence (Φ). The Grüneisen parameter is typically assumed to only vary weakly for different tissues (although this may not always be the case [31]), and therefore regarded as spatially invariant. Therefore, the spatial variations in the optical properties of the tissue sample, absorption and scattering, provide the main source of contrast. Absorption tends to dominate and is usually the main source of contrast in photoacoustic imaging, which makes the technique well suited to imaging anatomical features containing a high concentration of chromophores. Haemoglobin is one of the dominant chromophores in biological tissue, with an absorption coefficient at least one order of magnitude higher than any other major endogenous chromophore in the 400–900 nm wavelength range (see Fig. 3). This makes the technique an excellent fit to imaging the microvasculature. Other endogenous chromophores have also been imaged, such as melanin, used to visualise the retinal-pigmented epithelium [32] (RPE), or lipids, which can allow assessing the lipid content of atherosclerotic plaques [7]. Photoacoustic imaging has also been used to image genetically expressed reporters [33], as well as exogeneous targeted contrast agents [34] for molecular imaging. These contrast agents include organic dyes such as indocyanine green (ICG) and methylene blue, or nanoparticles, which typically absorb in the visible and NIR range of the spectrum. In addition, the unique spectral signature of these endogenous and exogenous chromophores can potentially enable quantifying the concentration of a specific chromophore via spectroscopic techniques. For example, by acquiring photoacoustic images at multiple wavelengths and knowing the spectral signatures of oxy- and deoxy-haemoglobin, it is possible to determine blood SO_2 by applying spectral inversion methods [2].

The imaging depth of photoacoustic tomography is fundamentally limited by the effects of optical and acoustic attenuation. Although acoustic attenuation is significicant for most soft tissues, optical attenuation is the dominant mechanism limiting penetration depth. Optical attenuation determines how deep the excitation light can penetrate into the biological tissue; this in turn is determined by the tissue's absorption and scattering properties, which are wavelength dependent. Typically, in order

T. J. Allen

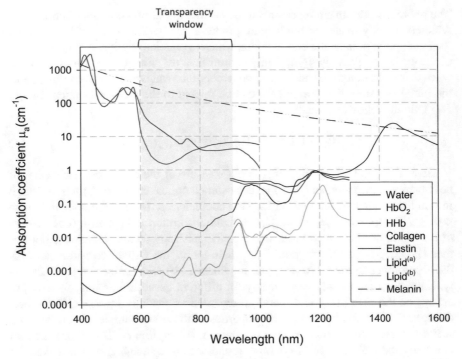

Fig. 3 Absorption coefficient spectra of endogenous tissue chromophores (oxyhaemoglobin (HbO$_2$), deoxyhaemoglobin (HHb) [35], water [36], lipid[a] [37], lipid[b] [38], melanin [39], collagen and elastin spectra [38])

to achieve a relatively large penetration depth (>5 mm), the emission wavelength of the excitation source is selected in the near-infrared (NIR) parts of the spectrum (600–900 nm) as here a transparency "window" exists due to the relatively low tissue scattering and the low absorption coefficients of water and blood.

Spatial resolution in photoacoustic tomography is directly related to the bandwidth of the recorded photoacoustic signal. In turn, the bandwidth is defined by the size of the optical absorber and can be very broad, with a frequency content extending to several tens of MHz, assuming the excitation pulse duration is selected to be short enough to not bandlimit the signal. As the photoacoustic signal propagates to the tissue surface, it suffers from frequency dependent acoustic attenuation in biological tissue, which bandlimits the signal. This effect fundamentally constrains the achievable spatial resolution, and is responsible for resolution decreasing with depth. For example, at a depth of one cm or more, the resolution is typically limited to the mm scale, for depths of several mm the resolution is typically on the order of a few hundreds of μm, whereas at a depth of a couple of mm or less, spatial resolution is limited to tens of μm. In addition to the above, resolution is also affected by other factors such as the frequency response and element size of the ultrasound detector, or the area over which the signals are recorded. These contributions are particularly relevant for shallow depths (<5 mm), as the detected signal is likely to be rather broadband, making it challenging to achieve an optimal detection bandwidth and spatial sampling requirements.

3 High-Power LEDs

An LED is a semiconductor device based on a p–n junction diode which can emit light when an electrical current flows through it. In essence, a P–N junction diode is formed by bringing two doped semiconductor materials in contact, an N-type material with an abundance of mobile electrons and few holes and a P-type material with an abundance of holes and few mobile electrons. When the junction is forward biased, by applying a voltage across the junction, holes (from the P-type material) and electrons (from the N-type material) are injected across the junction, resulting in an abundance of electrons and holes within the junction region. This high density of electrons and holes promotes strong electron–hole radiative recombination, resulting in a large number of photons being emitted. For LEDs, the radiative recombination process is based on spontaneous emission, where the injected electrons in a high energy state E_2 spontaneously make a downward transition to a lower energy state E_1 where they recombine with the injected holes; the energy difference ($hv = E_2 - E_1$) is then emitted as a photon. Spontaneous emission is random and photons may be emitted in any direction.

3.1 Characteristics of High-Power LEDs

3.1.1 Emission Wavelength

The choice of materials for designing an LED is primarily dictated by the desired emission wavelength, as the wavelength is determined by the energy bandgap ($E_g = E_2 - E_1$) of the materials involved. Therefore, by carefully selecting the semiconductor material, the emission wavelength of a high-power LED can be tailored to nearly anywhere in the visible to NIR range. The NIR range can be accessed using aluminum gallium arsenide ($Al_xGa_{1-x}As$), whose emission wavelength can be tuned over the 624–920 nm range by varying the mole fraction (x) of aluminium (Al). Aluminium gallium indium phosphide ($(Al_xGa_{1-x})_{0.5}In_{0.5}P$) can be used to access the 570–650 nm range of the visible spectrum; by increasing the mole fraction (x) of aluminium (Al), the emission wavelength is shifted from the longer to the shorter wavelengths. The 440–550 nm part can be accessed using indium gallium nitride (InGaN); as the content of indium (In) is increased, the emission wavelength shifts from the shorter to the longer wavelengths. However, the growth of high-quality InGaN becomes increasingly difficult with increasing indium content. As a result,

Table 1 Specifications of a range of commercially available high-power LEDs operating in CW mode

Model	Wavelength (nm)	Output power (W)	CW current rating (A)	Emitting area (mm^2)	Linewidth (FWHM) (nm)	Emission angle (°)	Typical lifetime (h)
M455D3 (Thorlabs Inc.)	455	1.445	1	1 × 1	18	80	>100,000
M530D3 (Thorlabs Inc.)	530	0.48	1	1 × 1	35	80	>100,000
M625D2 (Thorlabs Inc.)	625	0.92	1	1 × 1	17	80	>100,000
M850D3 (Thorlabs Inc.)	850	1.6	1.5	1 × 1	30	150	>10,000
SST-90 (Luminus Inc.)	623	4.2	6.3	9	19	100	>10,000

InGaN based devices are mainly used for ultraviolet (UV), blue and green LEDs, but rarely for achieving longer wavelengths.

3.1.2 Linewidth

The linewidth of an LED is relatively broad, on the order of tens of nm, and typically increases with the emission wavelength. For example, the linewidths of 455 nm and 850 nm devices are 18 nm and 30 nm, respectively (see Table 1, which shows the characteristics of a range of commercially available high-power LEDs). InGaN based devices show an exception to this behavior; they have a relatively broad linewidth when the indium content is high (e.g. green LEDs have a linewidth of around 35–39 nm). This can be attributed to the difficulties of growing the material, causing a greater incidence of defects and imperfections. For comparison, the linewidths of lasers commonly used for photoacoustic signal generation are on the order of a few nm or less.

3.1.3 Emission Angle

High-power LEDs have large emission angles, typically in excess of 80° (FWHM), as shown in Fig. 4. This, in combination with their relatively large (>1 mm^2) emitting areas, makes the emitted light difficult to collimate or to efficiently couple into an

Fig. 4 Typical spatial radiation distribution for a high-power LED (M850D3, Thorlabs Inc.) [40]

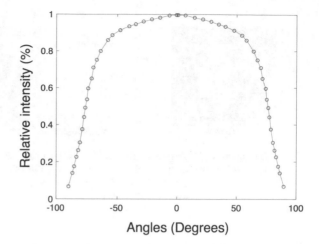

optical fibre. For photoacoustic applications, this typically requires the device to be placed in close proximity to the sample in order to achieve a relatively high fluence.

3.1.4 Output Power

High-power LEDs can provide an optical output up to a few Watts when operated in CW mode, depending on the emission wavelength and the size of the emitting element (see Table 1). The safe current rating is typically around 1 A or more, and is generally limited by the thermal damage threshold of the device, as the current causes heating within the substrate (junction temperature) due to Joule heating and non-radiative recombination processes. To maximise both the performance and lifetime of LEDs, they are typically mounted on a heat sink, removing some of the generated heat and, thus, lowering the temperature at the junction. Photographs of high-power LEDs are shown in Fig. 5, where Fig. 5a shows a device with a large emitting area (9 mm^2) and Fig. 5b shows a multi-wavelength device composed of 4 LED dies, each with a smaller emitting area (1 mm^2) and mounted on a metal-core printed circuit (MCPC) board for efficient heat removal.

3.1.5 Overdriving LEDs

Although high-power LEDs are typically specified only for CW operation, they can also be operated in pulsed mode. In pulsed mode, the amount of heat generated at the junction of the device depends not only on the peak current used to drive it, but also on its duty cycle. Therefore, if the duty cycle is kept low (e.g. <0.1%), it is possible to drive an LED with current pulses several times higher than their CW rating without the risk of exceeding the thermal damage threshold of the device. This allows achieving significantly higher output peak powers than what can be obtained

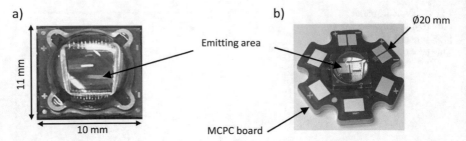

a)

11 mm

10 mm

Emitting area

b)

Ø20 mm

MCPC board

Fig. 5 Photographs of high-power LEDs; **a** high-power LED (SST-90) [41] with an emitting area of 9 mm^2; **b** high-power multi-wavelength LED (LZ4-00MC00, LedEngin, Inc.), composed of 4 LED dies emitting at 452, 520, 520, and 618 nm, each with a 1 mm^2 emitting area and mounted on a metal-core printed circuit (MCPC) board. These devices are encapsulated in spherical glass lenses

in CW mode. However, also in pulsed mode there is a limit as to how large the peak current pulses can be, which is a consequence of following three factors. First, the quantum efficiency of the device drops with increasing current. This is due to the fact that some of the injected carriers (electron and holes) are able to pass through the active region (junction) without recombining; the probability of this increases with current density. Eventually, as the drive current increases, the active region saturates, meaning that a further increase in injection current density does not increase the carrier concentration in the active region and therefore the emitted light. This drop in efficiency is illustrated in Fig. 6, which shows the output power of a pulsed high-power LED as a function of drive current (continuous blue curve) when operating at a duty cycle of 0.1%. It can be seen that the output power initially increases linearly, before increasing at a slower rate due to the drop in quantum efficiency. Second, operating an LED at excessively high drive currents can lead to faster ageing, as the high current density briefly increases the junction temperature, which strains

Fig. 6 Output power of a pulsed red LED (LXHL-PD09, Philips Lumileds) as a function of drive current (blue continuous curve) when operating at a duty cycle of 0.1%. The CW rating of the device is 1.4 A. (Data replotted from [42])

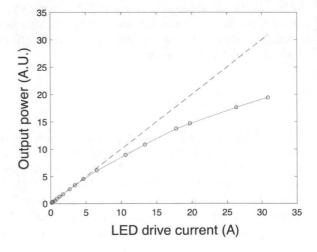

the crystal and causes defects within its lattice. This is an irreversible and gradual process, resulting in a reduction of light emission. Third, an excessive amount of current can cause the instantaneous failure of the device, such as the melting of the bond wires which connect the anode of the die to the voltage supply [42].

As mentioned above, LED manufacturers typically only specify the safe operational range of their devices for CW operation; little data is available for pulse mode operation, often leaving it to the user to empirically investigate the safe operational range of these devices when overdriven with current pulses that exceed their nominal rating. However, several studies [15, 16, 42, 43] have demonstrated the possibility of overdriving LEDs with current pulses ten times their rated current, leading to pulse energies of several µJ while operating a relatively low duty cycle (<0.1%), with no noticeable damage to the device.

To operate LEDs in pulsed mode, specific drivers are required. Pulsed LED drivers are most commonly composed of a capacitor, which is used as a storage element that discharges through the LED when a fast-switching device is activated [42–44]. Figure 7 shows a schematic representation of a simple driver circuit, composed of a capacitor C as a storage element and a fast-switching metal-oxide semiconductor field-effect transistor (MOSFET) T as the switching device. When the transistor is off, the capacitor charges to the voltage V_{cc} through the charging resistor r. When the transistor is turned on, via a MOSFET driver triggered by a signal generator, the capacitor discharges through the LED. It is also common practice to place a resistor R in series with the LED to limit the maximum peak current ($I = V_{cc}/R$) to avoid damage. To generate a quasi-squared pulse, the value of the capacitor is selected such that its exponential discharge is slow compared to the time delay between switching the transistor on and off. A current monitoring resistor with a low value (e.g. 0.01 Ω), when placed in series with the LED, can monitor the voltage drop across the

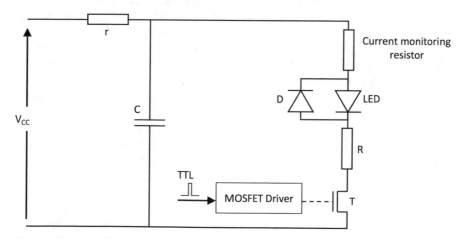

Fig. 7 Schematic of a typical LED driver for pulse operation mode. Vcc is the voltage provided by the power supply, T is a transistor used to switch the LED on and off, C is the storage capacitor, R is the limiting resistor, r is the charging resistor, D diode

resistor, hence to estimate the current that flows through the device. To protect the LED from potentially damaging negative transients due to the fast switching (ns rise time) of relatively large currents (>1 A), a diode D is connected across the LED. Generally, when designing LED driver circuits, it is important to keep any parasitic component introduced by the circuit components or physical layout to a minimum. For example, every inch of current carrying conductor adds approximately 20 nH of inductance to the circuit; this is problematic as it contributes to the rise time of the generated pulse.

4 Major Areas of Development in LED-Based Photoacoustic Imaging

In the following, four key areas of development in LED-based photoacoustic imaging will be discussed; single point measurements, tomography, spectroscopic imaging and novel excitation schemes. The aim of this section is to inform the reader of the major scientific achievements that have occurred, as well as provide an appreciation of how this field has developed within the past decade.

4.1 Single Point Measurements

The very first study [15] reporting the use of LEDs as a potential excitation source for the generation of photoacoustic signals for biomedical applications dates back to 2011. In this early experiment, a commercially available LED with an optical output power of 250 mW when driven at its nominal current rating (1 A) and emitting at a wavelength of 627 nm was used. The LED was overdriven by 40 times its nominal current, providing pulse energies of 400 nJ for a pulse duration of 60 ns and a PRF of 200 Hz. To generate and detect a photoacoustic signal with such a low pulse energy, a careful optimisation of the experimental setup was required. Importantly, the pressure amplitude of the generated photoacoustic signal was maximised by weakly focusing the light via an optical lens onto a highly absorbing gelatine based phantom. The generated photoacoustic signals were then detected using a focused ultrasound detector and signal averaged 50,000 times to improve the SNR. This study demonstrated, for the first time, the very possibility of generating photoacoustic signals using an LED. However, it was based on an unrealistic tissue phantom and hence did not yet show that LEDs could be used to generate photoacoustic signals in biological tissue. These initial limitations were overcome shortly afterwards via a study [16] that demonstrated the feasibility of generating a photoacoustic signal in a tissue mimicking phantom while illuminating a relatively large area (1 cm in diameter). In this experiment, a red (623 nm) high-power LED (CBT-120 from Luminus) was overdriven by 10 times its nominal current, providing pulse energies of 22 μJ for a

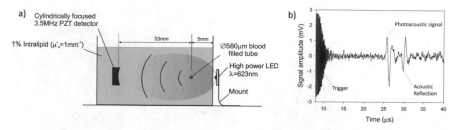

Fig. 8 Single point measurements in a tissue mimicking phantom [16]: **a** Experimental setup, composed of a single tube (0.58 mm in diameter) filled with human blood (35% hematocrit) and immersed to a depth of 5 mm in a solution of intralipid, mimicking the scattering ($\mu'_s = 1$ mm^{-1}) properties of biological tissue. **b** Recorded time domain signal. A reflection of the photoacoustic signal can be observed at t = 30 μs. This is caused by the mismatch in acoustic impedance at the interface between the intralipid and the wall of the tank. (Figures adapted with permission from [16])

pulse duration of 500 ns and a PRF of 200 Hz. The LED was used to illuminate the phantom from one side (see Fig. 8a), with an incident beam of approximately 1 cm in diameter. An ultrasound detector was placed on the opposite side of the LED to capture the generated photoacoustic signals. The detector was a cylindrically focused PZT detector (3.5 MHz, V383 Panametric) of focal length 33 mm, and was placed such that the blood-filled tube was in its focus. The detected photoacoustic signal, which is shown in Fig. 8b, was signal averaged 1000 times, leading to an SNR of approximately 13.

4.2 Photoacoustic Tomography

Following on from these early single point measurements, it was subsequently investigated whether high-power LEDs were suitable excitation sources for photoacoustic tomography of tissue mimicking phantoms [20]. To demonstrate this, a red (623 nm) high-power LED (SST-90 from Luminus) was overdriven by 20 times its nominal current, providing pulse energies of 9 μJ for a pulse duration of 200 ns, a PRF of 500 Hz. As shown in Fig. 9a, the LED was used to illuminate a tissue phantom composed of three tubes filled with human blood and immersed in a solution of intralipid, mimicking the scattering ($\mu'_s = 1$ mm^{-1}) properties of biological tissue. The beam incident on the phantom was approximately 1 cm in diameter. The generated photoacoustic signals were detected using a cylindrical focus 3.5 MHz PZT detector, which was located on the opposite side of the LED, i.e. operating in forward mode. The tubes were placed at the focus of the detector, whose focal length was 33 mm, and perpendicular to the imaging plane. The tubes were fixed in a mount which was rotated over a total range of 360 degrees in steps of 0.9° using a stepper motor. At each step, a photoacoustic signal was recorded and signal averaged 5000 times. In Fig. 9b, which shows the recorded time domain signals as a function of scan angle, it

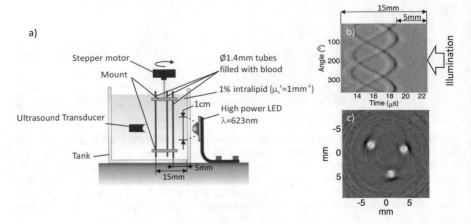

Fig. 9 Photoacoustic tomography of a tissue mimicking phantom [20]: **a** Experimental setup, consisting of three 1.4 mm diameter tubes filled with human blood (35% hematocrit) and immersed in a 1% solution of intralipid in order to mimicked the scattering ($\mu'_s = 1$ mm^{-1}) properties of biological tissue. **b** Recorded time domain signal. **c** Reconstructed photoacoustic image. (Figure adapted with permission from [20])

can be seen that photoacoustic signals generated at depth of up to 1.5 cm are clearly visible. Figure 9c shows the reconstructed photoacoustic image where all three tubes are visible.

Around the same time, further studies [17–19, 21, 22] into the possibility of using high-power LEDs for photoacoustic tomography were carried out by PreXion Corporation. These experiments were based on the development of two LED arrays, emitting at 850 nm, and placed on each side of an ultrasound linear array. Their aim was to a large extent to demonstrate that their custom-built source could generate detectable photoacoustic signals in high absorbers such as a biopsy needle [17–19], or in a relatively high concentration of ICG [21] injected in a young dead mouse. While these studies where somewhat limited in that they were based on absorption coefficients significantly higher that of the chromophores typically found in biological tissue, they have led to the development of a commercial LED based photoacoustic imaging system (AcousticX, initially developed by Prexion Corporation, Tokyo, Japan and since been bought by Cyberdyne Inc., Tsukuba, Japan). The system is composed of two LED bars, placed on each side of an ultrasound linear array. Each bar is composed of 144 LEDs (4 columns of 36 elements) and measures 12.4 mm × 86.5 mm. The pulse duration can be varied from 50 to 150 ns, and the PRF from 1 to 4 kHz. The pulse energy for each of the bars can be up to 200 μJ when operating at 4 kHz, a pulse duration of 100 ns and emitting at 850 nm. The bars can be designed to accommodate for a single wavelength or a combination of two wavelengths, and the wavelengths can be selected from a wide range (e.g. 470, 520, 620, 660, 690, 750, 820, 850, 940 or 980 nm). This commercial system has recently been characterized [24, 45–48], and its ability to acquire in-vivo images of the superficial microvasculature was demonstrated. For example, one such study demonstrated the possibility of imaging

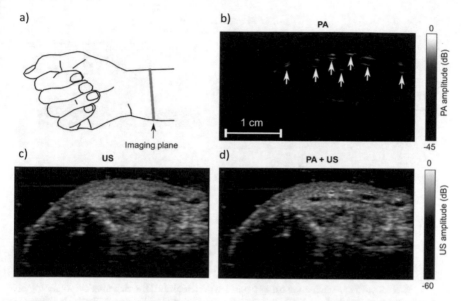

Fig. 10 In-vivo photoacoustic imaging of a human wrist [24]: **a** Schematic of a human wrist, indicating the region being imaged, **b** photoacoustic image (the white arrows indicating the presences of blood vessels), **c** ultrasound image and **d** a combined image showing both the photoacoustic and ultrasound image. (Figure adapted with permission from [24])

superficial blood vessels in a human wrist [24]. Figures 10 a, b shows respectively a schematic of the area being imaged and the achieved photoacoustic image, visualising a range of superficial blood vessels (indicated by arrows) with a depth penetration of at least 5 mm. The linear ultrasound array used to detect the photoacoustic signals can also operate in standard pulse echo mode, allowing to acquire not only a photoacoustic image but also a co-registered ultrasound image. Figure 10c, d show respectively the ultrasound images and the overlapping ultrasound and photoacoustic images.

4.3 Photoacoustic Spectroscopy

The fact that LEDs are available with a wide range of wavelengths (see Fig. 1) and the ability of creating compact arrays of devices emitting at different wavelengths make them well suited for photoacoustic spectroscopy applications. The ability of acquiring spectroscopic data with multi-wavelength LEDs was demonstrated first in a range of studies [16, 22, 49] based on generating photoacoustic signals in relatively high absorbers immersed in water, then via a study generating photoacoustic signals in a tissue mimicking phantom [20]. The latter was based on an LED composed of four wavelengths (Fig. 11a) to generate and detect photoacoustic signals in tubes (2.4 mm diameter) filled with human blood (35% haematocrit) immersed in a water

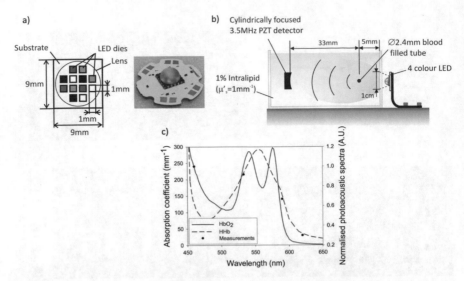

Fig. 11 Photoacoustic spectroscopy using a multi-wavelength LED: **a** Schematic and photograph of the device, composed of 12 LED dies emitting at 460, 523, 590, and 623 nm. **b** Schematic of the experimental setup. **c** Absorption spectra of oxy- and deoxyhaemoglobin overlaid with the photoacoustic spectra obtained from the blood-filled tube using the four wavelength device. (Figures adapted with permission from [20])

solution of intralipid. The experimental setup is shown in Fig. 11b. The phantom was illuminated sequentially by each of the four wavelengths. The peak amplitudes of the detected photoacoustic signals were normalised to their respective pulse energies and plotted as a function of wavelength alongside the absorption spectra of oxy- and deoxy-haemoglobin (Fig. 11c), showing that the relative trend in the measured photoacoustic spectra is broadly consistent with the absorption spectra of oxy- and deoxy-haemoglobin.

A more recent study explored the feasibility of measuring blood oxygenation levels in a superficial vessel of the index finger [45]. The experiment was conducted with the LED-based photoacoustic imaging system developed by Cyberdyne Inc.; this system is based on a pair of dual-wavelength LED bars that can emit at 690 and 850 nm. Photoacoustic images were acquired at both wavelengths and used to compute the blood oxygenation level in the finger, which was then compared to the oxygenation value obtained via a pulse oximeter, yielding a broad agreement. These studies can be considered a significant step towards LED-based photoacoustic spectroscopy.

4.4 Novel Excitation Schemes

So far, most studies investigating LEDs as an excitation source for photoacoustic imaging had been based on conventional, pulsed excitation methods. However, LEDs have the advantage over traditional excitation sources such as Q-switched Nd:YAG lasers that they can be arbitrarily modulated, allowing for a range of excitation schemes to be implemented, such as those that have been used with laser diodes [50–52]. One such example [20] consisted in a scheme by which several LEDs emitting at different wavelengths were first driven at once, so as to obtain a photoacoustic signal with an improved SNR, and then each of them was driven separately one after the other. The power spectrum of the photoacoustic signal generated by all wavelengths simultaneously was used to design a filter that, when applied to the sequentially acquired, individual wavelength signals, would improve the SNR of those. The validity of this approach is based on the assumption that the temporal shape of the photoacoustic signals generated at each individual wavelength are identical to each other, which is reasonable when the light penetration depth is significantly larger than the size of the absorber.

Another example is to use coded excitation schemes, which are based upon coded binary sequences (e.g. pseudo-random codes). Coded excitation has been shown to lead to an increased SNR compared to conventional single pulse excitation methods; specifically, it has been shown that the SNR of the photoacoustic signal increase as a function of $\sqrt{N}$, where N is the number of bits within the code. In addition, if two different coded binary sequences are selected that have a very small correlation between them, it is possible to interrogate the sample with LEDs emitting at two different wavelengths simultaneously, by driving the LEDs with these sequences. Besides improving SNR, this increases the acquisition speed of multi-wavelength data compared to sequential excitation approaches. The reduction of acquisition time was demonstrated experimentally [20], by showing that photoacoustic signals can be acquired at two wavelengths simultaneously when using orthogonal codes (so-called Golay codes) to drive two LEDs emitting at different wavelengths. In general, photoacoustic signals obtained via coded schemes must be decoded by taking the cross-correlation between them and the input binary sequence(s), before being suitable for further processing (e.g. tomographic reconstruction).

Although experimental studies on alternative, non-sequential excitation schemes are still in their infancy, they illustrate the driving flexibility that high power LEDs provide, and the potential this could bear for increasing SNR and, hence, the suitability of LEDs for photoacoustic applications.

5 Summary and Outlook

In the above—following a brief introduction to photoacoustic tomography and key characteristics of high-power LEDs—an overview of LED-based photoacoustic imaging has been given, covering four key areas of development, which are single point measurements, tomography, spectroscopy and novel excitation schemes.

In summary, early studies were based on single point measurements and focused on demonstrating the possibility of detecting photoacoustic signals generated by an LED, first in a high absorber, then in a tissue mimicking phantom. This was achieved by overdriving these devices by up to 40 times their rated current, and exploiting their high PRF to acquire and signal average many signals over a short period of time, in order to overcome the low peaks powers they provide. Following on from these results, a range of studies have demonstrated the possibility of using LEDs as an excitation source for widefield photoacoustic tomography. Experiments were first conducted in phantoms based on high absorbers before moving to tissue mimicking phantom and in-vivo studies. One example of the latter reported the ability of imaging the microvasculature of the human wrist or finger with depth penetrations of up to 5 mm [24]. The ability of using multi-wavelength LED arrays to acquire spectroscopic data was also investigated. One such study reported that by acquiring images at multiple wavelengths and employing spectroscopic processing techniques, the blood oxygenation levels in a superficial vessel of the index finger could be measured [45]. All of the above was complemented by the implementation of novel excitation schemes, taking advantage of the greater driving flexibility that LEDs provide. The first such study provided a strategy to improve the SNR of photoacoustic signals when acquiring spectroscopic data. The second study demonstrated the possibility to improve the acquisition speed of multi-wavelength imaging by using Golay codes to drive the LEDs (as opposed to conventional sequential acquisition of photoacoustic signals at multiple wavelengths).

Based on these results, and considering the characteristics of high-power LEDs in terms of their emission wavelength, pulse energies, and driving flexibility, it may well be concluded that these devices are suited to superficial vascular imaging where pulse energy requirements are relatively modest. In addition, the compact size of these devices as well as their low cost and reliability (they require practically no maintenance) make them well suited to achieving the long-sought translation of photoacoustic imaging from a laboratory technique to being deployable into a clinical environment. Compact, hand-held LED based systems could be developed for applications such as the imaging of arthritis, skin pathologies or oximetry type measurements.

Looking to the future, it is reasonable to anticipate that advances in high-power LED technology, which are mainly driven by the lighting industry, as well as novel signal processing schemes will increase the practical utility of LEDs in the context of photoacoustic imaging. Further progress may arise from the ability of optimising the light delivery. For example, it may be possible to weakly collimate the light on to the sample in order to avoid an excessively large illumination area. This would result in an improvement in the SNR of the generated photoacoustic signal due to the increase in fluence. It may also provide a greater flexibility in the design of photoacoustic imaging systems, as the LED would no longer need to be placed close to the tissue sample, which is currently required due to the large divergence of emitted light. In addition, if sufficient focusing can be achieved these devices may also be suitable for imaging modalities such as photoacoustic endoscopy or acoustic resolution photoacoustic microscopy [53], which require relatively modest pulse energies. On the other hand, for optical resolution photoacoustic microscopy, the use of LEDs is likely to be challenging, as it would be difficult to achieve the necessary micron scale diffraction limited spot sizes [54, 55].

Image reconstruction and enhancement techniques which are suitable for LED-based photoacoustic imaging are included in the next section of this book.

Acknowledgements The author would like to thank Olumide Ogunlade for his informative comments.

References

1. P. Beard, Biomedical photoacoustic imaging. Interface Focus **1**(4), 602–631 (2011)
2. B. Cox, J.G. Laufer, S.R. Arridge, P.C. Beard, Quantitative spectroscopic photoacoustic imaging: a review. J. Biomed. Opt. **17**(6), 061202 (2012)
3. P.J. van den Berg, K. Daoudi, W. Steenbergen, Review of photoacoustic flow imaging: its current state and its promises. Photoacoustics **3**(3), 89–99 (2015)
4. J. Brunker, P. Beard, Velocity measurements in whole blood using acoustic resolution photoacoustic Doppler. Biomed. Opt. Express **7**(7), 2789–2806 (2016)
5. I.V. Larina, K.V. Larin, R.O. Esenaliev, Real-time optoacoustic monitoring of temperature in tissues. J. Phys. D. Appl. Phys. **38**(15), 2633–2639 (2005)
6. M. Liu, Z. Chen, B. Zabihian, C. Sinz, E. Zhang, P.C. Beard, L. Ginner, E. Hoover, M.P. Minneman, R.A. Leitgeb, H. Kittler, W. Drexler, Combined multi-modal photoacoustic tomography, optical coherence tomography (OCT) and OCT angiography system with an articulated probe for in vivo human skin structure and vasculature imaging. Biomed. Opt. Express **7**(9), 3390 (2016)
7. T.J. Allen, A. Hall, A.P. Dhillon, J S. Owen, P.C. Beard, Spectroscopic photoacoustic imaging of lipid-rich plaques in the human aorta in the 740 to 1400 nm wavelength range. J. Biomed. Opt. **17**(6), 061209 (2012)
8. S.S.S. Choi, A. Mandelis, Review of the state of the art in cardiovascular endoscopy imaging of atherosclerosis using photoacoustic techniques with pulsed and continuous-wave optical excitations. J. Biomed. Opt. **24**(08), 1 (2019)
9. K.S. Valluru, J.K. Willmann, Clinical photoacoustic imaging of cancer. Ultrasonography **35**(4), 267–280 (2016)
10. A. Krumholz, L.L.V. Wang, J. Yao, Functional photoacoustic microscopy of diabetic vasculature. J. Biomed. Opt. **17**(6), 060502 (2012)
11. J.R. Rajian, X. Shao, D.L. Chamberland, X. Wang, Characterization and treatment monitoring of inflammatory arthritis by photoacoustic imaging: a study on adjuvant-induced arthritis rat model. Biomed. Opt. Express **4**(6), 900–908 (2013)
12. R.G.M. Kolkman, W. Steenbergen, T.G. van Leeuwen, In vivo photoacoustic imaging of blood vessels with a pulsed laser diode. Lasers Med. Sci. **21**(3), 134–139 (2006)
13. T.J. Allen, P.C. Beard, Pulsed near-infrared laser diode excitation system for biomedical photoacoustic imaging. Opt. Lett. **31**(23), 3462–3464 (2006)
14. K. Daoudi, P.J. van den Berg, O. Rabot, A. Kohl, S. Tisserand, P. Brands, W. Steenbergen, Hand-held probe integrating laser diode and ultrasound transducer array for ultrasound/photoacoustic dual modality imaging. Opt. Express **22**(21), 26365 (2014)
15. R. Skov Hansen, Using high-power light emitting diodes for photoacoustic imaging, in *Proceedings of SPIE*, vol. 7968 (2011), p. 79680A

16. T.J. Allen, P.C. Beard, Light emitting diodes as an excitation source for biomedical photoa-coustics, in *Proceedings of SPIE*, vol. 8581, 9 Mar 2013, p. 85811F
17. T. Agano, N. Sato, H. Nakatsuka, K. Kitagawa, T. Hanaoka, K. Morisono, Y. Shigeta, Develop-ment of environmentally friendly LED light source module for photoacoustic imaging system. Light. Diodes Mater. Devices Appl. Solid State Light. XIX, **9383**, 93831D (2015)
18. T. Agano, N. Sato, H. Nakatsuka, K. Kitagawa, T. Hanaoka, K. Morisono, and Y. Shigeta, Comparative experiments of photoacoustic system using laser light source and LED array light source. Photons Plus Ultrasound Imaging Sens. **9323**, 93233X (2015)
19. T. Agano, N. Sato, H. Nakatsuka, K. Kitagawa, T. Hanaoka, K. Morisono, and Y. Shigeta, Attempts to increase penetration of photoacoustic system using LED array light souce. Photons Plus Ultrasound Imaging Sens. **9323**, 93233Z (2015)
20. T.J. Allen, P.C. Beard, High power visible light emitting diodes as pulsed excitation sources for biomedical photoacoustics. Biomed. Opt. Express **7**(4), 1260 (2016)
21. T. Agano, N. Sato, Photoacoustic imaging system using LED light source, in *Conference on Lasers and Electro-Optics*, (2016), pp. 1–2
22. T. Agano, N. Sato, H. Nakatsuka, K. Kitagawa, T. Hanaoka, K. Morisono, Y. Shigeta, C. Tanaka, High frame rate photoacoustic imaging using multiple wave-length LED array light source. Photons Plus Ultrasound Imaging Sens. **9708**, 97084E (2016)
23. N. Sato, Y. Shigeta, M. Kuniyil Ajith Singh, T. Agano, Multispectral photoacoustic character-ization of ICG and porcine blood using an LED-based photoacoustic imaging system (2018), p. 129
24. W. Xia, M. Kuniyil, A. Singh, E. Maneas, N. Sato, A.E. Desjardins, Handheld real-time LED-based photoacoustic and ultrasound imaging system for accurate visualization of clinical metal needles and superficial vasculature to guide minimally invasive procedures. Sensors **18**(5), 1394 (2018)
25. T.J Allen, B.T. Cox, P.C Beard, Generating photoacoustic signals using high-peak power pulsed laser diodes, in *Proceedings of SPIE*, vol. 5697, (2005), pp. 233–242
26. O. Ogunlade, J.J. Connell, J.L. Huang, E. Zhang, M.F. Lythgoe, D.A. Long, P. Beard, In vivo three-dimensional photoacoustic imaging of the renal vasculature in preclinical rodent models. Am. J. Physiol. Physiol. **314**(6), F1145–F1153 (2018)
27. C. Hoelen, F. De Mul, Image reconstruction for photoacoustic scanning of tissue structures. Appl. Opt. **39**(31), 5872–5883 (2000)
28. B.E. Treeby, B.T. Cox, k-wave: MATLAB toolbox for the simulation and reconstruction of photoacoustic wave fields. J. Biomed. Opt. **15**(2), 021314 (2010)
29. K.P.K.P. Köstli, P.C. Beard, Two-dimensional photoacoustic imaging by use of fourier-transform image reconstruction and a detector with an anisotropic response. Appl. Opt. **42**(10), 1899–1908 (2003)
30. E. Zhang, J. Laufer, P. Beard, Backward-mode multiwavelength photoacoustic scanner using a planar Fabry-Perot polymer film ultrasound sensor for high-resolution three-dimensional imaging of biological tissues. Appl. Opt. **47**(4), 561–577 (2008)
31. D. Yao, C. Zhang, K. Maslov, L.V. Wang, Photoacoustic measurement of the Grüneisen parameter of tissue. J. Biomed. Opt. **19**(1), 017007 (2014)
32. T. Liu, Q. Wei, W. Song, J.M. Burke, S. Jiao, H.F. Zhang, Near-infrared light photoacoustic ophthalmoscopy. Biomed. Opt. Express **3**(4), 792–799 (2012)
33. J. Brunker, J. Yao, J. Laufer, S.E. Bohndiek, Photoacoustic imaging using genetically encoded reporters: a review. J. Biomed. Opt. **22**(7), 070901 (2017)
34. J. Weber, P.C. Beard, S.E. Bohndiek, Contrast agents for molecular photoacoustic imaging. Nat. Methods **13**(8), 639–650 (2016)
35. Absorption Coefficient of Oxyhaemoglobin and Deoxyhaemoglobin. Available, https://omlc. ogi.edu/spectra/hemoglobin/summary.html
36. G.M. Hale, M.R. Querry, Optical constants of water in the 200-nm to 200-microm wavelength region. Appl. Opt. **12**(3), 555–563 (1973)
37. R.L.P. van Veen, H.J.C.M. Sterenborg, A. Pifferi, A. Torricelli, E. Chikoidze, R. Cubeddu, Determination of visible near-IR absorption coefficients of mammalian fat using time- and

spatially resolved diffuse reflectance and transmission spectroscopy. J. Biomed. Opt. **10**(5), 054004 (2011)

38. C. Tsai, J. Chen, W. Wang, Near-infrared absorption property of biological soft tissue constituents. J. Med. Biol. Eng. **21**(1), 7–13 (2001)

39. Absorption Coefficient of Melanin. Available, https://omlc.ogi.edu/spectra/melanin/mua.html

40. Thorlabs Inc., *Technical Product Data Sheet, MCPCB-Mounted LED M850D3, 850 nm* (2018)

41. Luminus Devices Inc., *Technical Product Data Sheet, red LED SST-90* (2015), pp. 1–14

42. C. Willert, B. Stasicki, J. Klinner, S. Moessner, Pulsed operation of high-power light emitting diodes for imaging flow velocimetry. Meas. Sci. Technol. **21**(7), 075402 (2010)

43. C. Willert, S. Moessner, J. Klinner, Pulsed operation of high-power light emitting diodes for imaging flow velocimetry, vol. 21, no. 7 (2009), p. 075402

44. H. Zhong, T. Duan, H. Lan, M. Zhou, F. Gao, Review of low-cost photoacoustic sensing and imaging based on laser diode and light-emitting diode. Sensors **18**(7), 20–22 (2018) (Switzerland)

45. Y. Zhu, G. Xu, J. Yuan, J. Jo, G. Gandikota, H. Demirci, T. Agano, N. Sato, Y. Shigeta, X. Wang, Light emitting diodes based photoacoustic imaging and potential clinical applications. Sci. Rep. **8**(1), 9885 (2018)

46. E. Maneas, W. Xia, M. Kuniyil Ajith Singh, N. Sato, T. Agano, S. Ourselin, S.J. West, A.L. David, T. Vercauteren, A.E. Desjardins, Human placental vasculature imaging using an LED-based photoacoustic/ultrasound imaging system, in *Proceedings of SPIE*, vol. 10494 (2018)

47. J. Jo, G. Xu, Y. Zhu, M. Burton, J. Sarazin, E. Schiopu, G. Gandikota, X. Wang, Detecting joint inflammation by an LED-based photoacoustic imaging system: a feasibility study. J. Biomed. Opt. **23**(11), 1 (2018)

48. A. Hariri, J. Lemaster, J. Wang, A.S. Jeevarathinam, D.L. Chao, J.V. Jokerst, The characterization of an economic and portable LED-based photoacoustic imaging system to facilitate molecular imaging. Photoacoustics **9**, 10–20 (2018)

49. Y. Adachi, T. Hoshimiya, Photoacoustic imaging with multiple-wavelength light-emitting diodes. Jpn. J. Appl. Phys. **52**, 07HB06 (2013)

50. S.-Y. Su, P.-C. Li, Coded excitation for photoacoustic imaging using a high-speed diode laser. Opt. Express **19**(2), 1174–1182 (2011)

51. M.F. Beckmann, M.P. Mienkina, G. Schmitz, C.S. Friedrich, N.C. Gerhardt, M.R. Hofmann, Monospectral photoacoustic imaging using Legendre sequences, in *Ultrasonics Symposium (IUS), 2010 IEEE*, no. 2 (2010), pp. 386–389

52. M.P. Mienkina, C.-S. Friedrich, N.C. Gerhardt, M.F. Beckmann, M.F. Schiffner, M.R. Hofmann, G. Schmitz, Multispectral photoacoustic coded excitation imaging using unipolar orthogonal Golay codes. Opt. Express **18**(9), 9076–9087 (2010)

53. X. Dai, H. Yang, H. Jiang, In vivo photoacoustic imaging of vasculature with a low-cost miniature light emitting diode excitation. Opt. Lett. **42**(7), 1456 (2017)

54. T.J. Allen, O. Ogunlade, E.Z. Zhang, P.C. Beard, Large area laser scanning optical resolution photoacoustic microscopy using a fibre optic sensor. Biomed. Opt. Express **9**(2), 2117–2120 (2018)

55. T.J. Allen, J. Spurrell, M.O. Berendt, O. Ogunlade, S.U. Alam, E.Z. Zhang, D.J. Richardson, P.C. Beard, Ultrafast laser-scanning optical resolution photoacoustic microscopy at up to 2 million A-lines per second. J. Biomed. Opt. **23**(12), 1 (2018)

Image Enhancement
and Reconstruction Techniques

Deformation-Compensated Averaging for Deep-Tissue LED and Laser Diode-Based Photoacoustic Imaging Integrated with Handheld Echo Ultrasound

Michael Jaeger, Hans-Martin Schwab, Yamen Almallouhi, Celine Canal, Maike Song, Vincent Sauget, David Sontrop, Theo Mulder, Paul Roumen, Arno Humblet, Martin Frenz, and Peter Brands

Abstract Averaging is a fundamental necessity for deep photoacoustic (PA) imaging when using low-energy pulsed laser sources or LED's. Intrinsic (breathing, heartbeat…) or extrinsic (freehand probe guidance) tissue motion, however, leads to phase cancellation of the averaged PA signal when the axial displacement of tissue becomes larger than half the acoustic wavelength at the probe's centre frequency. Motion-compensated averaging (DCA) is a solution to this problem, and allows the detection of deep structures that are else not visible. In a combined PA and echo-ultrasound (US) system, tissue motion can be quantified in US images that are interleaved with PA images. In this chapter, we exemplarily illustrate the power of this technique when trying to image the optical absorption inside the carotid artery, using a fully integrated PA/US system based on a handheld clinical probe containing a miniaturised laser source. The key components of DCA are discussed and exemplified on volunteer data, and the influence of various parameters on image contrast is investigated. We demonstrate that DCA enables freehand PA detection of blood

M. Jaeger · M. Frenz (✉)
Institute of Applied Physics, University of Bern, Bern, Switzerland
e-mail: martin.frenz@iap.unibe.ch

M. Jaeger
e-mail: michael.jaeger@iap.unibe.ch

H.-M. Schwab
Cardiovascular Biomechanics, Eindhoven University of Technology, Eindhoven, The Netherlands

Y. Almallouhi · C. Canal
Lumibird, Les Ulis Cedex, France

M. Song
BrightLoop Converters, Paris, France

V. Sauget
SILIOS Technologies, Peynier, France

D. Sontrop · T. Mulder · P. Roumen · A. Humblet · P. Brands
Esaote Europe B.V., Maastricht, The Netherlands

© Springer Nature Singapore Pte Ltd. 2020
M. Kuniyil Ajith Singh (ed.), *LED-Based Photoacoustic Imaging*,
Progress in Optical Science and Photonics 7,
https://doi.org/10.1007/978-981-15-3984-8_3

47

vessels at a depth of 1.5 cm using only 2 mJ pulse energy, and give some guidelines for image interpretation.

1 Introduction

One of the promising application areas of photoacoustic (PA) imaging is its integration with clinical handheld ultrasonography [1, 2], to complement classical B-mode and colour flow imaging, and more recently elastography [3] and speed-of-sound imaging [4–8], with new valuable diagnostic information in a single multi-modal handheld system. For such a system being flexible and widely affordable, the pulsed light source is preferably integrated in the handheld probe itself. For this purpose, various groups and companies have developed light emitting diode (LED) and laser diode (LD) based miniaturised light sources [9–17]. So far, these systems have in common that the pulse energy is very low, compared to the more commonly used—but bulky and expensive—external solid-state lasers. For deep PA imaging where SNR becomes an important issue due to optical attenuation, the low pulse energy can partially be compensated for by increasing the pulse repetition frequency (prf) together with more extensive averaging. Laser safety regulations, however, limit the average irradiated power per unit area, so that—for a given total averaging time—the SNR (which is proportional to the *square-root* of the number of pulses) decreases with increasing prf due to the *linearly* decreasing maximum permissible pulse energy. Put differently, the lower the pulse energy, the longer the averaging time required to achieve a target SNR. This makes averaging substantially more important for low energy PA systems than for the ones using high-energy solid-state lasers.

Especially for deep imaging where longer averaging times are required than for superficial imaging, averaging becomes more challenging owing to tissue motion. On one hand, motion of tissue relative to the probe aperture occurs due to involuntary probe motion. On the other hand, the tissue exhibits intrinsic motion even when the probe is static, due to pulsating arteries, the beating heart or breathing, among others. With the 7.5–15 MHz centre frequencies that are typically used for high-resolution US imaging of a few cm depth range, the displacement magnitude of intrinsic tissue motion can easily exceed half an US wavelength. As a result, conventional averaging leads to phase cancellation of the PA signals, limiting the maximum averaging time up to which an SNR improvement is possible.

A solution to this problem is motion-compensated averaging, or—as previously named—displacement-compensated averaging (DCA) [18–21]. This technique takes benefit of the interleaved acquisition of pulse-echo data with PA data, which allows to estimate the tissue motion by tracking anatomical details in US images, and—subsequently—to motion-compensate PA images before averaging. DCA has originally been proposed for reducing clutter noise in PA imaging, along with other clutter-reduction techniques [22–27]. Clutter consists of PA echoes and out-of-plane PA signals, and it is a prominent noise source especially in reflection-mode PA imaging where it cannot be temporally separated from "real" direct in-plane PA signals [28,

29]. Since clutter is a systematic noise, it cannot be removed by conventional averaging, and thus poses an ultimate limit to imaging depth. DCA takes benefit of the fact that, upon tissue deformation, clutter behaves differently than the "real" signals, as the apparent reconstructed location of clutter does not coincide with the actual source location (else it would be "real" signal). Due to this different behaviour, clutter tends to decorrelate along a motion-compensated PA image sequence, thus the clutter intensity level can be reduced by averaging. In solid-state laser PA imaging where clutter is more prominent that thermal noise, we have demonstrated that DCA substantially improves contrast and imaging depth.

In low-pulse-energy deep PA imaging where thermal noise is more prominent than clutter, the main benefit of DCA is that the motion compensation allows for more extensive averaging and thus improved SNR by reducing the effect of phase cancellation. In an ideal case where the tracking is perfectly accurate and no out-of-plane motion occurs, at least the same SNR can be achieved as if tissue motion would be absent. In a more realistic scenario, however, decorrelation of US echoes and out-of-plane motion results in tracking and compensation errors. In addition, out-of-plane motion leads to decorrelation of PA image features so that phase cancellation can occur even with accurate motion compensation. For this reason, the optimum SNR is obtained in a trade-off between averaging time, tracking errors and out-of-plane motion. Depending on intrinsic tissue displacement magnitude and complexity of tissue structures (slipping boundaries and architectural anisotropy leading to 3D motion field), the sweet spot in this trade-off limits the achievable SNR. A further limitation to averaging time stems from the necessity of real-time feedback: the effective frame rate is given by the averaging time constant. Above a couple of seconds, the lag between freehand probe guidance and the effect on the DCA result makes it difficult for an operator to choose a probe placement that optimises the DCA result.

Along these lines, this chapter is dedicated to the elaboration of the various components and features of DCA and the investigation of their influence, exemplified on a specific implementation for an LD-based fully integrated handheld PA/US probe. In Sect. 2, we focus on the design of the main component of DCA, namely the motion tracking of US images. The tracking algorithm needs to be fast as well as robust, and its specific implementation is dictated by limitations of the specific acquisition system. In Sect. 3, we detail the experimental setup and processing steps, including a novel way of how to overlay the PA signal with the US image. In Sect. 4 we illustrate the various steps in the DCA processing in volunteer results, with a focus on the role of various parameters that influence its performance. To make the benefit of DCA most evident, we focus on the detection of the carotid artery which shows a large intrinsic pulsatile motion making it especially difficult to image. In addition we give more general experience on how to interpret DCA images, especially on how to identify real PA signal (as opposed to clutter) based on the US image. Our results demonstrate that detection of the carotid artery and other blood vessels at a depth of 1.5 cm is feasible using only 2 mJ pulse energy (80 Hz prf, 2.5 s averaging time), and they confirm that DCA is essential for achieving this imaging depth, not only by allowing effective averaging but also via its clutter-reducing effect. These results are especially important for LED-based PA imaging where the pulse energy is substantially lower.

2 DCA Prerequisites

2.1 US Image Quality

The motion tracking accuracy is crucial to the achievable DCA outcome. Tracking errors can stem on one hand from noise in the US images, on the other hand from imperfections of the tracking algorithm. First, we put a focus on the US image noise. Apart from thermal noise, US images contain clutter noise the same as PA images do. Clutter noise consists of system-related artifacts (such as side-lobes, grating lobes due to below-Nyquist sampling of the element-to-element pitch …) but also of tissue-related noise caused by higher-order echoes (multiply scattered US), which cannot be distinguished from the first-order echoes that make up a "clean" US image. Since the echogenicity (echo strength) of tissue varies on a large dynamic range (tens of dB), low echogenicity areas are easily dominated by clutter spreading from high echogenicity areas. The same as in PA, clutter noise in US tends to decorrelate with tissue deformation, leading to wrong detection of the echo shift in regions where clutter dominates. For this reason, a key point of attention in DCA is the optimisation of US image quality in terms of signal-to-clutter ratio (SCR).

In a classical line-by-line scan (LLS), US power is transmitted into a narrow collimated or slightly focused beam at a time, and the probe receives a dynamically focused signal (conventionally using delay-and-sum) from inside that beam. The time trace of the signal forms one image "line", and multiple lines are obtained when scanning the tissue with the beam and together form an image. With the advances of hardware development of the past decade, plane-wave (PW) (or ultrafast) US imaging has become popular [30], where a single plane US pulse is transmitted into the tissue and signals are digitized simultaneously on all elements on receive, allowing for the reconstruction of a large field-of-view (FoV) image in a single shot. While PW imaging has a great speed advantage over LLS, the big disadvantage is the much higher clutter noise level, making this type of image unsuitable for motion tracking. Figure 1a, b show an LLS and a PW image of the same region around the carotid artery, demonstrating the substantial difference in contrast. Especially note the increased apparent echogenicity inside the carotid in Fig. 1b.

The increased clutter noise in PW imaging as compared to LLS consists of diffuse 2nd order echoes. In the LLS, the US power is transmitted into a narrow beam. The 1st order echoes are generated only inside that beam, but 2nd order scattering leads to echoes that propagate back to the probe also from outside the beam. This is illustrated in Fig. 1c which shows an image when irradiating only one beam but reconstructing a large tissue area around the beam. The intensity follows the actual beam profile only near the transducer aperture (upper part of the image) and echoes

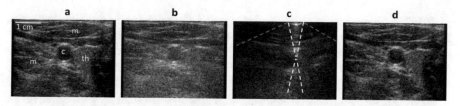

Fig. 1 **a** Line-by-line scan of tissue around carotid artery (*c*: carotid lumen; *th*: thyroid gland; *m*: muscle). **b** Plane-wave image of same area. **c** Image when irradiating a single line but reconstructing the full image area. The beam profile and the event horizon are indicated by *white and red dashed lines*, respectively. **d** Result of coherent plane-wave compounding. All images are displayed in the same dB scale covering 60 dB. Note that **c** was not taken at the exactly same position as **a**, **b**, **d**. These images were produced using a Vantage 64 LE research US system (Verasonics Inc. WA)

are reconstructed also outside the actual beam where echo intensity is dominated by 2nd and higher order echoes. These echoes are confined by an "event horizon" that marks the first possible arrival of an echo (of any order) at the different probe elements. In a line-by-line scan, the receive part is focused into the same area as the transmitted beam, i.e. only pixels inside that area are reconstructed. Therefore, the out-of-beam 2nd order echoes are less sensitively detected than the in-focus 1st order echoes. The sensitivity ratio determines the SCR in the image line. In a PW image, where a broad unfocused US pulse is transmitted into a large tissue region at once, 2nd order echoes are detected from inside the receiving beam area that originate from the irradiated tissue outside the receive area. Therefore, the relative contribution of 2nd order echoes is much higher than in a LLS.

An alternative to PW imaging for improving US quality is coherent plane wave compounding (PWC) [31]. In this technique, PW images are acquired with a variety of different PW transmit (Tx) angles. For each angle an image is reconstructed, and all images are coherently averaged (i.e. before envelope detection). The result is shown in Fig. 1d, an image that looks practically identical to the line-by-line scan in terms of contrast and spatial resolution. The observed SCR improvement can be understood from two perspectives: first, diffuse 2nd and higher order echoes decorrelate with varying Tx angle, so that averaging reduces the intensity of such echoes due to phase cancellation; second, coherent averaging of images obtained with different Tx angles corresponds to synthetic Tx focusing, similar to the (coherent) delay-and-sum (DAS) beamforming that synthetically focuses the transducer on receive (Rx). In that view, the Tx angle range in PWC is equivalent to the angular aperture in a LLS. For this equivalence to hold, the angle spacing should be chosen sufficiently small so that a hypothetical superposition of the plane pulses (with appropriate relative delays) could indeed result in a single focused or collimated beam within the size of the probe aperture. Then the SCR is similar or even better than the one of LLS. With a larger angle step, the hypothetical superposition would result in multiple parallel beams, the larger the step the more beams. This results in reduced SCR, equivalent to an increased clutter level that results from 2nd order echoes that couple from adjacent Tx beams into a Rx beam.

PWC is referred to as "ultrafast imaging" [30], referring to the fact that a full image can be reconstructed from a single or few PW acquisitions. One has, however, to keep in mind that to achieve an identical spatial resolution (identical angular aperture of Tx focusing) as well as an identical SCR (this was followed in Fig. 1), LLS and PWC require the same number of acquisitions (taking into account that dynamic Tx focusing can be achieved in a line-by-line scan by retrospective Tx beamforming [32]). A disadvantage of PWC compared to LLS can be the larger amount of data that needs to be transferred and processed, as for each angle, data are required for all probe elements and each pixel must be reconstructed. In LLS only the elements corresponding to a certain line need be active on receive and only the pixels inside the corresponding Tx beam area need to be reconstructed. A different practical difference between the two techniques is the different way in which motion affects the final image. In LLS, abrupt motion shows up as a relative shift of different parts of the image, and—in retrospective Tx beamforming—degrades SCR around the line that was acquired during the motion due to phase cancellation. In PWC, the phase cancellation due to motion equally degrades the SCR in the whole image, but the degradation is weaker than in LLS. Depending on the specific application, one or the other technique can be more advantageous. Ultimately, the system at hand will determine what type of data can be acquired (e.g. the specific system presented in this chapter only allows interleaved acquisition of PWC data with PA).

2.2   Motion Tracking Algorithm

Previously we proposed Loupas' phase correlation (LPC) [33] for estimating the tissue motion field based on quantifying the resulting phase shift of US echoes. Assuming that the array probes bandwidth (BW) is smaller than the centre frequency f_0 (this is typically the case for standard clinical US probes), it is practical to model a beamformed (e.g. using DAS) radio-frequency (RF) signal $s(z)$ (where z is the axial dimension) as the product of a "slowly" (given by BW) varying complex envelope $S(z)$ with a "quickly" (given by f_0) oscillating complex exponential carrier:

$$s(z) = S(z) \cdot \exp(2\pi i f_0 \cdot 2z/c) = S(z) \cdot \exp(2\pi i z/\Lambda) \text{ with } \Lambda = c/2f_0 \quad (1)$$

where c is the speed of sound. Note that Eq. 1 contains a factor 2 in the complex exponent that accounts for the two-way propagation in echo US. Assuming that no lateral nor out-of-plane motion occurs, and that the gradient of the motion along the axial direction z is small, image lines acquired before (s_n) and after (s_{n+1}) a motion step are identical apart from a z–dependent shift $\Delta z_{n,n+1}(z)$ and can therefore be modelled as:

$$s_{n+1}\big(z + \Delta z_{n,n+1}(z)\big) = s_n(z) \quad (2)$$

If, in addition, Δz is smaller than half the oscillation period Λ, Δz can be estimated from the point-wise Hermitian product $C_{n,n+1}$ between s_n and s_{n+1}:

$$
\begin{aligned}
C_{n,n+1}(z) &= s_n(z) \cdot [s_{n+1}(z)]^* = s_{n+1}(z + \Delta z_{n,n+1}(z)) \cdot [s_{n+1}(z)]^* \\
&= S_{n+1}(z + \Delta z_{n,n+1}(z)) \cdot S_{n+1}^*(z) \cdot \exp(2\pi i(z + \Delta z_{n,n+1}(z))/\Lambda) \cdot \exp(-2\pi i z/\Lambda) \\
&\cong S_{n+1}(z)^2 \cdot \exp(2\pi i \Delta z_{n,n+1}(z)/\Lambda)
\end{aligned}
\tag{3}
$$

$$
\rightarrow \widehat{\Delta z}_{n,n+1}(z) = \arg\{C_{n,n+1}(z)\}/2\pi \cdot \Lambda
\tag{4}
$$

For the last step of Eq. 3, one uses the assumption that $S(z)$ varies "slowly" compared to Λ and Δz and thus can be assumed constant over the distance Δz. Since the pre-factor to the complex exponential in Eq. 3 is thus real, Δz can be estimated based on the phase angle of C. Even though we assumed purely axial motion, LPC is not limited to axial motion: by acquiring US images with two (or more) different view directions through Tx and/or Rx beamsteering, a 2D-vector (or 3D for 2D arrays) field can be obtained.

Now, let's have a closer look at the various assumptions that were involved in the derivation of Eq. 4:

No lateral motion: Accepting some error, it can be slackened, saying that lateral motion has to be below the lateral resolution of the image. In case this assumption is not fulfilled, lateral motion causes decorrelation of S resulting in tracking errors. Such errors can be reduced by reducing the lateral resolution (e.g. by lateral spatial low-pass filtering [34]), but at the cost of lateral resolution of the motion field.

No out-of-plane motion: Out-of-plane motion can lead to decorrelation of S without any possible remedy. Therefore, this condition is crucial for any motion tracking algorithm to work. Also, out-of-plane motion can decorrelate the PA signal, making DCA useless. Therefore, real-time display of US images during DCA is a very important feedback for probe guidance, to help minimise out-of-plane motion.

Small axial gradient of motion field: The envelope S is the result of the interference (destructive and constructive) of echoes generated by reflectors that cannot be resolved by the axial impulse response (given by the BW) of the system. If the displacement magnitude changes by about 0.5Λ within the length of the axial impulse response, then the changing relative position of reflectors results in a changing interference of the echoes (destructive turns to constructive and vice versa) and thus to full decorrelation of S. Even for smaller gradients, partial decorrelation occurs [34].

Slowly varying envelope: The shorter the axial impulse response (the larger the BW), the more the complex envelope can vary with the displacement, and the pre-factor in the last line of Eq. 3 deviates from a real number so that the relation between the phase angle and the displacement magnitude becomes inaccurate. Even for a broadband signal, it is possible to enforce a narrower BW and thus a more slowly varying envelope by bandpass filtering. The increased length of the axial impulse response, however, leads in turn to more decorrelation of S according to the previous paragraph. For this reason, the correlation length together with the axial gradient of the motion field determine a minimum decorrelation rate.

The net effect when the above conditions are only partially fulfilled is thus decorrelation of S, which in turn results in tracking noise. To reduce tracking noise, one employs a convolution of C with a typically 2D window function (or 3D for matrix arrays) before calculating the phase angle. Similar to the axial impulse response length, an increasing axial length of this tracking window induces errors when the gradient of the motion field is not zero. The choice of the tracking window length thus deserves attention, and depends on the application.

The main limitation of LPC is that the displacement magnitude has to be smaller than 0.5Λ (0.075 mm at 5 MHz) to avoid phase aliasing. One could argue that it is possible to enforce this condition by properly choosing f_0 via bandpass filtering. In practise however, this approach is limited by the bandwidth of the system. A way to track large tissue displacements is to make sure the condition is fulfilled between successive US acquisitions, and accumulate the motion field over time [34]. With externally induced tissue motion, the motion between successive US acquisitions can be controlled to be sufficiently small. In carotid imaging, however, the intrinsic pulsatile motion of the artery wall can easily lead to a total displacement on the order of several Λ in a fraction of a second. In such a case, one can in principle choose the framerate fast enough to capture sufficiently small motion intervals. Depending on the US system at hand, however, such a high frame rate may not be possible, either due to limited data transfer speed and/or due to limited processing speed. With the system used for this exemplary study, the frame rate was limited by the transfer and processing speed to about 10 fps. This made a type of tracking algorithm necessary that is capable of accommodating displacement magnitudes of several Λ length.

In the US elastography literature, block-matching (BM) is often used for tracking large displacement magnitudes [35–37]: the similarity of image patches of successive images is quantified using a similarity measure (e.g. cross-correlation) for a variety of test displacements ("search approach"), resulting in a map of the value of the similarity measure. The displacement is then estimated as the one that optimises the similarity measure. In comparison to the BM techniques, LPC has several advantages: (i) it is fast because it is based on a point-wise calculation whereas BM requires a time-consuming search approach; (ii) it is more robust because it can accurately determine phase shift even in low SNR situations where BM fails when the noise modifies the amplitude distribution ("peak-hopping") [38]; and (iii) it directly gives an accurate continuous-valued result of displacement magnitude, whereas an error-prone interpolation is required in BM to determine fractional displacements from the discrete search area. To avoid the interpolation, some authors have proposed to combine BM with LPC (BM-LPC) [39, 40], where BM is used for a rough estimate of the displacement and LPC is used for fine-tuning.

For the system proposed in this chapter we designed a different approach that is more robust and faster than BM-LPC: similar to some BM techniques, this approach makes use of the envelope of the complex RF-mode image, but instead of BM it consequently employs the LPC concept for improved speed and accuracy. As mentioned above, the limitation to the use of LPC is the limited bandwidth of the RF signal, so that the low frequencies that would be required for detecting large displacements are not available. The envelope, on the other hand, can have far lower spatial frequencies, but also a much larger fractional bandwidth so that the prerequisite for LPC is not fulfilled. To solve this problem, we bandpass-filter the envelope, to obtain synthetic RF data to which LPC can be readily applied. For each motion step n, the bandpass-filtered envelope at the bandpass frequency k, $u_{n,k}$, is defined as:

$$u_{n,k} = K_k * \left(|s_n|^2\right) \qquad (5)$$

Note that the envelope is defined here as the *squared* absolute value of the RF signal. The reasoning behind this definition as opposed to the absolute value itself is as follows: the absolute value can have sharp edges at the locations where adjacent echoes interfere to zero amplitude. These edges contain artificially high spatial frequencies above the actual spatial resolution given by the probe bandwidth. The squared absolute value, on the other hand, contains only truly resolvable spatial variations. It is moreover reasonable in a physics sense, as it corresponds to an actual physical quantity, i.e. energy density, whereas the absolute value itself doesn't.

The filtering and tracking are done in a multi-stage approach: In a first stage, the bandpass centre frequency is chosen sufficiently small so that the largest experienced displacement magnitude is smaller than half the wavelength Λ_k of the kth bandpass filter. LPC of successive frames $u_{n,k}$ and $u_{n+1,k}$ results in a first displacement estimate $\widehat{\Delta z}_{n,n+1,k}$, albeit at a low spatial resolution. To increase the spatial resolution, this displacement estimate is then used to motion-compensate the frames $u_{n,k+1}$ and $u_{n+1,k+1}$ at the next higher bandpass frequency. After motion compensation, the residual displacement is ideally smaller than Λ_{k+1} so that LPC can be applied without phase aliasing on this stage, resulting in an estimate of the residual displacement. This estimate is added to $\widehat{\Delta z}_{n,n+1,k}$ resulting in a refined estimate of the displacement, $\widehat{\Delta z}_{n,n+1,k+1}$. This procedure is repeated for all chosen filter stages. In the end, a final residual displacement estimate can be obtained from LPC of the motion corrected RF signals s_n and s_{n+1}.

3   Combined Handheld PA and US System

3.1   Acquisition System

For illustration of its benefit for low energy handheld PA imaging, we exemplarily show results of the implementation of DCA on a system that was developed within the H2020 project *Cvent*. The goal of *Cvent* is an improved diagnosis of plaque vulnerability using PA detection of blood clots inside carotid plaque. The system contains a fully integrated hand-held probe, based on a pre-existing commercial linear-array probe (7.5 MHz centre frequency, 5 MHz bandwidth, 0.245 mm element pitch) that was re-engineered to contain a built-in multi-wavelength diode laser source for PA imaging. Figure 2 shows a picture of the probe. Probes with various combinations of optical wavelengths in the near infrared were produced. The results presented in the next section were obtained using a single wavelength at 808 nm, irradiating the skin alongside the linear array through an elongated area of 1.5 cm^2 with pulses of 60 ns duration and 2 mJ pulse energy. An average pulse repetition rate of 80 Hz was used in this study, resulting in 100 mW/cm^2 time average irradiance well below the safety limit of 330 mW/cm^2 according to IEC 60825-1. The probe is connected to a commercial portable ultrasound system (MyLab™ One, Esaote Europe B.V., NL) for data acquisition. The limited on-board memory of the system allows to acquire a maximum of 9 PW data frames (for US imaging), each covering a depth range of 10 mm, and 10 PA data frames covering a depth range of 20 mm. After filling the on-board memory, the data (a "burst") is transferred via USB to a PC, where processing is performed on graphical processing units (GPU). For the presented results, an Acer Aspire E 15 laptop (Intel core i7-6500U, 2.5 GHz) was used with a built in NVIDIA GeForce 940 MX graphics card. With this PC, the over-all speed (transfer and processing) allowed for processing 8 bursts per second, allowing real-time imaging.

Fig. 2 Handheld PA/US probe containing the integrated diode laser source. The laser light exits the probe through the glass window alongside the acoustic lens (*arrowhead*) covering the linear array transducer

3.2 Image Reconstruction

Both the US and the PA images were reconstructed using conventional DAS algorithms. The carotid artery of the volunteer was located at a depth of 10–15 mm and had about 5 mm diameter, and thus could be easily covered by the 20 mm depth range of the PA acquisitions. The US acquisitions, however, only cover a range of 10 mm. For visual inspection of potential echo clutter in the PA images, the US images must show the tissue located between the probe and the carotid. This allows identifying PA signal as real signal or clutter based on the absence or presence of strong echoes at roughly half the depth. At the same time, the US images must also contain the carotid to allow motion tracking of the tissue at the location where DCA is most important. An US depth range covering superficial and deep tissue is also desired by the clinicians as it helps interpretation of the anatomical context during freehand probe guidance. A 20 mm US depth range was therefore achieved via the spatial distribution and superposition of the 9 US patches (Fig. 3a). The distribution was done in a way that PWC with 7 different angles ($-3°$ to $3°$ in $1°$ steps) was achieved in an area covering the upper part of the carotid, where US image quality was most important for motion tracking. The areas above and below the artery as well as at the lateral edges of the image contained less angles, resulting in reduced image quality. This was regarded acceptable as these regions were only needed for identification of the anatomy. For the demonstration of imaging depth using DCA, however, our desire was to achieve the maximum possible motion tracking accuracy resulting in maximum possible contrast-to-noise ratio (CNR) given the system limitations. For this purpose, we designed an additional mode where all 9 US acquisitions were placed on top of the location of the carotid

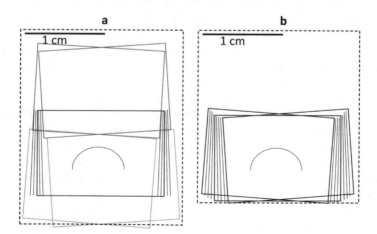

Fig. 3 Arrangement of the 9 US patches (indicated by *solid rectangles and lines*) acquired with various different angles for plane-wave compounding, covering an extended depth range (**a**) and a reduced depth range (**b**). The full image area is denoted by a *dashed rectangle*, and the upper edge of the carotid lumen is indicated by *solid arcs*

artery (Fig. 3b) so that PWC could be performed using 9 ($-4°$ to $4°$ in $1°$ steps) instead of only 7 angles, as we noticed that the extra angles resulted in a visible reduction of the tracking noise.

Both the US and the PA images were reconstructed in complex RF-mode with a pixel resolution of 0.15 mm (laterally) by 0.083 mm (axially). The 10 PA acquisitions of each burst were averaged before image reconstruction to reduce numerical cost. This could be done without motion compensation, as tissue motion during a burst was negligible.

3.3   DCA Details

Motion tracking was performed as described in Sect. 2. The bandpass convolution kernels were implemented as a truncated sinc function multiplied with an exponential carrier

$$K_k(z) = \frac{\sin(\pi kz)}{\pi kz} e^{2\pi ikz} (-1 < kz < 1) \tag{6}$$

for k corresponding to virtual centre frequencies [0.25, 0.5, 1.0] MHz. The local averaging of the complex correlation prior to determining the phase angle was implemented as a successive convolution in lateral and axial direction, with hamming windows with length $1/k$ corresponding to [3.0, 1.5, 0.75] mm. For the last stage, the tracking of the actual RF signal, the local averaging was implemented the same way but with a window length of 1.5 mm.

Motion compensation was performed via interpolation of the complex envelope of the complex RF-mode images (the IQ-data), followed by a phase correction to account for deviations of the continuous-valued displacement map from the discrete pixel grid. The reason is that the axial grid required for interpolation of the IQ-data can be sampled with the resolution given by the probe bandwidth, whereas interpolation of RF-data would require a higher resolution given by the (larger) centre frequency and, thus, increased computing time for image reconstruction.

Apart from the tracking and motion compensation algorithm, an important detail of a DCA implementation is what motion information is extracted from the US images and how this information is used for motion compensation: motion tracking can either be performed between successive US images or relative to a fixed reference. When determining displacement relative to a fixed reference, the disadvantage is that the increasing decorrelation of echoes with time (e.g. due to out-of-plane motion) leads to increasing displacement quantification errors. The advantage is that, for any displacement, such errors are only made once. When determining displacement between successive US frames, the quantification errors are on average smaller, but when accumulating displacement maps over larger time intervals, errors accumulate. Depending on the statistics of echo correlation/decorrelation, the accumulated errors can become larger than the errors of a large but single tracking step. Motion compensation can either be performed in a forward or in a backward way. In forward compensation, one option is to compensate the DCA result of time step $n - 1$ for the displacement from US frame $(n - 1)$ to frame n, and update with the PA frame from time step n. This implementation is computationally efficient because only one motion compensation is required per time step, but it is only applicable if a moving average with exponential weights is desired/acceptable. Alternatively, a number of past PA frames from time steps $(n - m)$ to $(n - 1)$ can be compensated for the displacement between US frames $(n - m)$ to $(n - 1)$ and the US frame at step n, and averaged with the PA frame at step n using arbitrary averaging weights. The disadvantage of this approach is that it is computationally more expensive as m motion compensation operations are required at each time step. In forward compensation, tissue motion is visually preserved in the DCA result, which can be considered an advantage. In backward compensation, on the other hand, PA frames from time step n are compensated for the displacement between US frame n and the US frame at a constant time step $n_0 < n$. Averaging can be performed with arbitrary averaging weights, even though only one motion compensation operation is required per time step. This combination of flexibility and computational efficiency is a big advantage of backward in comparison to forward compensation. In addition, backward compensation of US images in parallel to PA can serve as a useful feedback for assessing tracking quality: a static backward-compensated US image indicates perfect tracking.

Given the rather low framerate of the *Cvent* system, it turned out that a combination of fixed reference motion tracking and backward compensation worked best for imaging down to the depth of the carotid artery. For visual feedback on tracking accuracy, DCA was applied not only to PA but also to US images. The US images were thus not only backward compensated, but also averaged over time. In addition to allowing a coarse assessment of tracking accuracy based on the absence of motion in the back-compensated image, averaging of motion-compensated RF-mode US images provides additional feedback as small tracking errors show up as phase cancellation artifacts.

A moving average with exponential weights $w(n)$ was used to define the DCA result $P_{dca}(n)$ because this could be efficiently implemented in a recursive way:

$$w(n) = \frac{1}{\sum_{n'=0}^{\infty} e^{-n'/T}} e^{-n/T} \tag{7}$$

$$P_{dca}(n) = \sum_{n'=-\infty}^{n} w(n - n') P(n') = w(0) P(n) + (1 - w(0)) P_{dca}(n - 1) \tag{8}$$

Thereby, $P(n)$ is the nth motion-compensated frame (either PA or US). For PA, a time constant T of 20 bursts (i.e. w decreases to $1/e$ over 20 bursts) is used in practise, corresponding to a 2.5 s average delay between acquisition and display, but we will also show the influence of different time constants in the next section. For US feedback on tracking quality, a shorter time constant of 5 bursts was used, as this was sufficient to reveal phase cancellation due to tracking errors but provided a more immediate feedback (0.6 s) which was helpful for adjusting probe guidance.

3.4 Image Display

The software of the *Cvent* system was programmed to display two different images side-by-side: on the left, image 1 displays a "high quality" (HQ) US image, to help guide the radiologist within the anatomical context when looking for plaque. This US image is not motion-compensated nor averaged. Standard US image post-processing was implemented (envelope detection, logarithmic compression, speckle filtering) to match, as far as feasible (given the limited amount of data), standard B-mode US image quality. On the right, image 2 provides an overlay of the DCA PA image with the DCA US image. Again, envelope detection and logarithmic compression was used for the PA image, but no speckle filtering. Depth-gain compensation (TGC) was applied to both to US and PA, to reduce intensity variations caused by optical and ultrasound attenuation.

Combining PA and US data in a single image has the advantage that the PA signal can be identified within the anatomical context given by the echo texture in US. Various different ways of combining PA and US data are possible. One way would be to simply blend the colour maps of the two modalities. When using clinical probes this is, however, not a good approach: the limited bandwidth of clinical probes cuts off low spatial frequencies of PA signals. High spatial frequency detail in PA images is mainly found as sparsely distributed small blood vessels and surfaces of larger blood vessels or tissue interfaces (e.g. between fat and muscle). Large areas of the PA image thus do not contain useful information, and simply blending colour maps would unnecessarily tone the US image in these areas. A simple and popular way of combining PA and US data is thus by displaying PA data in colour scale only where PA intensity exceeds a certain threshold which is chosen above the expected noise intensity level. The disadvantage of this technique is that noise not only consists of thermal noise but also contains clutter noise. The latter can vary, depending on the imaging location but also depending on variations in the light coupling efficiency, so that the threshold needs to be adapted in an unpredictable way.

To avoid this problem, we have chosen a different approach based on the coherence factor (CF) concept. Conventionally, the CF quantifies the coherence of an US or PA signal across the probe aperture, as the squared coherent sum (phase preserved) normalised to the incoherent sum (of the intensity, no phase information) of the signal (after applying the same time delays as in conventional DAS). A perfectly coherent signal yields the largest CF value, whereas a perfectly incoherent signal yields a comparably small value due to phase cancellation in the coherent sum. We adapted this concept to measure the coherence along the sequence of bursts, per pixel in the reconstructed (after DAS) and motion-compensated PA images. Clutter and noise decorrelate along the burst sequence, thus being "incoherent", whereas true PA signal is correlated and thus shows a higher "coherence". We therefore defined the

CF as:

$$CF(x, z, n) = \frac{\left[\sum_{n'=-\infty}^{n'=n} w(n - n') P(x, z, n')\right]^2}{\sum_{n'=-\infty}^{n'=n} w(n - n') \sum_{n'=-\infty}^{n'=n} w(n - n')[P(x, z, n')]^2} \qquad (8)$$

The sum in the numerator equals the conventional DCA result, whereas the sum in the denominator is an incoherent version of DCA. For perfectly correlated PA signal, this CF becomes one, independent of the magnitude of the PA signal, whereas it takes on small values in noise- or clutter-dominated areas. The CF provides a more practical way of identifying "real" PA signal than thresholding the PA intensity because it does not depend on the PA signal magnitude but only on the coherence. We use this CF for the combined PA/US display in the following way: first, the intensity (squared envelope) of the PA image is logarithmically compressed and the result is coded into an RGB colour map. Then, the colour channels are multiplied with the CF in each pixel. This approach opens up the freedom to code the intensity of PA signals as colour hue alone, whereas the colour brightness is determined by the "coherence". This allows to distinguish different signal intensity levels based on hue while the brightness emphasizes "real" signal independent of signal intensity. For the results shown in the next section, a "blackbody" colour map was chosen. In a last step, the PA colour image is blended with the grayscale US DCA image, using (1-CF) as spatially dependent transparency value, after multiplying the US grayscale by 0.5 to provide better colour contrast between PA and US. To make the effect of the CF on the final image more pronounced and independent of depth-dependent optical attenuation, the CF is logarithmically compressed to a range of 7.5 dB, starting from −5 dB and including a TGC of 3.5 dB/cm.

4　Results

This section presents results of detecting blood vessels at the depth of the carotid artery in a healthy volunteer. In a first part, the influence of the various steps of the signal processing are illustrated on a single example data set. In a second part, various different imaging examples are shown, to underline the reproducibility of the achieved imaging depth and to provide some recommendations on image interpretation.

4.1　Illustration of Processing Steps

Figure 4a shows right away the final side-by-side display of the "conventional" HQ B-mode US image and the PA/US overlay. For this and following results, the limited US depth range configuration was used to achieve maximum tracking accuracy. The HQ B-mode US image on the left reveals anatomical detail of the tissue surrounding

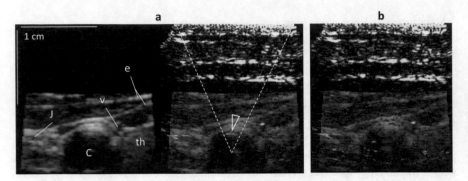

Fig. 4 **a** Side-by-side display of B-mode US (left) and PA/US overlay (right). The US image shows a transversal section through the left carotid artery (*C*: carotid lumen; *J*: internal jugular vein; *v*: small vessel; *th*: thyroid gland and *e*: muscle epimysia). The overlay of PA (blackbody colour scale) with US (grayscale) shows strong PA signal at the upper surface of the carotid lumen (*arrowhead*), around the jugular vein and around the small vessel. The lateral extension of the carotid lumen signal is limited by the receive angular aperture, indicated by *dashed lines*. **b** For comparison, when averaging without motion compensation, the carotid signal is missing. Intensity display ranges were: 50 dB for US (5 dB/cm TGC), 25 dB for PA (12.5 dB/cm TGC)

the carotid artery, including the carotid lumen, thyroid gland, small vessels and muscle epimysia. The PA/US overlay clearly shows PA signal emanating from the upper surface of the carotid lumen. The appearance of this signal is typical for PA imaging using conventional clinical US probes: only the lumen surface is seen because the limited BW suppresses low spatial frequencies of optical absorption (the light penetration depth itself is few mm), and the signal is laterally limited due to the limited probe aperture: from the cylindrical transient generated by the vessel lumen, only the part that propagates within the sector indicated by lines is detected by the probe. Apart from the carotid lumen, PA signal is observed at small vessels that are indicated in the US image. Note that the signal intensity is similar in spite of the substantially different vessel sizes. This is expected, as the signal amplitude is proportional to the absorption contrast at the vessel boundaries, which is similar as all vessels are situated at roughly the same depth. The averaging time constant was 20 bursts, corresponding to 2.5 s. Figure 4b shows the PA/US overlay when averaging the PA signal over the same 20 bursts but without prior motion compensation. As a consequence of phase cancellation, the carotid signal is not discernible from the background noise. The results in Fig. 4 thus clearly demonstrate that DCA is a requirement for deep imaging, especially in low energy PA imaging where long averaging times are required. Apart from enabling the detection of the carotid signal, a comparison of the PA results reveals a reduced background noise level inside and around the carotid lumen in Fig. 4a compared to Fig. 4b. This indicates that part of the noise was systematic noise that persisted in Fig. 4b but was reduced in Fig. 4a due to the motion correction. Such noise can be explained by clutter stemming from reverberations within the superficial tissue layers, of PA transients that are generated in or just below the skin where laser fluence is largest.

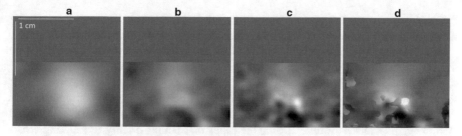

Fig. 5 Axial displacement field detected at maximum carotid motion, based on multi-stage bandpass-filtered envelope tracking: **a** stage 1 at 0.25 MHz bandpass centre frequency; **b** stage 2 at 0.5 MHz; **c** stage 3 at 1.0 MHz. **d** Phase correction at 7.5 MHz smoothens the result of stage 3. The colour scale spans ± 0.25 mm, where white and black indicate motion towards and away from the transducer, respectively

Figure 5 illustrates the multi-stage motion tracking process, exemplified on the axial displacement map detected at the peak of the carotid wall motion. A set of three envelope bandpass stages was chosen at bandpass centre frequencies (0.25, 0.5 and 1.0 MHz). In a first step, the 0.25 MHz are used for tracking. Figure 5a shows the resulting displacement map. Note that displacement values are available only within the region of interest (RoI) that is covered by the US image. The 0.25 MHz were chosen so as to provide the most robust motion-compensation of the US images (by visual inspection). This displacement map, however, has very low spatial resolution, so that short-scale variations of displacement magnitude are missed. As a result, only the upward motion of the upper carotid wall is captured because the upper carotid wall gives the strongest echo within the spatial resolution. This leads to an overestimation of the displacement magnitude in a large area outside the vessel lumen. To avoid such errors, the displacement map is refined in a second step, by phase-tracking the 0.5 MHz-filtered US envelope, resulting in Fig. 5b. As a result of the refinement, this displacement map shows an improved spatial resolution, so that it is able to capture also the downwards motion of the lower carotid wall. At the same time the displacement magnitude is decreased above the upper carotid wall because the weaker echoes from the tissue overlying the carotid can now be spatially separated from the strong carotid wall echo, resulting in more accurate estimation of the displacement magnitude in this area. The displacement map is further refined in the third tracking stage based on the 1.0 MHz-filtered envelope (Fig. 5c). While the spatial resolution is again markedly improved, the spatial distribution shows an increased level of bumpy short-scale spatial variations. In a last step (Fig. 5d), the displacement map is refined based on phase tracking the RF-mode US images as opposed to the bandpass filtered envelope. After this phase correction, the spatial distribution has become smooth over a large area, as one would expect from a real displacement field. The difference between Fig. 5d and c underlines the earlier made statement that RF phase tracking is more robust than envelope tracking: the increased noise in Fig. 5c can be assigned to decorrelation of the envelope due to a changing interference of echoes. Only few small areas can be identified in Fig. 5d with unrealistic sharp discontinuities that result from phase aliasing. Comparison to the HQ US image in Fig. 4 reveals that these areas are found in regions of low echogenicity, mainly inside the hypoechoic carotid lumen. This is expected, as such low echogenic areas can be dominated by higher order echoes that move differently than 1st order echoes.

With the chosen backward compensation, the DCA PA image is static and thus the overlay with US also requires a static US image. Therefore, motion compensation is applied to US the same way as to PA. As already mentioned, the motion compensated US image provides a useful real-time feedback on tracking quality, as tracking errors can be identified based on residual motion. In addition, when averaging the motion-compensated US images, even small errors can become visible when they lead to phase cancellation. This is illustrated in Fig. 6: phase cancellation shows up as fluctuating small areas where echo intensity is transiently reduced. Based on the visual assessment of residual echo motion and phase cancellation areas, freehand probe motion can be adapted in real-time to minimise the frequency of tracking errors, to provide an optimum data set to which DCA can be most successfully applied.

Apart from choosing between forward or backward motion compensation, one main decision to be made when implementing DCA is between accumulative and fixed-reference motion tracking. As already mentioned, both approaches can have advantages and disadvantages. For our system we use fixed-reference tracking because accumulative tracking turned out to be less robust. This is illustrated in Fig. 7: accumulative and fixed-reference tracking led to similar contrast-to-noise ratio (CNR), but accumulative tracking resulted in an increasingly distorted DCA image already after 34 bursts (4 s averaging time) due to accumulation of tracking

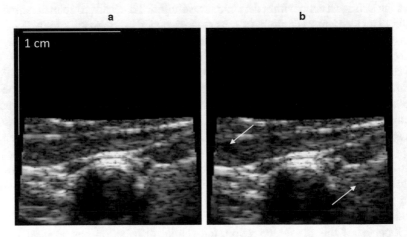

Fig. 6 Illustration of using the DCA US image for real-time tracking quality feedback. This figure shows the US image underlying the PA/US overlay, but without the PA data. **a** Perfect tracking results in an US image that is static and shows a stable intensity distribution. **b** With less-than-perfect tracking, the image may still look static, but small tracking errors result in small fluctuating areas of decreased intensity due to phase cancellation (*arrows*). 50 dB (5 dB/cm TGC)

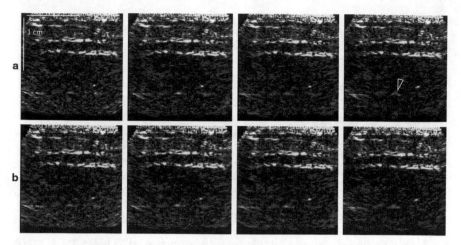

Fig. 7 DCA result for accumulative (**a**) and fixed reference tracking (**b**), at burst number 14, 24, 34, 50 (left to right). 20× averaging time constant, 25 dB intensity range (12.5 dB/cm TGC). Whereas contrast is similar between the two techniques, accumulative tracking leads to geometrical distortions, best visible at the lower edge of the image or at the curvature of the carotid signal (*arrowhead*)

errors. In comparison, fixed-reference tracking is limited in tracking accuracy, but no error accumulation occurs so that the DCA result remains robust over time.

A further decision to be taken is between coherent averaging (of motion-compensated RF-mode PA images) and incoherent averaging (of the motion-compensated PA envelope). Incoherent averaging can have the advantage that tracking inaccuracies cannot lead to phase cancellation as in coherent averaging. On the downside, incoherent averaging is less efficient in terms of noise reduction: the convergence rate of averaging the square of a Gaussian distributed random number is weaker than when averaging the random number itself. In addition, averaging the square converges to a positive number instead of zero, which adds a disturbing bias to the signal intensity distribution. This is illustrated in Fig. 8, where the DCA results using the two different techniques can be compared. Note that both images are displayed in identical dB scales. Within the first cm depth range where SNR is large, the colour hue of strong PA signals is coded in the same colour in both images. In areas where the signal is dominated by noise, the colour reveals an increased intensity level in Fig. 8b compared to Fig. 8a due to the biased convergence limit. Due to the TGC, the colour indicates larger intensity values with increasing depth. Since the SNR of e.g. the carotid signal is quite small, the positive average noise adds an offset so that the carotid signals colour hue indicates a higher intensity in 8b than in 8a. Note that the CNR is visibly decreased, e.g. around the jugular vein, as the noisy background achieves the same colour hue as the signal from the jugular vein. We decided for coherent DCA because it has an over-all improved CNR (better convergence property, not markedly sensitive to phase cancellation errors) compared to incoherent DCA.

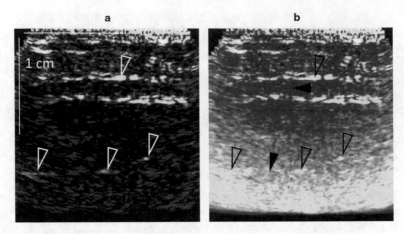

Fig. 8 Coherent (**a**) and incoherent (**b**) DCA. 25 dB intensity range (12.5 dB/cm TGC). Note the similar intensity level of superficial and deep PA signals (*empty arrowheads*), but due to the positive value of the average square noise (*full arrowheads*) incoherent DCA has reduced CNR compared to coherent DCA. 25 dB intensity range

Finally, an important parameter for DCA is the averaging time constant. All PA DCA results shown so far were based on a time constant of 20 bursts. The time constant has to be chosen in a trade-off between SCR and real-time feedback. For comparison of SCR, Fig. 9 shows DCA results for various different constants, i.e. 10, 20, 40 and 80 bursts, corresponding—for an 8 Hz burst rate—to 1.25, 2.5, 5 and 10 s averaging time, respectively. As one can see, SCR markedly increases from 10 to 20, but converges above 40 bursts. The reason for this convergence as opposed to a continuous increase is that part of the noise consists of persistent clutter. Even though part of clutter can be reduced due to the carotid motion, this reduction is limited: since the carotid displacement magnitude is limited and the motion is periodic, only a limited number of statistically independent clutter realisations can be averaged regardless of how long the averaging time is chosen. Even though SCR slightly improved from 20 to 40 bursts, we considered 5 s a too long averaging time regarding real-time feedback. Therefore, we decided for a time constant of 20 bursts

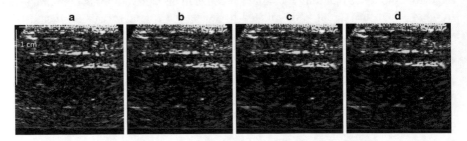

Fig. 9 Influence of averaging time on DCA constrast: **a** 10×; **b** 20×; **c** 40×; **d** 80×. 25 dB intensity range (12.5 dB/cm TGC). CNR converges between (**b**) and (**c**)

corresponding to 2.5 s, which at the same time provides close-to-maximum SCR with a still acceptable delay between acquisition and display.

The last step in the DCA processing chain is the combination of the PA with the US image. The way this is done has an important influence on the visibility and interpretation of the PA signal. As mentioned earlier, we decided for an overlay of the two images based on the DCA coherence factor (CF), as the CF provides an amplitude-independent measure of the significance of a PA signal. The way the CF is used for that purpose is illustrated in Fig. 10. The first step is the choice of the base colour map that defines the colour hue for displaying the PA signal intensity. We decided for a "blackbody" colour map because it resulted in the visually most pleasant blend with the grayscale of the US image. The blackbody colour map starts with black for zero signal intensity, over red and yellow, to white for the highest intensity. Therefore, not only the colour hue depends on signal intensity but also the colour brightness. Alternatively, one could choose a colour map that codes signal intensity entirely in colour hue. As a step into this direction, we also show results for

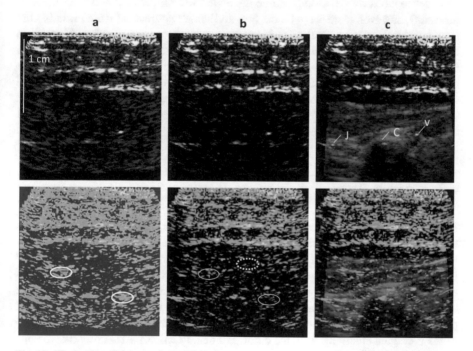

Fig. 10 Illustration of the use of the DCA coherence factor (CF) for colour-coding PA data and for the PA/US overlay, departing from two different base colour maps, blackbody (top) and modified jet (bottom): **a** PA data without CF fading. **b** Fading of PA colour map using CF. **c** Overlay of faded PA colour map with US grayscale map using (1-CF) as transparency value. To improve visibility of the PA signal, the US colour values were multiplied with 0.5 before overlay. The signals from carotid lumen (*C*), jugular vein (*J*) and the small vessel (*v*) maintain their colour value with CF fading. Some medium intensity features (e.g. *solid circles*) keep their colour hue but are attenuated by CF fading. The low intensity background (blue in modified jet map) becomes black by CF fading, apart from some pixels (e.g. *dotted circle*)

a modified "jet" colour map. The original jet colour map starts from dark blue over bright blue, etc. to bright red and dark red. To reduce brightness variation, this map was truncated to range from bright blue to bright red. Figure 10 exemplarily shows results for both the blackbody and the modified jet map. Figure 10a shows the result of coding the PA intensity into the respective colour map. Assuming that only coherent signal (coherent in the sense that we defined earlier) is "real" and worth displaying, the CF is then used to highlight "coherent" and supress "incoherent" signal, by coding the CF into the colour brightness value. The result is shown in Fig. 10b. In both colour maps, some features retain the original colour hue and brightness. In the lower half of the image, these are notably the features that can be assigned (based on the US image) to jugular vein, carotid, and a small vessel. Relative to these, the brightness of the background is reduced. In case of the blackbody colour map, it is difficult to determine in the CF-faded image alone whether the darkness is due to a dark value in the original colour map or due to the CF fading. In the modified jet colour map, on the other hand, it is clear that mostly areas that were initially blue (low intensity) were set to black by CF fading. Some areas that were initially green (intermediate intensity), however, were set to black, too, and some features that were initially blue (low intensity) remain so in the CF-faded image, demonstrating that the CF provides complementary information to signal intensity. In a last step, the CF is used to set the local transparency value in the overlay of the PA onto the grayscale US image (Fig. 10c): in pixels where the CF is high, transparency is set low, so that the pixel colour is determined by the PA signal, whereas transparency is set high in pixels where the CF is low, to show the anatomical context in areas where the PA signal can be assumed not to contain valuable information.

4.2 Further Results

The remainder of this section is dedicated to showing and discussing further results and provide experience on how to interpret images.

As mentioned before, the limited bandwidth of the clinical probe acts like a spatial bandpass filter that allows to detect only rapid spatial variations of optical absorption, e.g. only the surface of large blood vessels. In addition, the limited aperture size allows only part of the surface to be seen, i.e. the part from which the PA transient propagates into the probe aperture. This leads to the typical arclet-shaped appearance of the PA signal emerging from the carotid lumen. In case of a 1D array like the one we are using in this study, the limited aperture of the probe has a further implication not discussed so far: to provide a well-defined imaging plane, the array aperture is by design focused in elevation (the dimension perpendicular to the imaging plane). For that reason, a main requirement for detecting the carotid signal is that the imaging plane has to be perpendicular to the lumen surface. By that the transients emanating from the lumen surface and hitting the transducer have propagated parallel to the imaging plane and are detected with maximum elevation sensitivity. With increasing deviation from a perpendicular orientation, the same transients arrive at an increasing elevation and sensitivity rapidly decreases. For reasons that are yet unclear (but will be discussed in the next section) it can be difficult to detect the carotid signal in a transverse section even with perfectly perpendicular orientation. The chance to at least partially detect the carotid is higher in a longitudinal section: then the carotid signal appears as a line (Fig. 11a) that covers a larger number of pixels and thus provides a richer statistics for identification of this signal. It makes it also more practical to adjust the probe orientation: by only slightly tilting or moving the probe, the angle between imaging plane and lumen surface changes rapidly due to the surface curvature, thus it is possible to optimise the sensitivity without substantially changing the imaging plane position. For a transversal section, optimising the angle between imaging plane and lumen surface requires a search over a large probe tilt angle range, and the location of the section area within the lumen changes together with the tilt angle, so that two probe orientation parameters need be simultaneously optimised for detecting a desired location.

In our experience, it is much easier to catch the signal emanating from the internal jugular vein (and from smaller vessels) than from the carotid itself. This is illustrated in Fig. 11b–d where the signal from the internal jugular vein can be clearly identified even when it is located near the lower edge of the image. Note that, in Fig. 11c, the carotid lumen is visible on US at the same depth as the jugular vein (but no PA signal is detected due to the difficulties mentioned before). These results therefore further underline the ability of the presented system to detect optical absorption in the blood at the depth of the carotid artery.

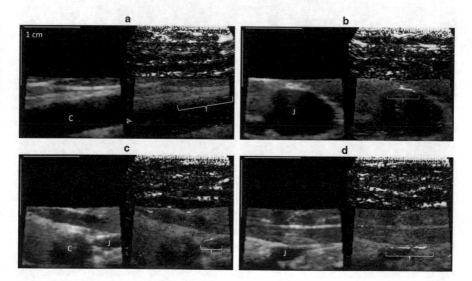

Fig. 11 Further results demonstrating imaging depth. **a** longitudinal section of left carotid artery; **b–d** transversal sections of right internal jugular vein. In **c** and **d** the jugular vein appears flattened due to compression of the tissue (*C*: carotid lumen; *J*: jugular vein lumen). In all images, the lateral extent of the PA signal is indicated by *brackets*. Intensity range: 50 dB for US (5 dB/cm TGC), 25 dB for PA (12.5 dB/cm TGC)

So far, all results were based on a limited RoI US image. As mentioned in the previous section, the US RoI depth range was limited to 1 cm in order to achieve maximum angle coverage for PWC and thus maximum tracking quality within the hardware limitations of the available system. Apart from showing the anatomical context of PA signal and serving for motion tracking, however, the US image is needed for identification of PA reflection artifacts based on the presence of echogenic structures seen in US. For this purpose, the US image needs to show structures that are located at half the depth of the PA signal of interest. Therefore, we implemented a software version where a larger depth range is shown on the US image, at the cost of giving up on angle coverage and thus on tracking quality and DCA performance. Figure 12 shows some results using this software, to illustrate how the US image can be used for evaluating the authenticity of PA signal: in Fig. 12a (as already in Fig. 4) PA signal is visible which can anatomically be related to the surface of the left carotid lumen. At half its depth, no reflecting structure is seen on US (dashed line). These two observations together give confidence that the PA signal is actually from the lumen, not a reflection artifact. In Fig. 12b, the left internal jugular vein is seen on US. It appears as a line because it is fully collapsed due to the static pressure exerted by the probe. Next to the jugular vein, a small vessel is visible. PA signal is visible inside the jugular vein as well as in the small vessel. The authenticity of the

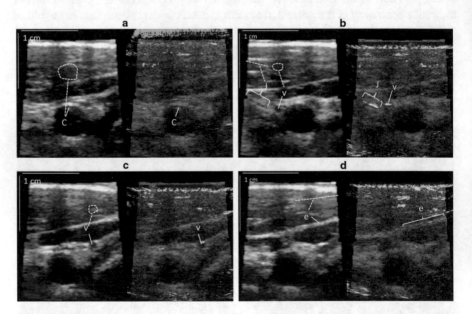

Fig. 12 a–d Evaluation of authenticity of PA signal is based on US in two ways: by the correspondence of location of PA signal to anatomical features (indicated by *solid white lines* or *brackets*; *C*: carotid lumen, *J*: internal jugular vein, *v*: vessel, *e*: epimysium), and by the absence of strongly reflecting structures on US image at half the depth of the PA signal (indicated by *straight dashed lines* and *dashed circumscribed areas*. Intensity range: 40 dB for US (5 dB/cm TGC), 25 dB for PA (12.5 dB/cm TGC)

PA signal is again suggested by the correspondence of the PA signal to the anatomy seen on US, together with the fact that no reflectors are seen in US at half the depth. In Fig. 12c, PA signal is visible at the location of a vessel seen on US. At half the depth, a structure is visible on US which might cause a reflection artifact at the depth of the PA signal. However, one would then expect further artifacts with similar or higher intensity in near vicinity corresponding to other and more intense structures visible in the US image at similar depth. Moreover, based on the fact that the same PA signal is often found at the location of this vessel (as in Fig. 4), one can safely assume that this is real signal. Apart from blood vessels, epimysia often exhibit a distinct PA signal as in Fig. 12d. Again, the authenticity is confirmed by the anatomical correspondence and by the absence of a reflector at half the depth. The horizontal stripes of high PA signal intensity that have been present in all images shown so far can partially be interpreted as echo artifacts: they do not show anatomical correspondence with the US image, but can be interpreted as reverberations between skin surface (outside the RoI shown in the US images) and the muscle surface and horizontal muscle layers (visible on US).

5   Discussion and Conclusion

The results presented in this chapter demonstrate that DCA is a key requirement for deep PA imaging using low energy (LED or LD based) PA systems, and we have shown that this technique allows detecting the PA signal of the carotid artery using a compact fully-integrated PA/US probe in a freehand approach.

The results show the known limitations of using a clinical linear array probe for PA imaging, namely that the limited probe bandwidth and aperture allow only to detect sharp boundaries, e.g. at the upper (and sometimes lower) edge of a vessel lumen. Moreover, sensitivity depends on the relative orientation angle between the PA signal sources and the imaging plane. This certainly puts limits to how much can be interpreted from PA images, and—in the worst case—renders quantitative interpretation of PA signal amplitude impossible. To solve this problem, one approach has been to use concave arrays that are better matched to e.g. the neck or breast geometry to increase angle coverage [41, 42], and a large bandwidth (more) suitable to capture a tomographic section. Such a system has, however, the disadvantage of providing a less well defined imaging plane: the size of the elevation focus which defines the thickness of the imaged tissue slice depends on the acoustic wavelength in relation to the transducer element size. Below a certain frequency limit, the focusing capability is lost, and if these frequencies are not suppressed by the transducer's frequency response (or by the successive signal filtering), signals from anywhere in the 3D tissue sample are projected onto the same 2D image. This becomes a problem in PA when signals from below the skin surface but outside the imaging plane are orders of magnitude stronger than the signals coming from deep inside tissue due to the large difference in laser fluence. Therefore, the lowest detected frequency at the same time defines the minimum imaging plane thickness. If the wish is to detect the inside of the lumen of the carotid artery, for example, the imaging plane thickness would implicitly have to be larger than the diameter of the artery, i.e. around 10 mm. This would preclude such a system from the combination with conventional US images where a much better elevational resolution is desired (typically below 1 mm). A way to avoid the ambiguity of the PA imaging plane is to use a 2D array [43] for Rx beamsteering in azimuth *and* elevation. To achieve a sufficient tomographic coverage, however, the aperture size of such an array must be large in both dimensions, so that it becomes again impractical for an integration with standard handheld US.

We therefore foresee that, while a dedicated broadband and curved array can have specific clinical applications, a conventional clinical probe can have advantages in applications where the limited bandwidth is not a substantial problem and where PA can add important diagnostic information to conventional US: if, for example, the goal is to quantify blood oxygen saturation (SO_2) in small vessels, reconstructing the inside of the vessel lumen is not required (SO_2 can be regarded to be uniform across the lumen) and quantitative absolute values of PA signal amplitude are not required (SO_2 can be determined from relative variations of PA signal amplitude as function of optical wavelength [44]). In the *Cvent* project, the goal is to detect blood clots inside plaque. In the previous section we mentioned that detecting the signal from the carotid lumen requires a high level of experience in probe guidance as the detection sensitivity depends on a perpendicular orientation of the imaging plane relative to the lumen surface. This is, however, less of a problem when detecting blood clots: the optical absorption by clots is distributed non-uniformly inside plaque so that we expect it to act like a collection of small independent absorbing centres [45]. Figure 12 illustrates in a phantom the expected difference of the PA signal inside plaque compared to the signal from a healthy artery. For this purpose, a uniformly absorbing but acoustically transparent cylinder was embedded inside a background medium that is acoustically scattering. In one position, the cylinder contains a small echogenic volume in which graphite powder was mixed, mimicking plaque containing blood clots. When imaging the phantom in a "healthy" area (Fig. 13a), it looks similar as in a healthy volunteer: the cylinder appears as hypoechoic area on US, and the transversal section shows a PA signal in the shape of an arclet at the upper surface of this area, and the longitudinal section shows a line-shaped signal. When imaging at the position of the "plaque" (Fig. 13b), the plaque appears as a collection of diffuse PA speckle. This diffuse type of signal can be detected independent of the orientation of the imaging plane because it acts like a collection of independent and isotropically radiating sources.

In our study we made the interesting observation that detecting the PA signal from the carotid artery is substantially more difficult than the one from adjacent small vessels or from the internal jugular vein at the same depth. The 808 nm optical wavelength used for this study is very near the isosbestic point of the optical attenuation spectra of oxy- and deoxyhaemoglobin, so the absorption contrast is expected to be identical for carotid and other vessels. A possible explanation for the observed difference in signal intensity may, however, be found in the different morphology of the carotid artery wall compared to surrounding vessels: it contains a substantially thicker muscle cell layer (tunica media) that is perfused by capillaries (vasa

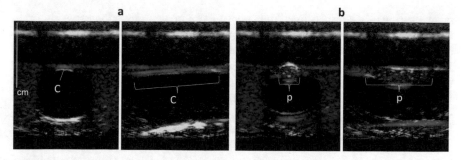

Fig. 13 Phantom mimicking a healthy carotid artery (**a**) and an artery containing plaque with haemorrhage (**b**). C: "healthy carotid" signal; p: "plaque" signal. Note that the apparent PA signal at the lower edge of the "carotid" lumen are echo artifacts, caused by an impedance mismatch between the background and the cylinder medium

vasorum). The depth profile of optical absorption in haemoglobin may thus resemble more a staircase than a single-step function, so that the optical contrast is blurred towards low spatial frequencies that may be less well detected. At the same time, the intensity of the light reaching the lumen interior is reduced by the thicker tunica media.

The most important component of DCA is the tracking algorithm. The goal of the presented study was to demonstrate that sufficient imaging depth could be achieved to detect the PA signal from blood vessels at the depth of the carotid artery. For this purpose, we relayed on an easily implementable, robust and real-time capable *ad-hoc* algorithmic solution. The advantage of the chosen algorithm compared to the commonly used block-matching (BM) technique is the lower numerical cost: Fig. 4 indicates that the peak displacement magnitude of the carotid wall motion was roughly 0.25 mm, corresponding to 2.5 wavelengths (0.1 mm) of the oscillations of the RF-mode image at the 7.5 MHz centre frequency. A BM technique would thus require at least 5 test displacements (2.5 in positive and in negative axial direction) but preferably more, to retrieve the optimum value of the block-matching criterion (e.g. correlation coefficient) with sufficient resolution. In comparison, only three filter stages were needed in our approach. At each stage, the displacement is directly estimated from the correlation phase (thus not requiring a search approach), and multiple filter stages are only used for refining the spatial resolution of the displacement map. A displacement map with slightly reduced quality could even be obtained with only two stages.

Even though the chosen tracking algorithm was sufficient for the demonstration of imaging depth in the presented volunteer results, it has potential for improvement: so far, we used only axial motion tracking and compensation, as it is the axial motion that leads to phase cancellation of the average PA signal if not accounted for. Lateral motion, on the other hand, can laterally blur the average PA image, thus the SNR of the DCA result can be further improved by lateral motion tracking and compensation. As previously mentioned, a 2D displacement vector field can be obtained for this purpose by acquiring two US images with different view directions (via Tx and/or Rx beamsteering). One-dimensional tracking of these images along the respective view direction results in projections of the displacement vector onto the different directions, and the displacement vector field can be reconstructed from these projections. An advantage of this approach is that it is substantially faster than a BM approach that requires a 2D search area. A disadvantage is the reduced lateral resolution if Rx beamsteering is used (as the full Rx angular aperture must be split into different view directions), or the increased data size if Tx beamsteering is used (due to the larger number of acquisitions). The envelope-based LPC technique proposed in this chapter is a practical alternative which combines the advantage of BM (full resolution without increasing data size) with one-dimensional tracking (low computational cost): phase tracking of the bandpass-filtered squared envelope can be applied to the lateral dimension equally well as to the axial dimension. This directly results in the lateral component of the displacement field with only a factor 2 increase in computational cost. This approach is very similar to spatial quadrature [46], where tracking is based on the complex RF signal and a lateral oscillation is achieved via Rx apodisation. Apart from increasing the dimensionality of the displacement field, multi-dimensional motion tracking has been shown to improve the accuracy of each dimension over a single-dimensional tracking [35]. Further ideas for improvement are found in literature on US strain imaging [36, 47–49].

As previously mentioned, the accuracy of the motion tracking relies on the US image quality. In the presented results, the US image quality was good in the sense that the intensity level of higher-order echo clutter was lower than the intensity of first-order echoes in most of the image area. Preliminary experience from an ongoing clinical study, however, reveal that motion tracking is more difficult in a large part of cases. Anatomy and acoustic properties of the neck vary substantially between subjects. Fat in and between the musculature above the artery can lead to reverberations of ultrasound that obscure the artery lumen so that the detected displacement is determined by the motion of the superficial tissue from where the reverberations originate, rather than by the actual artery wall motion. Similarly, calcifications inside plaque lead to reverberations that obscure the lower artery wall, such that the tracking result at the lower wall is determined by the motion of the upper wall. To enable reliable results independent of anatomy, the tracking algorithm thus must be able to (better) discriminate between superposing first- and higher-order echoes. This might be achieved via identification of different statistical features of the RF signal, via the different motion speed using a blind signal separation technique, via deep learning, or via a combination of these.

The presented results were obtained with an LD-based system providing 2 mJ pulse energy, using an average prf of 80 Hz and an averaging time constant of 2.5 s. As previously mentioned, the resulting average irradiance at the skin surface was a factor 3 below the safety limit (for 808 nm). With a faster data transfer and processing speed, the prf could thus have been increased by a factor 3 up to 240 Hz. Maintaining the 2.5 s averaging time constant, this would have led to a factor 1.7 increase in SNR (amplitude). By irradiating the skin on two sides of the linear probe instead of only one, the total increase in SNR would augment to a factor of 3.5. This indicates that identical results as the presented ones could have been achieved with—by a factor of

3.5—reduced pulse energy, i.e. only 0.7 mJ. This is a promising result, as it suggests that imaging the carotid artery is within the reach of the performance of LED-based systems.

Acknowledgements This project has received funding from the European Union's Horizon 2020 research and innovation programme under grant agreement No. 731771, Photonics Private Public Partnership, and is supported by the Swiss State Secretariat for Education, Research asnd Innovation (SERI) under contract number 16.0160. The opinions expressed and arguments employed herein do not necessarily reflect the official view of the Swiss Government.

References

1. J.J. Niederhauser, M. Jaeger, R. Lemor, P. Weber, M. Frenz, Combined ultrasound and optoacoustic system for real-time high-contrast vascular imaging in vivo. IEEE Trans. Med. Imaging **24**(4), 436–440 (2005). https://doi.org/10.1109/TMI.2004.843199

2. M.K.A. Singh, W. Steenbergen, S. Manohar, Handheld probe-based dual mode ultrasound/photoacoustics for biomedical imaging, in *Frontiers in Biophotonics for Translational Medicine* (Springer, Singapore, 2016), pp. 209–247

3. J.-L. Gennisson, T. Deffieux, M. Fink, M. Tanter, Ultrasound elastography: principles and techniques. Diagn. Interv. Imaging **94**, 487–495 (2013). https://doi.org/10.1016/j.diii.2013.01.022

4. M. Jaeger, G. Held, S. Peeters, S. Preisser, M. Grünig, M. Frenz, Computed ultrasound tomography in echo mode for imaging speed of sound using pulse-echo sonography: proof of principle. Ult. Med. Biol. **41**(1), 235–250 (2015). https://doi.org/10.1016/j.ultrasmedbio.2014.05.019

5. M. Jaeger, M. Frenz, Towards clinical computed ultrasound tomography in echo-mode: dynamic range artefact reduction. Ultrasonics **62**, 299–304 (2015). https://doi.org/10.1016/j.ultras.2015.06.003

6. M. Jaeger, E. Robinson, H.G. Akarcay, M. Frenz, Full correction for spatially distributed speed-of-sound in echo ultrasound based on measuring aberration delays via transmit beam steering. Phys. Med. Biol. **60**, 4497–4515 (2015). https://doi.org/10.1088/0031-9155/60/11/4497

7. P. Stähli, M. Kuriakose, M. Frenz, M. Jaeger, *Forward Model for Quantitative Pulse-Echo Speed-of-Sound Imaging.* arXiv: 1902.10639v2 [physics.med-ph]

8. M. Imbault, M.D. Burgio, A. Faccinetto, M. Ronot, H. Bendjador, T. Deffieux et al., Ultrasonic fat fraction quantification using in vivo adaptive sound speed estimation. Phys. Med. Biol. **63**, 215013 (2018). https://doi.org/10.1088/1361-6560/aae661

9. A. Hariri, J. Lemaster, J. Wang, A.S. Jeevarathinam, D.L. Chao, J.V. Jokerst, The characterization of an economic and portable LED-based photoacoustic imaging system to facilitate molecular imaging. Photoacoustics **9**, 10–20 (2018). https://doi.org/10.1016/j.pacs.2017.11.001

10. A. Hariri, E. Zhao, A.S. Jeevarathinam, J. Lemaster, J. Zhang, J.V. Jokerst, Molecular imaging of oxidative stress using an LED-based photoacoustic imaging system. Sci. Rep. **9**, 11378–11410 (2019). https://doi.org/10.1117/12.2509204

11. J. Jo, G. Xu, Y. Zhu, M. Burton, J. Sarazin, E. Schiopu et al., Detecting joint inflammation by an LED-based photoacoustic imaging system: a feasibility study. J. Biomed. Opt. **23**(11), 110501 (2018). https://doi.org/10.1117/1.JBO.23.11.110501

12. W. Xia, M.K.A. Singh, E. Maneas, N. Sato, Y. Shigeta, T. Agano et al., Handheld real-time LED-based photoacoustic and ultrasound imaging system for accurate visualization of clinical metal needles and superficial vasculature to guide minimally invasive procedures. Sensors **18**, 1394 (2018). https://doi.org/10.3390/s18051394

13. Y. Zhu, G. Xu, J. Yuan, J. Jo, G. Gandikota, H. Demirci et al., Light emitting diodes based photoacoustic imaging and potential clinical applications. Sci. Rep. **8**, 9885 (2018). https://doi.org/10.1038/s41598-018-28131-4

14. K. Daoudi, P.J. van den Berg, O. Rabot, A. Kohl, S. Tisserand, P. Brands, W. Steenbergen, Handheld probe integrating laser diode and ultrasound transducer array for ultrasound/photoacoustic dual modality imaging. Opt. Express **22**(21), 26365–26374 (2014). https://doi.org/10.1364/OE.22.026365

15. A. Fatima, K. Kratkiewicz, R. Manwar, M. Zafar, R. Zhang, B. Huang et al., Review of cost reduction methods in photoacoustic computed tomography. Photoacoustics**15**, 100137 (2019). https://doi.org/10.1016/j.pacs.2019.100137

16. M. Erfanzadeh, Q. Zhu, Photoacoustic imaging with low-cost sources: a review. Photoacoustics **14**, 1–11 (2019). https://doi.org/10.1016/j.pacs.2019.01.004

17. K. Sivasubramanian, M. Pramanik, High frame rate photoacoustic imaging at 7000 frames per second using clinical ultrasound system. Biomed. Opt. Exp. **7**(2), 312–323 (2016). https://doi.org/10.1364/BOE.7.000312

18. M. Jaeger, L. Siegenthaler, M. Kitz, M. Frenz, Reduction of background in optoacoustic image sequences obtained under tissue deformation. J. Biomed. Opt. **14**(5), 054011. https://doi.org/10.1117/1.3227038

19. M. Jaeger, S. Preisser, M. Kitz, D. Ferrara, S. Senegas, D. Schweizer, M. Frenz, Improved contrast deep optoacoustic imaging using displacement-compensated averaging: breast tumour phantom studies. Phys. Med. Biol. **56**, 5889–5901 (2011). https://doi.org/10.1088/0031-9155/56/18/008

20. M. Jaeger, D.C. Harris-Birtill, A. Gertsch, E. O'Flynn, J. Bamber, Deformation compensated averaging for clutter reduction in epiphotoacoustic imaging in vivo. J. Biomed. Opt. **17**(6), 066007 (2012). https://doi.org/10.1117/1.JBO.17.6.066007

21. M. Jaeger, K. Gashi, H.G. Akarcay, G. Held, S. Peeters, T. Petrosyan et al., Real-time clinical clutter reduction in combined epi-optoacoustic and ultrasound imaging. Photonics Lasers Med. **3**(4), 343–349 (2014). https://doi.org/10.1515/plm-2014-0028

22. M. Jaeger, J.C. Bamber, M. Frenz, Clutter elimination for deep clinical optoacoustic imaging using localised vibration tagging (LOVIT). Photoacoustics **1**, 19–29 (2013). https://doi.org/10.1016/j.pacs.2013.07.002

23. T. Petrosyan, M. Theodorou, J. Bamber, M. Frenz, M. Jaeger, Rapid scanning wide-field clutter elimination in epi-optoacoustic imaging using LOVIT. Photoacoustics **10**, 20–30 (2018). https://doi.org/10.1016/j.pacs.2018.02.001

24. M.K.A. Singh, M. Jaeger, M. Frenz, W. Steenbergen, Photoacoustic reflection artifact reduction using photoacoustic-guided focused ultrasound: comparison between plane-wave and element-by-element synthetic backpropagation approach. Biomed. Opt. Exp. **8**(4), 2245–2260 (2017). https://doi.org/10.1364/BOE.8.002245

25. M.K.A. Singh, M. Jaeger, M. Frenz, W. Steenbergen, In vivo demonstration of reflection artifact reduction in photoacoustic imaging using synthetic aperture photoacoustic-guided focused ultrasound (PAFUSion). Biomed. Opt. Exp. **7**(8), 2955–2972 (2016). https://doi.org/10.1364/BOE.7.002955

26. M.K.A. Singh, W. Steenbergen, Photoacoustic-guided focused ultrasound (PAFUSion) for identifying reflection artifacts in photoacoustic imaging. Photoacoustics **3**(4), 123–131 (2015). https://doi.org/10.1016/j.pacs.2015.09.001

27. H.-M. Schwab, M.F. Beckmann, G. Schmitz, Photoacoustic clutter reduction by inversion of a linear scatter model using plane wave ultrasound measurements. Biomed. Opt. Exp. **7**, 1468–1478 (2016). https://doi.org/10.1364/BOE.7.001468

28. G. Held, S. Preisser, H.G. Akarcay, S. Peeters, M. Frenz, Effect of irradiation distance on image contrast in epi-optoacoustic imaging of human volunteers. Biomed. Opt. Exp. **5**(11), 3765–3780 (2014). https://doi.org/10.1364/BOE.5.003765

29. S. Preisser, G. Held, H.G. Akarcay, M. Jaeger, M. Frenz, Study of clutter origin in in-vivo epi-optoacoustic imaging of human forearms. J. Opt. **18**, 094003–94009 (2016). https://doi.org/10.1088/2040-8978/18/9/094003

30. M. Tanter, M. Fink, Ultrafast imaging in biomedical ultrasound. IEEE Trans. Ult. Ferr. Freq. Cont. **61**(1), 102–119 (2014). https://doi.org/10.1109/TUFFC.2014.6689779

31. G. Montaldo, M. Tanter, J. Bercoff, N. Benech, M. Fink, Coherent plane-wave compounding for very high frame rate ultrasonography and transient elastography. IEEE. Trans. Ult. Ferr. Freq. Cont. **56**(3), 489–506 (2009). https://doi.org/10.1109/TUFFC.2009.1067

32. S. Freeman, P.-C. Li, M. O'Donnell, Retrospective dynamic transmit focusing. Ult. Imag. **17**, 173–196 (1995). https://doi.org/10.1006/uimg.1995.1008

33. T. Loupas, J.T. Powers, R.W. Gill, An axial velocity estimator for ultrasound blood flow imaging, based on a full evaluation of the Doppler equation by means of a two-dimensional autocorrelation approach. IEEE Trans. Ult. Ferr. Freq. Cont. **42**(4), 672–688 (1995). https://doi.org/10.1109/58.393110

34. M. O'Donnell, A.R. Skovoroda, B.M. Shapo, S.Y. Emelianov, Internal displacement and strain imaging using ultrasonic speckle tracking IEEE Trans. Ult. Ferr. Freq. Cont **41**(3), 314–325 (1994). https://doi.org/10.1109/58.285465

35. E. Konofagou, J. Ophir, A new elastographic method for estimation and imaging of lateral displacements, lateral strains, corrected axial strains and poisson's ratios in tissues. Ult. Med. Biol. **24**(8), 1183–1199 (1998). https://doi.org/10.1016/s0301-5629(98)00109-4

36. P. Chaturvedi, M.F. Insana, T.J. Hall, 2-D companding for noise reduction in strain imaging. IEEE Trans. Ult. Ferr. Freq. Cont. **45**(1) (1998). https://doi.org/10.1109/58.646923

37. W.-N. Lee, C.M. Ingrassia, S.D. Fung-Kee-Fung, K.D. Costa, J.W. Holmes, E.E. Konofagou, Theoretical quality assessment of myocardial elastography with in vivo validation. IEEE Trans. Ult. Ferr. Freq. Cont. **54**(11), 2233–2245 (2007). https://doi.org/10.1109/TUFFC.2007.528

38. E. Weinstein, A.J. Weiss, Fundamental limitations in passive time-delay estimation—part II: wide-band systems. IEEE Trans. Acoust. Speech Signal Process **32**(5), 1064–1078 (1984). https://doi.org/10.1109/TASSP.1984.1164429

39. M.A. Lubinski, S.Y. Emelianov, M. O'Donnell, Speckle tracking methods for ultrasonic elasticity imaging using short-time correlation. IEEE Trans. Ult. Ferr. Freq. Cont. **46**(1), 82–96 (1999). https://doi.org/10.1109/58.741427

40. T. Shiina, N. Nitta, E. Ueno, E., J.C. Bamber, Real time tissue elasticity imaging using the combined autocorrelation method. J. Med. Ult. **26**(2), 57–66. https://doi.org/10.1007/BF02481234

41. G. Diot, S. Metz, A. Noske, E. Liapsis, B. Schroeder, S.V. Ovsepian et al., Multispectral optoacoustic tomography (MSOT) of human breast cancer. Clin. Cancer Res. **23**(22), 6912–6922. https://doi.org/10.1158/1078-0432.CCR-16-3200

42. A. Dima, V. Ntziachristos, Non-invasive carotid imaging using optoacoustic tomography. Opt. Express **20**(22), 25044–25057 (2012). https://doi.org/10.1364/OE.20.025044

43. X.L. Dean-Ben, D. Razansky, Functional optoacoustic human angiography with handheld video rate three dimensional scanner. Photoacoustics **1**, 68–73 (2013). https://doi.org/10.1016/j.pacs.2013.10.002

44. K.G. Held, M. Jaeger, J. Ricka, M. Frenz, H.G. Akarcay, Multiple irradiation sensing of the optical effective attenuation coefficient for spectral correction in handheld OA imaging. Photoacoustics **4**(2), 70–80 (2016). https://doi.org/10.1016/j.pacs.2016.05.004

45. M.U. Arabul, M. Heres, M.C.M. Rutten, M.R. van Sambeek, F.N. van de Vosse, R.G.P. Lopata, Toward the detection of intraplaque hemorrhage in carotid artery lesions using photoacoustic imaging. J. Biomed. Opt. **22**(4), 041010 (2017). https://doi.org/10.1117/1.JBO.22.4.041010

46. M.E. Anderson, Multi-dimensional velocity estimation with ultrasound using spatial quadrature. IEEE Trans. Ult. Ferr. Freq. Cont. **45**(3), 852–861 (1998). https://doi.org/10.1109/58.677757

47. L. Chen, G.M. Treece, J.E. Lindop, A.H. Gee, R.W. Prager, A quality-guided displacement tracking algorithm for ultrasonic elasticity imaging. Med. Imag. Anal. **13**(2), 286–296 (2009). https://doi.org/10.1016/j.media.2008.10.007

48. Y. Petrank, L. Huang, M. O'Donnell, Reduced peak-hopping artifacts in ultrasonic strain estimation using the Viterbi algorithm IEEE Trans. Ult. Ferr. Freq. Cont. **56**(7), 1359–1367 (2009). https://doi.org/10.1109/TUFFC.2009.1192
49. H. Rivaz, E. Boctor, P. Foroughi, R. Zellars, G. Fichtinger, G. Hager, Ulstrasoundelastography: a dynamic programming approach. IEEE Trans. Med. Imaging **27**(10) (2008). https://doi.org/10.1109/TMI.2008.917243

Ultrasound Receive-Side Strategies for Image Quality Enhancement in Low-Energy Illumination Based Photoacoustic Imaging

Sowmiya Chandramoorthi and Arun K. Thittai

Abstract PAT (Photoacoustic Tomography) is a hybrid noninvasive imaging modality that provides functional cum structural information about the underlying tissue medium. Conventional PAT employs bulky and expensive solid-state pulsed lasers as an illumination source, however, Light Emitting Diodes (LED) and Pulsed Laser Diodes (PLD) have been recently explored as a suitable alternative that are portable and inexpensive. However, its depth of penetration is relatively lower than that of solid-state lasers due to lower energy per pulse. Averaging of multiple frames is usually employed as a common practice in high PRF LED/PLD systems to improve the PAT image SNR. Recently an approach of sub-pitch translation of ultrasound linear array was demonstrated to contribute to improvement in SNR of PAT image, with just fewer number of frame averaging. In this chapter, the various methods proposed in literature for improving the achievable image SNR in low energy LED/PLD based PAT systems are described. Specifically, details of the simulation and experimental studies conducted using sub-pitch translation approach are provided. Overall, this chapter briefly summarizes the reports that demonstrate feasibility of achieving improvement in image quality by employing novel methods at receive-side (ultrasound data acquisition and beamforming) while using low energy sources of illumination in PAT.

1 Introduction

Photoacoustic Tomography (PAT) is an emerging biomedical imaging modality that combines contrast from pure optical imaging and spatial resolution from ultrasound (US) imaging. In PAT, the tissue medium under investigation is illuminated with a

S. Chandramoorthi · A. K. Thittai (✉)
Department of Applied Mechanics, Indian Institute of Technology, Madras, Chennai 600036, India
e-mail: akthittai@iitm.ac.in

S. Chandramoorthi
e-mail: sowmiyachandramoorthy@gmail.com

© Springer Nature Singapore Pte Ltd. 2020
M. Kuniyil Ajith Singh (ed.), *LED-Based Photoacoustic Imaging*,
Progress in Optical Science and Photonics 7,
https://doi.org/10.1007/978-981-15-3984-8_4

short-pulsed laser beam, which causes increase in temperature of specific tissue chromophores due to absorption of the incident electromagnetic energy. This absorption causes heating resulting in thermal expansion of the absorber. This leads to the generation of ultrasonic waves [1, 2]. Typically, high energy solid state lasers, such as Nd: YAG, are most widely used as excitation sources for PAT systems. These sources are capable of delivering high energy of ~100 mJ that produces sufficient signal-to-noise ratio (SNR) required for performing deep photoacoustic imaging [3–9]. However, these lasers are bulky, expensive and have low pulse repetition rate (PRF) (<10 Hz), which limits its utility for performing real-time photoacoustic imaging in clinical environment and point-of-care applications. With this motivation in mind, several researchers have explored the use of low-cost, low-energy pulsed light sources that can emit pulses at high PRF in the order of kHz as a potential alternative.

Light Emitting Diode (LED) and Pulsed Laser Diode (PLD) are two kinds of light sources that have been successfully demonstrated as an alternative to Nd: YAG in recent years [10–16]. But the maximum achievable energy per pulse while using LED/PLD as source is relatively lower than that of Nd: YAG and other solid state sources. This directly affects its depth of penetration in the tissue medium, due to lower signal strength of the ultrasound signal travelling from deep-seated optical absorbers. Modifications/improvisations to the illumination-side or ultrasound receive side factors could potentially lead to improved signal-to-noise ratio (SNR). In this chapter, the emphasis will be on the receive-side ultrasound data acquisition schemes and beamforming methods that can contribute to achieving improved resolution, contrast and depth of penetration in LED/PLD based PAT systems.

Averaging of multiple frames is employed as a common practise to improve SNR and to extend the achievable imaging depth by exploiting the high PRF of LED/PLD laser sources. However, averaging over a large number of frames, typically thousands, reduces the effective frame rate of the photoacoustic (PA) images. Further, the improvement in SNR due to averaging process is only proportional to the square root of the number of frames averaged. Hence recently, development of novel methods of US data acquisition and beamforming to compensate for low SNR while keeping the number of frame averaging low in LED/PLD based PAT imaging system are being explored widely. Some of the studies that employ standard signal processing approaches includes usage of adaptive denoising [17], empirical mode decomposition [18], wavelet transform [19], Wiener deconvolution [20], deep neural network [21], filtered delay multiply and sum [22], double stage delay multiply and sum [23] or short lag spatial coherence [24]. Most of them are software-based approaches that are primarily focused on the beamforming algorithm or post processing in frequency domain. Methods adapting corrective measures in hardware to improve image quality in the context of LED/PLD PAT systems are quite few.

Most of the common PAT reconstruction techniques assume ideal conditions, such as, (1) homogeneous sound speed, (2) full-angle view, (3) impulse excitation, (4) wideband detection, (5) point detector measurement, (6) continuous sampling, amongst others [1]. However, it is not practically feasible to satisfy all the above-mentioned conditions. For example, inhomogeneous acoustic properties in the medium blurs an image reconstructed assuming uniformity, because the variations in speed of sound causes significant changes in the time of flight of sound from the source to detectors. In addition, a finite-sized aperture with partial detection view, finite-sized detector elements instead of a point detector and discontinuously sampled array that causes gaps in collected raw RF data are typically employed. The above violations cause degradation in practical spatial resolution obtained, as compared to that predicted theoretically. Improvements in image quality by compensating for each of these violations has been demonstrated using either software or hardware based techniques [25–33].

However, methods to address violation due to spatial discrete sampling of the transducer are very few in linear array PAT literature. Here, the use of a simple and inexpensive way of increasing the discrete spatial sampling of the array is described. This is done by employing a strategy of actuator-assisted translation of the linear array transducer for acquiring data from sub-pitch locations. This method was found to better track the discontinuities in optical absorption at sub-λ level resulting in an overall improvement in image quality. This has been demonstrated through simulation and experiments [34, 35]. In fact, this is the only work that has been reported on the effect of increasing the discrete spatial sampling of the detector elements on the achievable image quality in low energy PAT systems.

Further, due to the weak light absorption-to-ultrasound conversion efficiency for LED and PLD sources, the recorded photoacoustic raw radiofrequency (RF) data has lower ratio of signal content versus noise levels. The main sources of external noises include laser induced electronic noise, jitter, EMI (electromagnetic interference) from surrounding equipment, etc. These noise factors have pronounced detrimental effect on the quality of the reconstructed PA images resulting in further lowering of the SNR values. Here, it is demonstrated that by strategically employing a noise-reduction filter on the raw data prior to reconstruction stage yields significant improvement in SNR (~61%) compared to mere averaging in PLD-based PAT systems.

This chapter is organized in the following manner. In Sect. 2, a review of some of the key image reconstruction/ ultrasound reception-side strategies proposed in literature for improving the image quality of PAT systems has been summarized. In Sect. 3, the sub-pitch translation approach as a methodology of improving the resolution, contrast and depth of penetration for low-SNR PAT settings is described. Further, results obtained from simulation and experiments conducted with varying source/medium and receive side parameters is also reported. In Sect. 4, a frequency domain filtering method to remove the effect of EMI from raw RF data is described. The results obtained from employing the said method are summarized subsequently.

2 Review of Image Enhancement Strategies for PAT Systems

In this section, a review of the articles published in this context of achieving improved image quality in PAT systems using ultrasound receive-side innovations is summarized. Firstly, the methods reported on LED-PAUS system are discussed. Later, a few relevant ultrasound hardware-based techniques reported using other sources in PAT literature are also discussed.

2.1 LED-PAUS Imaging System

In this subsection, two such strategies reported to achieve improved PA image quality while employing the commercially available LED-PAUS imaging system (Acous-ticX, Cyberdyne Inc., Ibaraki, Japan) is presented. This scanner is capable of acquiring interleaved PA and US B-mode images in real time. This system uses LED arrays for PA excitation and a linear US transducer for ultrasonic excitation and detection. For B-mode PA/US data acquisition, two LED arrays are positioned on either side of the US probe as shown in Fig. 1. Each of these LED arrays consists of four rows of 36 1 mm × 1 mm LED elements. These LED arrays are capable of delivering a maximum optical energy of 200 µJ per pulse and can be driven with a repetition rate of 1–4 kHz with pulse duration of 30–150 ns. The ultrasound probe is a lead zirconate titanate (PZT) 128-element linear array transducer having a pitch of 0.3 mm and total length of 38.4 mm. The central frequency of the transducer is 7 MHz and the measured −6 dB bandwidth is 75%. The ultrasound and photoacoustic modalities

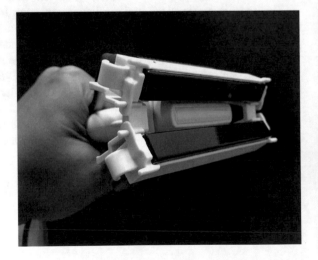

Fig. 1 Commercial LED-PAUS probe with two LED arrays affixed on both sides of the linear array transducer (provided by Cyberdyne, Inc.)

have sampling rates of 20 and 40 MHz, respectively. This LED-PAUS system has been tested for utility in several phantom as well as in vivo studies [15], [36–39].

2.1.1   Evaluation of DS-DMAS on LED-PAUS Imaging Dataset

Signal processing algorithms can improve SNR in photoacoustic data. The most common beamforming algorithm in linear-array PAT is delay-and-sum (DAS). Delay-multiply and Sum (DMAS) has been demonstrated to yield superior image quality compared to DAS for linear-array PAT [22]. However, it still suffers from low contrast when noise is present in the dataset. In order to reduce the effects of off-axis signals in reconstructed images, while retaining resolution, the DMAS is combined with eigenspace minimum variance beamformer [40]. In this work reported by Mozaffarzadeh et al. [23] a Double Stage-DMAS (DS-DMAS) has been shown to further improve image quality in comparison to DAS and DMAS while using the LED-based scanner.

The equation corresponding to the conventional DAS beamformer is as follows:

$$y_{DAS}(k) = \sum_{i=1}^{M} x_i(k - \Delta_i) \tag{1}$$

where, $y_{DAS}(k)$ is the output of beamformer, k is time index, M is the number of array elements, and $x_i(k)$ and Δ_i are detected signals and corresponding time delay for detector i, respectively. One of the algorithms introduced to improve the DAS beamformer performance is DMAS, which is written as follows

$$y_{DMAS}(k) = \sum_{i=1}^{M-1} \sum_{j=i+1}^{M} x_i(k - \Delta_i)x_j(k - \Delta_j) = \sum_{i=1}^{M-1} \sum_{j=i+1}^{M} x_{id}(k)x_{jd}(k) \tag{2}$$

$$
\begin{aligned}
&= [x_{1d}(k)x_{2d}(k) + x_{1d}(k)x_{3d}(k) + \cdots + x_{1d}(k)x_{Md}(k)] \\
&+ [x_{2d}(k)x_{3d}(k) + x_{2d}(k)x_{4d}(k) + \cdots + x_{2d}(k)x_{Md}(k)] + \cdots \\
&+ \left[x_{(M-2)d}(k)x_{(M-1)d}(k) + x_{(M-2)d}(k)x_{Md}(k)\right] \\
&+ \left[x_{(M-1)d}(k)x_{Md}(k)\right]
\end{aligned} \tag{3}
$$

where, $x_{id}(k)$ and $x_{jd}(k)$ are delayed detected signals for element i and j($x_i(k - \Delta_i)$ and $x_j(k - \Delta_j)$), respectively. This DMAS is a non-linear algorithm and uses a correlation process to form a high contrast photoacoustic image, however, DMAS is still insufficient when high level of imaging noise is present in the data. Hence, another stage of correlation process inside the DMAS was suggested to suppress noise and artifact unmitigated by the DMAS in a method named, Double Stage Delay Multiply and Sum (DS-DMAS). The formula of DS-DMAS is as follows

$$y_{DS-DMAS}(k) = \sum_{i=1}^{M-2} \sum_{j=i+1}^{M-1} x_{it}(k)x_{jt}(k) \qquad (4)$$

where, x_{it} and x_{jt} are the ith and jth term shown in [3]. This DS-DMAS beamformer was evaluated on LED-PAUS imaging system experimentally using point targets, as well as a hair and a rabbit eye. To illustrate the effect of this method, example results obtained by the authors from hair phantom and point target phantom is shown in Figs. 2 and 3, respectively.

This algorithm has been demonstrated to compensate for the low SNR of LED-based systems and offer better lateral resolution of about 60% and 25% when compared to DAS and DMAS, respectively. In addition, it results in higher contrast ratio

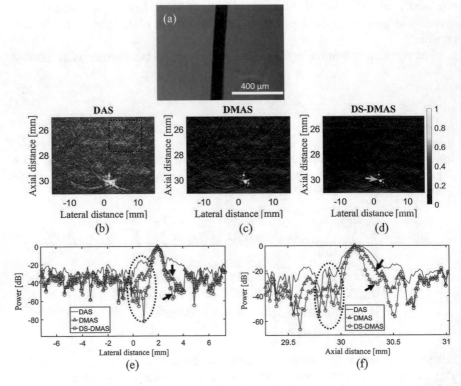

Fig. 2 **a** The microscopy image of a hair. Reconstructed photoacoustic images using the data generated by the hair using (**b**) DAS, **c** DMAS, and **d** DS-DMAS algorithms. All images are shown with a dynamic range of 40 dB (for better evaluation). 100 frames of the detected photoacoustic signals were averaged to have a higher SNR. The arrow points to the target of imaging. The dashed square shows the region that was used for contrast evaluation. DS-DMAS suppresses the sidelobes about −38 dB and −23 dB, in comparison with DAS and DMAS, respectively. **e** and **f** are the lateral and axial variations of the images, respectively. The arrows and circle show the level of sidelobes where the superiority of DS-DMAS is proved. (reprinted with permission from [23])

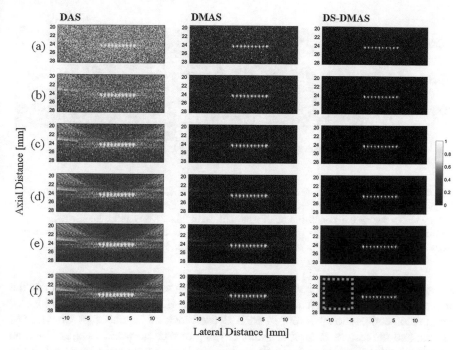

Fig. 3 Reconstructed photoacoustic images of the point-target phantom. DAS, DMAS, and DS-DMAS were used for generating the first, second, and third columns of the images, respectively. All images are shown with a dynamic range of 60 dB. The images obtained by averaging **a** 1 frame, **b** 3 frames, **c** 10 frames, **d** 20 frames, **e** 50 frames, and **f** 100 frames of the detected photoacoustic signals are shown. The background noise is suppressed (darker background) using a higher number of frames. The dotted square is used for CR calculation. (reprinted with permission from [23])

of about 97% and 34% than DAS and DMAS, respectively. Further, DS-DMAS offers this using a smaller number of frames (only 2% of all the frames).

2.1.2 Deep Neural Network for LED-PAUS Imaging System

In the work by Anas et al. [21] a deep neural networks based image enhancement approach to improve the quality as well as reduce the scanning time of LED-based PA images has been reported. The proposed architecture uses convolutional neural networks (CNN) to extract the spatial features and recurrent neural networks (RNN) to leverage the temporal information in PA images. The CNN is built upon a densenet-based architecture that uses series of skip-connections to enhance the image content. For the RNN component, a convolutional variant of short-long-term-memory is used to exploit the temporal dependencies in a given PA image sequence.

Phantoms containing wire & magnetic nanoparticle target and in vivo human fingers have been scanned using the LED-PAUS system. All of the experimentally acquired data is divided into training, validation and test sets. The training set is used

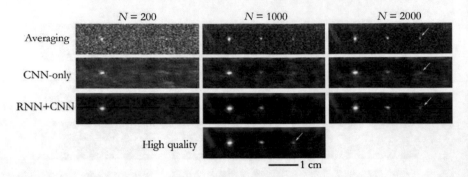

Fig. 4 Comparative performance of the simple averaging, CNN-only and RNN + CNN methods in detection of point target objects with different concentrations. A successive decrease in signal quality with a decrease in concentration of nanoparticle in tubes 3–5 is noticeable. In addition, the improved performance of RNN + CNN method can be observed (marked by arrow) with respect to the two other methods. (reprinted with permission from [21])

for optimizing the network parameters; the validation set, in contrast, is used to fix the hyper-parameters of the architecture that mainly include number of dense blocks, number of convolutional layers in each dense block and number of ConvLSTM layers; and the test set is used to evaluate the proposed network. Results obtained from optical contrast analysis on the nanoparticle phantom embedded with tubes containing decreasing nanoparticle concentration from tube 3–5 (left to right) is shown in Fig. 4. Comparison among three competing techniques namely; Averaging, CNN only and RNN + CNN for three different values of number of frames averaged is shown. The PA images in this figure indicate a successive decrease in the image quality with a decrease in the concentration of the optical absorber. A better recovery of the target object for the proposed RNN + CNN method (shown by arrows) when compared to the other two methods is observed.

2.2   Effect of Hardware-Based Improvisations Implemented on Ultrasound Transducer for PAT Application

Some of the key hardware-based methods implemented on the ultrasound transducer for PAT application includes usage of a negative acoustic lens reported in the work by Manojit et al. [33], placing an acoustic reflector in the work by Bin Huang et al. [41] and increasing array element density in curved array PAT by Dima et al. [42].

2.2.1 Negative Acoustic Lens and Modified DAS

In the work by Manojit et al. [33] an improvement in tangential resolution is achieved by attaching an acoustic concave lens, made of acrylic in front of the flat detector surface.

Briefly, in planar circular scanning geometry mode of PAT, for a given transducer bandwidth, the aperture size of the detector affects the tangential resolution greatly when the object of interest is near the detector surface. In the work by Manojit et al. this issue of deteriorating tangential resolution was overcome by attaching an acoustic concave lens, made of acrylic in front of the flat detector surface. Figure 5A shows the photographs of the transducers with and without the negative cylindrical lens. Figure 5B shows the reconstructed images of the needle with a flat detector when the needle was placed at different distances from the scanning center.

From the results it is evident that, the tangential resolution is poor when the target object is far from the scanning center and it is improved significantly with the use of the negative lens. However, there are some practical challenges with this method. Attaching the in-house made acoustic lens to the detector surface without the formation of any air bubbles was found to be difficult. Also, absorption of ultrasound signal and the impedance mismatch between the negative lens and the acoustic coupling medium (water/mineral oil) resulted in loss of signal. Manufacturing a custom-made detector with curved piezo surface or a negative lens attached to the piezo surface inside the transducer are very expensive. Hence the author later proposed a modified

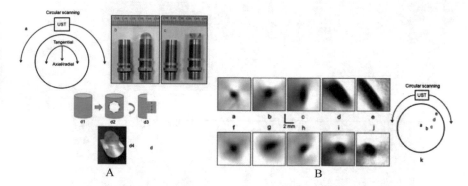

Fig. 5 **A**: **a** Diagram showing how radial and tangential resolution is defined in planar circular scanning configuration. **b** and **c** Photographs of the flat ultrasonic transducer and the ultrasonic transducer glued to a negative cylindrical lens made of acrylic. The active area of the detector was completely covered by the lens. Minor ticks of scale in **b** and **c** corresponds to 1 mm. **d** Step-by-step schematic of how the negative cylindrical lens is made from an acrylic cylindrical rod. **B**: Reconstructed TAT images, using the flat ultrasonic detector, of a needle (18 gauge, 1 mm in diameter) inserted inside a pork fat base placed at a distance of **a** ~4 mm, **b** ~14 mm, **c** ~32 mm, **d** ~50 mm, and **e** ~64 mm from the scanning center. Corresponding TAT images obtained with the negative lens detector are shown in (**f**), (**g**), (**h**), (**i**), and (**j**), respectively. **k** Location of the needle inside the scanner is shown (reprinted with permission from [33])

DAS method to improve tangential resolution by removing point detector assumption and accounting for the actual detector surface dimension during beamforming [32, 43]. In conventional DAS, the backprojection term is given by,

$$b(\vec{r_0},t) = 2p(\vec{r_0},t) - 2ct\frac{\partial p(\vec{r_0},t)}{\partial t} \tag{5}$$

where, $b(\vec{r_0},t)$ is the backprojection term and $p(\vec{r_0},t)$ is the measured acoustic signals. In Eq. (5), large-aperture detectors are considered as a point detector, typically at the center of the transducer surface. This introduces artifacts in the reconstructed image. Hence a modified delay-and-sum reconstruction algorithm, where the entire surface area of the detector is considered during backprojection was proposed. In modified DAS, the recorded pressure signal at r_0 is represented as a surface integral over the detector aperture given by,

$$p'(\vec{r_0},t) = \iint p(\vec{r_0'},t) W(\vec{r_0'}) d^2 \vec{r_0'} \tag{6}$$

This is achieved by considering many small segments on the detector from which the recorded PA signal $p'(\vec{r_0},t)$ is backprojected instead of a single center point of the line. The results obtained from simulation is shown in Fig. 6.

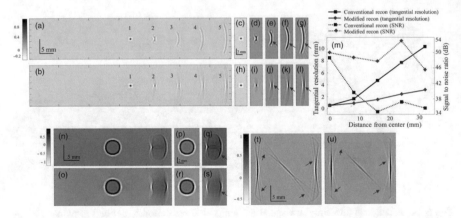

Fig. 6 a–l Simulation results for point source phantoms. **a** Conventionally reconstructed PAT images of 5-point targets. **b** Reconstructed using modified delay-and-sum reconstruction algorithm. **c–g** Zoomed in point targets 1–5 in (**a**). **h–l** Zoomed in point targets 1–5 in (**b**). **m** Comparison of the tangential resolution and SNR between conventional and modified reconstruction algorithm as a function of distances from the scanning center. **n–s** Simulation results for circular shaped numerical phantom. **n** Conventionally reconstructed PAT images of two circles at different distances from scanning center. **o** Reconstructed modified reconstruction algorithm. **p** and **q** Zoomed-in individual circles in (**n**). **r** and **s** Zoomed in individual circles in (**o**). **t** Conventionally reconstructed PAT image of N-shaped blood vessel numerical phantom. **u** Reconstructed using modified reconstruction. Red arrow points to the places where improvements are clearly visible. (reprinted with permission from [32])

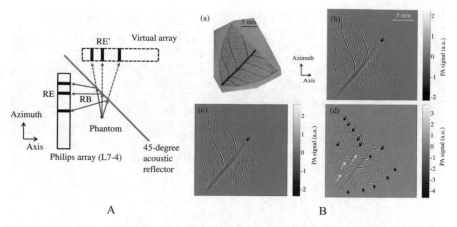

Fig. 7 A: Top view of the experimental setup. RB, reflected beam; RE, physical receiving element; RE′, virtual receiving element. The commercial Philips array operated in B-mode to collect photoacoustic signals, a 45-deg acoustic reflector (glass) formed a virtual array, and a laser illuminated light orthogonally to the drawing from the top (not shown in the figure). **B**: PAT images of a leaf skeleton phantom. **a** Photo of the phantom. **b** Image of the leaf skeleton phantom acquired without the presence of the acoustic reflector. **c** Image of the phantom acquired with the presence of the reflector, but reconstructed without incorporating data from the virtual array. **d** Image of the phantom acquired with the presence of the acoustic reflector and reconstructed with data from the virtual array incorporated (reprinted with permission from [41])

2.2.2 Acoustic Reflector

In the work by Huang et al. [41] a simple way of handling limited view problem of linear array PAT with the use of a 45° acoustic reflector is proposed. The Fig. 7A shows the experimental setup in a water tank in top view and Fig. 7B shows the results obtained using proposed method on phantom consisting of a leaf skeleton embedded in agar.

Figure 7B b) is an image acquired without the presence of the reflector, and Fig. 7B c) is an image acquired with the presence of the acoustic reflector, but reconstructed without incorporating the virtual array. In both images, the major skeletons on the lower-right side of the leaf were missing due to limited view. The images shown in 7B d) clearly indicate that by incorporating the virtual array, those missing skeletons can be recovered, as indicated by the arrows.

2.2.3 Large Number of Detector Elements

In the work by Dima et al. [42], improvements in PAT image quality was demonstrated by utilizing a densely-packed larger number of detector elements implemented on Multispectral Optoacoustic Tomography (MSOT) having 64, 128 and 256 elements on curved array transducers. This imaging study was done on phantoms and animals imaged under similar conditions. Figure 8 shows the experimental results obtained

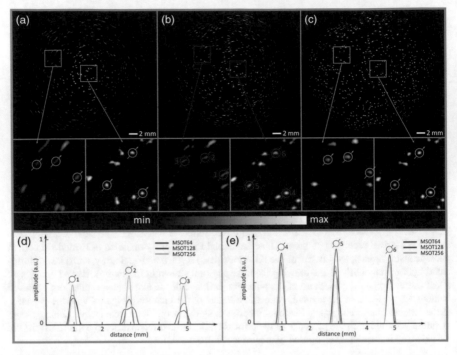

Fig. 8 Top row shows experimental phantom reconstruction results from **a** MSOT64, **b** MSOT128, and **c** MSOT256. For each image, negative values were set to zero and the remainder normalized to 1. Furthermore, images were segmented based on their local maxima. Details at the periphery and the center, framed ($4 \times 4 \text{ mm}^2$) and color-coded by array, were magnified for display at the bottom left and right, respectively. Graphs below depict diagonal cross-sections of individual microspheres (circled blue, red, and green) for **d** the periphery (numbered 1 to 3) and **e** the center (numbered 4 to 6). Individual cross-sections were artificially spaced to allow sufficient distance for comparison. (reprinted with permission from [42])

from black polyethylene microsphere phantom while using the three different arrays. The results demonstrated higher PA sensitivity for MSOT128 compared to MSOT64 and the highest sensitivity for MSOT256.

3　Sub-pitch Translation for Improving PAT Image Quality

Previous studies that attempted at increasing the array sampling density, did so by manufacturing an array containing densely packed elements [30, 31]. However, manufacturing a dense array transducer having smaller inter-element spacing and smaller element width is complex and hence expensive. Such special types of transducers are not a default in standard clinical US scanners that are currently in use. On the other hand, usage of a conventional linear array transducer with a reasonable footprint that comes as a default in most standard US clinical scanners seem to be a more preferred choice for PAT. A typical linear array is normally sampled by pitch distance that is equal to λ, corresponding to the center frequency of that particular linear array. In cases where the target dimension is much smaller than λ, a regular λ-pitch transducer may not be sufficient to reconstruct the actual target dimension and may also result in low SNR. For example, microvasculature comprises of small feed arteries, arterioles, venules and capillaries. The arterioles typically range in diameter from about 5 to 100 μm [44] and small feed arteries immediately upstream from the arterioles have typical diameter ranging between 100 and 400 μm [45].

In such scenarios a denser sampling of the array and acquisition of new raw RF lines at sub-pitch locations may be crucial to provide better image contrast and resolution in the reconstructed images. However, manufacturing a transducer with "sub-λ" pitch, especially at higher frequencies (>10 MHz) is very expensive and makes the system complex. Hence, there is a need for alternate inexpensive ways of increasing discrete array sampling in PAT.

Recently, the possible improvements to image resolution, contrast and depth of penetration in a PLD-based PAT system by employing an actuator-assisted sub-pitch translation of the linear array transducer was investigated and reported [35]. This method was found to better track the discontinuities in optical absorption at sub-λ level resulting in an overall improvement in image quality. This has been demonstrated through simulation and experiments.

In the following section a detailed description of the theory and methodology behind sub-pitch translation approach is given.

3.1 Theory and Methodology

3.1.1 Effect of Discrete Spatial Sampling on Image Quality: Theory

The image contrast and resolution improvement of PAT images (which is an ultrasound signal at receive) due to spatial sampling is majorly based on two factors: Nyquist criterion and the PSF, respectively, which are described below in detail.

(i) *Nyquist criterion (half wavelength rule)*:

In many imaging systems, it is not practically feasible to satisfy the Nyquist criterion in the spatial domain. Particularly in ultrasound imaging, there is a challenge of manufacturing physical array elements of the necessary size, which prevents the system from adhering to the sampling rule even at the clinical ultrasound wavelength corresponding to the center frequency of the array. This directly impacts PA signal acquired using ultrasound array detectors.

The photoacoustic signal measured at the boundary is sampled in space with a series of detector elements placed at discrete spatial positions and in time with a temporal sampling frequency. According to the Nyquist criterion, to accurately reconstruct the measured signal, the sampling frequency must be greater than at least twice the maximum frequency of the signal. In the spatial domain, if the incident PA signal has a spatial frequency, k, then the bandlimited temporal frequency, ω_t, should equal ck, where, c is the speed of ultrasound in the medium. To resolve all the spatial information, the spatial sampling frequency k_s should satisfy the condition $k_s > 2$ k and hence the signal must be sampled with a spatial sampling interval δ_s < 1/2 k [46].

$$\delta_s < \frac{c}{2\omega_t} \tag{7}$$

$$\delta_s < \frac{\lambda}{2} \tag{8}$$

Therefore, if the spatial sampling period is less than $\lambda/2$, then aliasing related to spatial discrete sampling can be significantly reduced contributing to improvement in image quality. However, in a regular linear array the spatial sampling interval is restricted to λ instead of the necessary $\lambda/2$.

(ii) *PSF Versus Spatial Sampling*:

System PSF can be defined as the spatial integration of signal from point source over the individual detector elements in a periodic linear array. Several factors affect the PSF of an imaging system, namely, detection view angle, imaging depth, element width and spatial sampling. In a linear array transducer the limited detection view angle (α) has a significant effect on the lateral resolution [47]. The lateral resolution of the imaging system (P) is inversely proportional to the numerical aperture sin α. Hence the empirical relationship between α and P is given by [47]

$$P(\alpha) = \frac{r_1}{\sin \alpha} \tag{9}$$

where, $r_1 = 0.3$. According to the equation [7], the resolution deteriorates with decrease in the detection view angle α. This detection view angle in turn decreases with increase in imaging depth, which is given by the relation, $\tan \alpha = \frac{D}{2z}$, where z is the imaging depth and D is the transducer array length. In the case of a single element scanning based measurement system for a planar recording geometry, the effect of element width on the PSF has been derived in detail in literature [48, 49]. Wherein, it is stated that the lateral extension of the PSF is approximately equal to the element width. Further, it has been shown that when the element is scanned with a spacing less than or equal to half the diameter of the detector element there is significant reduction in aliasing artifacts [50]. Here, this is further extended by increasing the discrete spatial sampling to smaller sub-λ steps for the case of a periodic linear array scanning based measurement system and its impact on PAT image quality was studied while keeping the other transducer parameters such as element width, detection view angle and the imaging depth constant.

Rigorous derivation of PSF for a periodic linear array in ultrasound imaging has been reported recently [47]. The final expression of PSF with respect to PA imaging context is reiterated here for convenience of the readers. Let us assume a point target located at q, image field point r and array of detector elements denoted by u. Then the general form of the PSF for a periodic linear array is given by:

$$P(r,q) = \int F_0(\omega)\{ \int E(u) W(u,r) D(\hat{q}';\omega) B(q';\omega) e^{ik(|r'|-|q'|)} du \}^2 d\omega \quad (10)$$

where $r' = r-u$ and $q' = q-u$. $F_0(\omega)$ is the combined frequency response of the PA input signal and the receiver array element. And,

$$E(u) = \sum_{a=-\infty}^{\infty} \delta(u - u_0 - ap\hat{p}) \quad (11)$$

is the sampling function that describes the position of element centers in a 1D linear array, u_0 is the position of a reference element, p is the pitch, $\hat{p}$ is the unit vector and $W(u, r)$ is the weighting function. The other factors in equation [9] are as follows: $D(\hat{q}';\omega)$ is the directivity of array elements, $B(q';\omega)$ is the beam spread associated with wave propagation, and the exponent is the phase shift associated with wave propagation and imaging algorithm.

At the pre-staggered resolution level, the span of the spatial detector integration will not be able to exceed the detector element pitch. However, on the upsampled sub-λ grid, where the PSF is defined, the spatial integration from each detector can now span multiple high-resolution samples. This can be noted from the PSF expression in equation [9] that is analytically dependent on a finer/courser sampling function, weighting factor and element directivity to an incoming PA signal.

3.1.2 Sub-pitch Translation Methodology in PAT

In PAT, acoustic signals are generated by optical absorption of pulsed light by an endogenous optical absorber present in the medium. These photoacoustic signals may be received either by using a scanned single element or a multi element array. The detected signal at each element is later used for reconstructing the initial pressure distribution generated from the optical absorber, which acts as the photoacoustic source. As mentioned earlier, a densely sampled array using point acoustic detectors having a wide detection view angle provides the best possible image resolution and contrast in PAT with little time for data acquisition. However, it is not practical to achieve such a receive-setup.

In the approach described below, the focus is on enhancing image quality while using a typical λ-pitch (0.3 mm) linear array transducer having 128 elements at receive for PAT. Here, an actuator-assisted sub-pitch linear translation approach is used for increasing the density of acquired raw RF data in the lateral direction. In this approach, the acoustic signal that is propagating from the photoacoustic source for one light excitation pulse, or an average over several pulses, is first received over

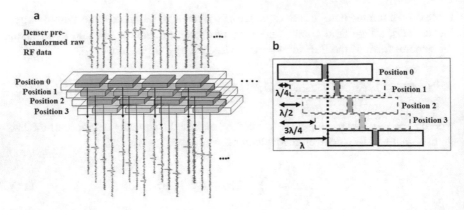

Fig. 9 **a** Schematic representation of the method adopted for sub-pitch translation of linear array in PAT, **b** A zoomed version of only two elements and their relative positions with respect to initial position is shown for better clarity

the transducer array kept in its original position. Thereafter, the transducer array is translated by sub-pitch amount to acquire photoacoustic signals generated for the next excitation pulse(s) and additional RF line data from locations in-between the original positions of the detector elements are recorded. The RF data from sub-pitch locations are augmented to the RF line data from original position to create a densely-sampled frame data. In this study, sub-pitch translation of $\lambda/4$, $\lambda/2$ and $3\lambda/4$ from initial position were considered. A schematic representation of the various translated positions (0, 1, 2 and 3) of the array transducer is shown in Fig. 9. The acquired sets of dense raw RF frame data are then beamformed using standard delay and sum (DAS) reconstruction method. This reconstructed image from denser data is compared with a single reconstructed image from λ-separated RF line data in terms of resolution, image contrast and depth of penetration.

3.1.3 Simulation

Simulations were performed using k-Wave Toolbox in MATLAB®(The MathWorks, Inc., MA, USA) [51, 52]. This simulation platform computes the time evolution of an acoustic wave field within homogeneous or heterogeneous media using the equations of linear acoustics. Photoacoustic source/optical absorbers were simulated using numerical PSF phantom and numerical blood vessel phantom whose initial pressure distribution is given by a grid-based image representative of a series of circular optical absorbers (Fig. 12a) and vasculature (Fig. 13d), respectively. On the detector side, a typical diagnostic ultrasound linear array probe was simulated having parameters as listed in Table 1.

In simulation, sub-pitch translation was done by moving the phantom instead of moving the array. First, position 0 data were acquired without moving the transducer and phantom. The phantom was then translated by $\lambda/4$ (0.075 mm) and position 1

Table 1 Parameters used in k-Wave simulation

Parameters	Diagnostic ultrasound linear array
Transducer width	38.4 mm (1536 grid points)
Element width	0.275 mm(11 grid points)
Element spacing (kerf)	0.025 mm (1 grid point = dx)
Element height	4 mm (2 grid points, dz = 2 mm)
Element pitch	0.3 mm (12 grid points = λ)
Number of elements	128
Sound speed	1540 m/s

data were acquired. In a similar manner, position 2 and position 3 data were also acquired by translating the phantom by $\lambda/4$ from previous positions, respectively. Position 0 and position 2 data were staggered next to each other to form the $\lambda/2$-pitch configuration. All the individually acquired RF lines from position 0, 1, 2 and 3 were staggered next to each other to form the $\lambda/4$-pitch configuration. These two sets of densely staggered raw RF data were reconstructed and compared against the image reconstructed using unstaggered configuration in terms of improvements in resolution and contrast.

3.1.4 Experiment

A pulsed laser diode (PLD) module (LaserComponents GmbH, Germany) of 905 nm, 226 W peak power/pulse, 120 ns pulse width operated at 5 kHz PRF was used as source of light excitation. The PLD was integrated with a Sonix Touch Q + Ultrasound scanner (Analogic Ultrasonix®, MA, USA) by external triggering with function generator. The ultrasound scanner was made to operate in passive receive mode by reprogramming the code in Texo SDK platform using *texosetsyncsignals( )* command and by appropriately rearranging the BNC connectors in the PCI card to receive the external trigger [53]. The ultrasound probe attached to the scanner is a typical 5 MHz, 128 element linear array probe with 0.3 mm pitch (all the parameters used in experiments were same as given in Table 1 under simulation).

Phantoms were made with 6% gelatin derived from acid-cured porcine skin (Sigma-Aldrich Corp., St. Louis, MO) and 1% by weight dilution of 20% Intralipid® solution (Sigma-Aldrich Corp., St. Louis, MO) for PAT imaging. The PLD module generates an elliptical beam with divergence of 12° along the slow axis and 20° along the fast axis. The energy per pulse was 23.5 μJ. Three kinds of phantoms were prepared for evaluating improvements in resolution, contrast and depth of penetration. All the phantoms were manufactured using a custom-made mould of either 50 × 50 × 30 mm or 80 × 80 × 40 mm dimensions, which can be opened on the source and detector side for easy assembly on the stage for performing PAT experiment in transmission mode. A schematic of the setup is shown in Fig. 10. The linear array probe was mounted using a high precision 3D motion controller translation stage that has a

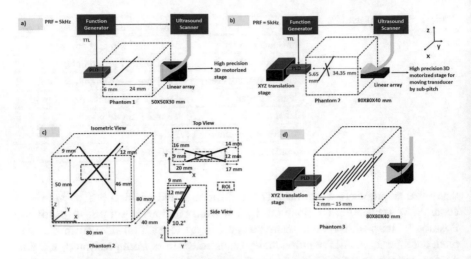

Fig. 10 An illustration of the setup used for performing **a** 2D PAT on phantom 1 with single point target **b** 3D PAT on phantom 2 for contrast study, **c** shows a detailed schematic of the phantom 2 with measurements, **d** shows phantom 3 prepared for depth of penetration study containing 9 leads present at depths ranging from 2 to 15 mm from the PLD

pitch (distance covered by one 360° rotation) of 4 mm and can provide resolution of up to 2.5 μm (i.e. 1600 steps/rotation). The 3D motion controller was translated in steps of 0.075 mm separation and four such sub-λ position raw RF pre-beamformed data sets were acquired and stored as illustrated in Fig. 9. Post-acquisition, these data sets were staggered appropriately and beamformed using traditional DAS reconstruction technique to obtain photoacoustic images corresponding to λ (128 lines with 0.3 mm pitch), λ/2 (256 lines with 0.15 mm pitch) and λ/4 (512 lines with 0.075 mm pitch) configurations. The various phantoms prepared for each study is described below.

Phantom for resolution study (Phantom 1)

A phantom containing a single lead inclusion of either 0.2 mm diameter or 0.5 mm diameter (phantom 1) embedded at 6 mm from the source-side surface of the phantom and 24 mm from the transducer-side surface of the phantom was utilized to study the improvement in PSF. The cross-section of the lead was imaged by keeping the transducer orthogonal to the target as shown in Fig. 10a. The improvement in PSF was demonstrated in phantom 1 by analyzing the lateral profile taken across the lead inclusion from the photoacoustic image obtained for the λ, λ/2 and λ/4 configurations.

Phantom for contrast study (Phantom 2)

The image quality improvement in terms of contrast was studied using a phantom (phantom 2) with 2 leads crossing each other and inclined at an angle of 10.2° with the source side surface having dimensions of 80 × 80 × 40 mm as shown in the schematic in Fig. 10b. Detailed measurements of phantom 2 is shown in Fig. 10c. Firstly, the position 0 data was acquired and stored, following which the array was translated by $\lambda/2$ in the x-direction and the sub-pitch data was acquired. Later, the transducer was scanned in the z-direction orthogonal to the plane of imaging in steps of 2 mm. Seven such imaging planes were stacked one on top of another to complete a 3D PAT data acquisition. The PLD source was scanned along z direction for each plane using a manual XYZ translation stage. An illustration of the setup is shown in Fig. 10b. In each plane both the λ and $\lambda/2$ position data were acquired, stored, staggered and beamformed to obtained staggered and unstaggered configurations. In addition, to compare the effect of averaging vis-à-vis sub-pitch translation on the improvement in contrast, same number of frames as the number of sub-λ position data (in this case, 2 frames) at λ position were acquired without translating the transducer and averaged. 2X averaging of λ pitch data was compared against $\lambda/2$ pitch data in order to maintain comparable acquisition time. 2D maximum intensity projection (MIP) image along the XY plane of the obtained 3D data was used to evaluate the overall improvements in image quality.

Phantom for depth of penetration study (Phantom 3)

The results obtained from contrast study suggested that there is a possibility for this improvement in contrast to translate as increased sensitivity of photoacoustic signal propagating from deeper located targets. This is due to the increased signal strength and noise cancellation stemming from the coherent summation of sub-pitch data. This phenomenon was analyzed using phantom containing targets embedded at different depths from the surface. 9 leads of 0.7 mm diameter was embedded with increasing depth from the source-side surface of the phantom and a cross section was imaged by keeping the transducer orthogonal to the targets. A schematic of the setup used is shown in Fig. 10d and a schematic of the cross section of the phantom containing leads at depths ranging from 2 mm to 15 mm along with the ROI chosen for calculating Contrast Ratio (CR) is shown in Fig. 11.

3.1.5 Performance Evaluation

The image contrast improvements in both simulation and experiment was analyzed by measuring Contrast ratio (CR) $= 20\log_{10}(\mu_1/\mu_2)$ [54]. Where, μ_1 is the mean intensity within the high amplitude resolution lines at each level where the targets were present and μ_2 is the mean intensity of background region. For the depth of penetration study, the target was considered to be detectable if the measured contrast ratio, CR, was > 1 and CR in dB, was > 0 dB. In addition to evaluating the detectability of the deeper located targets based on threshold on the contrast value, the improvement in contrast of the last visible target in the 64 frame-averaged, λ-pitch configuration was also analyzed and compared against its corresponding $\lambda/2$ and $\lambda/4$ pitch images by plotting of their axial and lateral profiles.

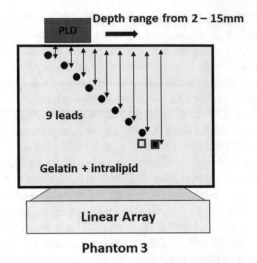

Phantom 3

Fig. 11 This figure shows a cross section of the phantoms prepared for depth of penetration study (phantom 3) shown in Fig. 10d. The red box indicates the ROI chosen for calculating contrast values

3.2   Results and Discussion

In this section, the results obtained from simulations and experiments performed on the various phantoms for evaluation of PSF, contrast and depth of penetration is described in detail.

3.2.1   Simulation Results

(i)   *Evaluation of PSF improvement*:

The improvement in PSF was first evaluated through simulation on a numerical phantom having series of six-point absorbers with decreasing diameter ranging from 0.5 mm to 0.05 mm. The initial pressure distribution of the input image grid is shown in Fig. 12a. Reconstructed images obtained from λ, $\lambda/2$ and $\lambda/4$ configurations are shown in Fig. 12 b, c and d, respectively. The lateral profiles from images obtained for each configuration are plotted in Fig. 12e along with the ground truth. A zoomed image of 0.5 mm, 0.2 mm and 0.1 mm targets is shown below the lateral profile.

The results shown in Fig. 12 clearly demonstrate the improvements in resolution when using a denser sampled array obtained by staggering of raw RF data acquired at sub-pitch locations. Compared to λ-pitch array, LR obtained by augmenting true RF A-lines at sub-pitch locations improved by 16.66% for 0.5 mm and by 42.85% for 0.2 mm while using a $\lambda/2$ pitch array, respectively. Further, it can also be noted that the magnitude of side lobes is considerably reduced in the lateral profile obtained

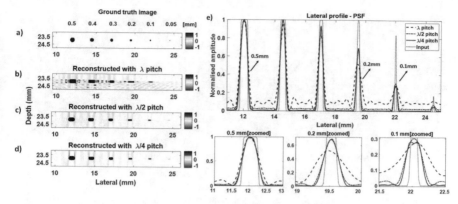

Fig. 12 Simulation results obtained from PSF target is shown in this figure. Ground truth image is shown in (**a**). Reconstructed images obtained using data acquired with **b** λ, **c** λ/2 and **d** λ/4 pitch separation between the detector elements of the linear array is shown. **e** shows the lateral profiles from the images obtained using the three configurations along with zoomed versions of 0.5 mm, 0.2 mm and 0.1 mm targets

using data from λ/2 and λ/4 pitch when compared to using data from only λ-pitch array.

Blood vessel numerical phantom

The results obtained in simulations when imaging the numerical blood vessel phantom are shown in Fig. 13. Gaussian noise was added such that the signal to noise ratio of the data was 40 dB.

The images obtained using sub-pitch translation clearly demonstrates a better resolution compared to that obtained using a λ-pitch array. Specifically, the thin vessel branch located at 6 mm depth and around 16 mm laterally appear more continuous in Fig. 13b, c than in Fig. 13a. The initial pressure distribution of representative vasculature image is shown in Fig. 13d. The data corresponding to the location marked by dark blue dashed line running across 4 vessels in Fig. 13d is plotted in Fig. 13e. It is clearly seen that the last two vessels that are closely spaced appear as one peak in image reconstructed from λ-pitch data. On the contrary, two separate peaks are clearly seen on images reconstructed using both λ/2 and λ/4 pitch data (marked with black arrow in Fig. 13e). It can be noted that some of the vertical vessels in the phantom were not clearly visible in the reconstructed image due to limited view problem. Several strategies have been reported in literature to overcome this issue [55–59], however, the focus in this section was discuss approaches that exploit the effect of increasing lateral array sampling and therefore this aspect is not described here.

From the above simulation results, it is evident that denser array created by sub-pitch translation mechanism is able to provide better LR without the need for increasing the number of elements. In addition, there is a visible reduction in background noise while using λ/2 pitch, which further decreases in reconstructed image using λ/4 pitch data (compare background noise present within the black box in 13a, b, c.

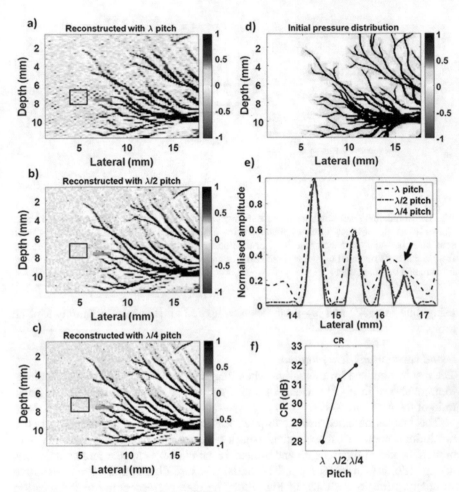

Fig. 13 Simulation results obtained using blood–vessel numerical phantom are shown in this figure. PAT images reconstructed from data acquired using **a** λ, **b** λ/2 and **c** λ/4 pitch separation between the detector elements of the linear array is shown. Initial pressure distribution of the input image is shown in **d**. **e** shows the lateral profiles along the dark blue dashed line marked in 13(**d**) from the images obtained using the three configurations and **f** shows the CR calculated from a, b and c (black box denotes background noise ROI and light blue solid linein **a**, **b** and **c** denotes the profile along which the signal peak values were taken)

This is also evident in the CR values plotted in Fig. 13f. The above results indicate that, in the case of thin vasculature in real tissue, it may be crucial to have a dense array sampling to obtain better image resolution.

Experiment results
Evaluation of PSF improvement (phantom 1)
Experiments using a point target phantom (phantom 1) using setup shown in Fig. 10a with a 0.2 mm lead and 0.5 mm lead as targets were performed. The scanner

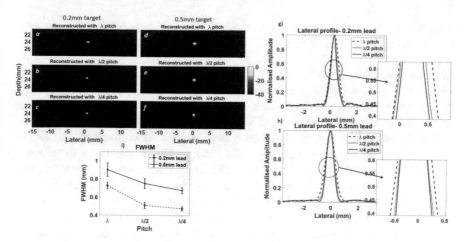

Fig. 14 Photoacoustic images, lateral profiles and FWHM values obtained from 0.2 mm lead and 0.5 mm lead as targets is shown in this figure. PAT images reconstructed with **a** λ pitch, **b** λ/2 pitch and **c** λ/4 pitch for a 0.2 mm lead is shown. The results from 0.5 mm lead target is presented in **d–f** in the same order. **g** and **h** shows the lateral profiles of the three configurations for a 0.2 mm and 0.5 mm lead respectively. A zoomed version of the profile is shown alongside Fig. 14g, h. The plot of their corresponding FWHM values is shown in (**i**)

was coded to acquire an average of 128 frames obtained from 128 individual laser firings pulsed at 5 kHz PRF for better SNR. Figure 14 shows the photoacoustic images, lateral profiles and FWHM values obtained using data from λ, λ/2 and λ/4-pitches, respectively. The images shown in Fig. 14 a–f were obtained by averaging 128 frames.

The results shown in Fig. 14 clearly demonstrate the improvement in LR obtained by using the proposed sub-pitch translation mechanism for achieving high density array. It is noticeable from Fig. 14a–c and d–f that as the sampling increases the point target looks tighter. Also, in the lateral profile plotted in Fig. 14g, h the red solid line has a narrower width than blue dotted line and the black dashed line. This is evident in the zoomed version shown alongside Fig. 14g h for better clarity. Figure 14i shows a plot of the FWHM values computed for the two cases of 0.2 and 0.5 mm. The LR improves by 29.66% and 16.66% while employing λ/2-pitch array configuration when compared to that of conventional λ-pitch array while using a 0.2 mm lead and 0.5 mm lead, respectively. Similarly, LR improves by 34.48% and 25% when a λ/4-pitch array was used as opposed to a λ-pitch array for the 0.2 mm and 0.5 mm case, respectively. The error bar shown in Fig. 14i was obtained from 5 independent realizations acquired at different parallel planes orthogonal to the axis of the lead target. In addition, it can be noticed that the pixelation within the point target decreases with increasing the array sampling. Similar to the results obtained from simulation, experimental results also showed higher percentage improvement with 0.2 mm target than with 0.5 mm target. Thus, demonstrating that sub-pitch translation is likely to improve resolution of sub-λ targets better.

Contrast Study (phantom 2)

Figure 15a, b shows stacked 3D visualization of the reconstructed photoacoustic images from 7 planes, obtained using λ-pitch without and with averaging, respectively. Figure 15c shows the stacked 3D visualization of images obtained by using λ/2-pitch reconstructed data. A zoomed version of their corresponding 2D MIP images along the XY plane is shown in Fig. 15d, e and f. A picture of the phantom used during experiments is shown in Fig. 15g. The pixels corresponding to the solid white line shown in Fig. 15g taken from the MIP images obtained using data from λ-pitch, λ-pitch (averaged) and λ/2-pitch configurations, i.e., Figure 15d, e and f is plotted in Fig. 15h. In addition, a side view of the 3D stack obtained with staggering is showed in Fig. 15i to indicate the relative source-lead-transducer distance at each plane. All images are displayed in 40 dB dynamic range.

Comparing Fig. 15a, b and c it is evident that Fig. 15c has a better contrast between background and lead target. This trend is also visible in Fig. 15d, e and f

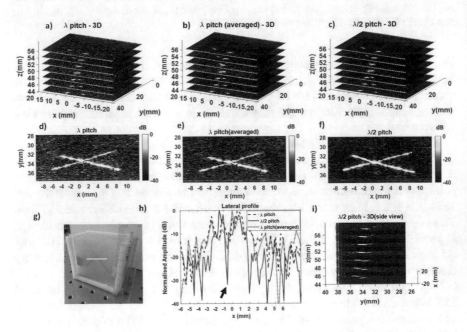

Fig. 15 Photoacoustic images obtained from a phantom with embedded lead targets is shown in this figure. Stacked arrangement of 2D reconstructed PAT images with **a** λ pitch **b** λ pitch (averaged) and **c** λ/2 pitch obtained from 7 planes between 4.4 and 5.6 cm from the bottom surface at 2 mm separation from each other is shown. 2D MIP images of the 3D stack on the XY plane that were reconstructed with **d** λ pitch **e** λ pitch (averaged) and **f** λ/2 pitch is shown. Also shown in **g** is a picture of the phantom used in the experiment. The lateral profiles taken along the white solid line marked in Fig. 15(**g**) from the images obtained for the three configurations is shown in (**h**). A side view with relative source-lead-transducer distance at each plane is shown in Fig. 15i

where the edges of the lead target have a better definition in image from $\lambda/2$-pitch configuration than that from both the λ-pitch configurations. It can be observed from the plots shown in Fig. 15h that the separation between the two leads is demarcated with a sharper dip (marked with black arrow) in the image data obtained using $\lambda/2$-pitch configuration, while they are hardly separable in the image data obtained from both the averaged and unaveraged λ-pitch configurations. This may be an important value addition when it comes to real situation of resolving extremely thin vasculature in tissue that lie close to each other.

The improvement in contrast was quantified using Contrast Ratio as a metric whose values obtained at 7 different y-levels on the MIP image is tabulated in Table 2. The ROI chosen for calculating μ_1 and μ_2 is marked in Fig. 16. For each y-level numbered from 1 to 7, the mean intensity of target (μ_1) was chosen within the solid black boxes and the mean intensity of background (μ_2) was taken from the solid white boxes at the same level in the background.

Depth of penetration Study

The results obtained by employing sub-pitch translation on phantom 3 and the corresponding percentage improvement in depth of penetration with variation in parameter settings are reported below.

Figure 17a shows the PAT B-mode images reconstructed using λ, $\lambda/2$ and $\lambda/4$-pitch configurations obtained from averaging of 2, 16, 32 and 64 frames, respectively. All the images were displayed in 30 dB dynamic range. Figure 17b shows plots of axial profiles along the 9th lead from the λ-pitch PAT image obtained by averaging 64 frames, $\lambda/2$-pitch image averaged over 32 frames and $\lambda/4$-pitch image averaged over 16 frames respectively. Figure 17c shows the plot of the contrast values of the

Table 2 CR values obtained in decibels for the different Y-levels

CR in dB	1	2	3	4	5	6	7
λ pitch	14.96	19.22	21.84	20.08	18.92	19.99	20.95
λ pitch (avg)	16.65	19.92	22.97	20.34	19.80	20.49	22.96
$\lambda/2$ pitch	21.72	23.39	26.21	24.26	22.55	23.51	24.35

ROI for CR Calculation

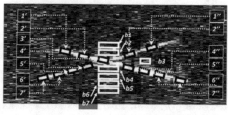

Fig. 16 Region of Interest (ROI) chosen for calculation of Contrast Ratio [CR $= 20\log_{10}(\mu_1/\mu_2)$] for 7 regions occupying different y positions on the MIP image is shown and Table 2 tabulates the CR values obtained from each region

same 9th lead with respect to the increasing sampling density. These values plotted are not in dB and hence $C = 1$ level in the graphs marks the detectability threshold. This is denoted by the grey solid line marked in Fig. 17c

It is clear from the images shown in Fig. 17a that the depth of penetration increases from 12.42 mm (location of 8th lead) to 14.3 mm (location of 9th lead) by performing sub pitch translation as opposed to performing mere averaging. This is evident by comparing the bottom-right corner-most image to the top-left corner-most image in Fig. 17a, wherein the visibility of the 9th lead present at 14.3 mm is much brighter than the same lead present in any of the four λ-pitch images displayed along the first row of the figure. Further the profile plots showed in Fig. 17b shows the decrease in background noise level with increasing sub-pitch sampling despite using lower number of frames for averaging. The Contrast, C, measured from 9th lead target for the

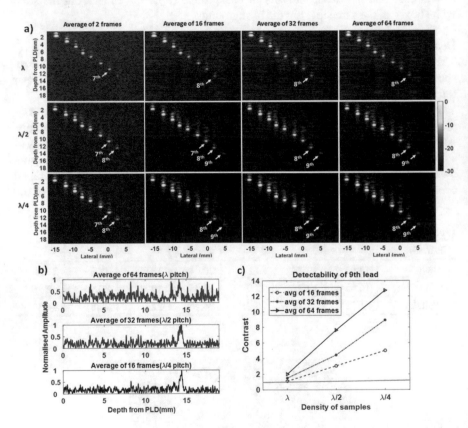

Fig. 17 Photoacoustic images reconstructed using **a** λ, λ/2 and λ/4 pitch configurations obtained by averaging 2, 16, 32 and 64 frames. **b** shows plots of the axial profiles of the 9th lead from λ-pitch (average of 64 frames), λ/2-pitch (average of 32 frames) and λ/4-pitch (average of 16 frames). **c** shows a plot of the contrast with respect to sample density for the 9th lead in the PAT images obtained with average of 16, 32 and 64 frames

different settings (average of 16, 32 and 64 frames) as a function of increasing sub-pitch sampling density is plotted in Fig. 17c. The contrast values obtained using $\lambda/2$ translation approach with 16 frame averaging demonstrate a 63.9% improvement over λ-pitch configuration obtained from average of 16 frames, 50.9% improvement over λ-pitch configuration obtained from average of 32 frames and 34.57% improvement over that obtained from average of 64 frames. Further, the contrast values obtained using $\lambda/4$ translation approach with 16 frame averaging demonstrate a 39.2% improvement over $\lambda/2$-pitch configuration obtained from average of 16 frames and 11.6% improvement over $\lambda/2$-pitch configuration obtained from average of 32 frames respectively.

An increase in depth of penetration by about 15% to 14.3 mm in the PLD based PAT setup by employing just $\lambda/2$ translation was obtained. Further improvement in contrast was achieved by increasing the sampling to $\lambda/4$. Overall, this study demonstrated the feasibility of achieving increased sensitivity of the photoacoustic signal travelling from deep-seated targets without incurring additional cost or system complexity.

4 Removal of EMI in Low SNR PAT Images

Most of the photoacoustic receive setups, including commercially available ultrasound scanners and data acquisition boards, are highly sensitive to electromagnetic interference (EMI) and other electronic noises, unless proper shielding methods are used. Even with the use of metallic shielding it is only possible to reduce the impact of EMI and it may not be possible to remove its effect completely. In practice, photoacoustic signals generated with the use of low energy sources, such as a PLD or LED, travelling from deep-seated optical absorbers inherently suffer from low SNR, due to interference from such noises. In this situation, where the photoacoustic signal may be corrupted due to the presence of some external electromagnetic field, mere averaging of multiple frames may not be sufficient to suppress the noise source. This noise may also limit the depth of penetration achievable with the light source. Hence, methods to de-noise and improve the signal strength of low SNR PA signals can be of particular importance in several applications.

In this section, investigation on the use of frequency domain filtering on post-beamformed and pre-beamformed ultrasound RF data, respectively, is reported and corresponding improvement in SNR of PA images is evaluated. Further, their performance is compared with that obtained from mere averaging of multiple frames.

The results suggest that filtering the raw RF data before beamforming results in better noise reduction than filtering the post beamformed photoacoustic image. It should also be emphasized that this methodology is of importance, especially, when dealing with deeper located targets. In this work we have explored a depth range of 18–22 mm, which is considered deep for the energy level of the pulsed laser diode source that was utilized.

In the following subsection, the steps adopted for implementing the algorithm is described.

4.1 Methodology

Frequency Domain Filtering

Ultrasound signals received using commercial scanners are mostly bandlimited signals having center frequency of 2–15 MHz. Band-pass filters are most commonly used for de-noising and improving SNR of photoacoustic signal. However, regular low pass/high pass/band pass filters cannot be applied on the Fourier transformed ultrasound raw RF data when noise and signal share overlapping frequency spectrum [60, 61]. In this work, a thresholding-based digital notch filtering approach is employed to remove significant noise components from the spectrum. Other similar filtering approaches that have been demonstrated for PAT are cited [18, 20, 60, 62–65].

The raw data was transformed to Fourier domain by performing 2D FFT,

$$F(k, l) = \sum_{x=0}^{N} \sum_{y=0}^{M} f(x, y) e^{i2\pi\left(\frac{kx}{N} + \frac{ly}{M}\right)} \tag{12}$$

A threshold value σ was calculated empirically as $0.8*M$ after evaluating the range of values in F, where M is the maximum absolute value of the 2D spectrum.

A binary image of the spectrum was obtained using thresholding as follows

$$g = \begin{cases} 1 \text{ if } |F| \geq \sigma \\ 0 \text{ if } |F| \leq \sigma \end{cases} \tag{13}$$

The binary image g is clustered into signal and noise components, and a digital notch filter was implemented to subtract the noise components in g from the original spectrum F. The 2D IFFT of the resultant spectrum resulted in the noise free image data back in the spatial domain. The steps performed are detailed in the flowchart shown in Fig. 18

In Fig. 18a–d, the step by step filtering process employed is reported. Figure 18a shows an image of the raw RF data before being filtered. Figure 18b shows the 2D spectrum of the data and Fig. 18c shows its corresponding thresholded binary image. Figure 18d is the final filtered raw RF data before reconstruction. The image clearly shows the removal of noise while leaving the photoacoustic signal intact.

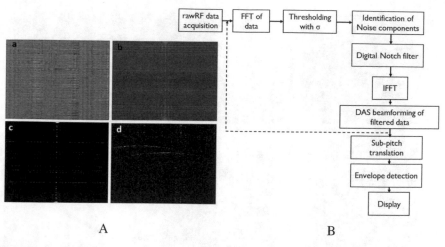

A   B

Fig. 18 **A** This figure shows the step by step process of filtering. Image **a** shows the noisy raw RF data, **b** is the magnitude image of the 2D spectrum F, **c** shows the binary image, **g** and **d** is the filtered raw RF data before beamforming. Alongside is shown the flowchart of the algorithm

4.2 Results

The effect of EMI removal on rawRF data using the proposed method in comparison to regular band pass filtering is shown in Fig. 19.

The results shown in Fig. 19 clearly show that the proposed methodology is able to remove the noise in the rawRF data due to EMI without affecting the signal when compared to regular band pass filtering of the data. In Fig. 20 we show the corresponding reconstructed images.

Figure 20a–c shows the photoacoustic images reconstructed after averaging over 128 frames. Figure 20a shows the PA image obtained without any filtering on either pre-beamformed or post-beamformed RF data. Figure 20b shows the PA image reconstructed from filtered post-beamformed RF data. Figure 20c shows the PA image

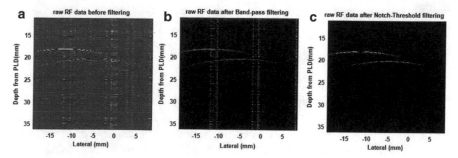

Fig. 19 This figure shows the photoacoustic rawRF data **a** without any filtering, **b** with regular band pass filtering and **c** with the proposed thresholding based digital notch filtering

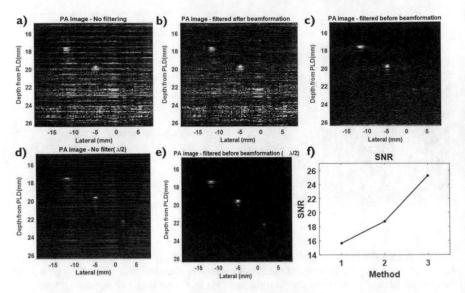

Fig. 20 Photoacoustic images obtained using **a** average of 128 frames and reconstructed without any filtering, **b** average of 128 frames and filtered after reconstruction and **c** average of 128 frames and filtered before reconstruction, **d** λ/2 pitch image averaged over 128 frames and reconstructed without any filtering, **e** λ/2 pitch image averaged over 128 frames and filtered before reconstruction and **f** is the plot of the SNR of target located at 20 mm depth in images **a**, **b** and **c**

reconstructed from filtered pre-beamformed raw RF data. Figure 20d shows the λ/2 pitch PA image obtained without any filtering on either pre-beamformed or post-beamformed RF data and Fig. 20e shows the λ/2 pitch PA image reconstructed from filtered pre-beamformed raw RF data. It is clear from the images shown in Fig. 20 that by performing filtering before reconstruction it is possible to significantly reduce the electronic/EMI noise without causing any blurring or loss of the photoacoustic signal. This is also evident in the values of SNR plotted in Fig. 20f. The SNR of target located at 20 mm depth increases by 61.4% and 20.05% by performing filtering before and after beamforming, respectively, as opposed to regular averaging of 128 frames.

5 Conclusion

This chapter discussed the various ultrasound receive-side innovations that has been demonstrated to improve the SNR and image quality in low energy LED/PLD based PAT systems. First, a brief discussion on the beamforming schemes and ultrasound hardware-based techniques that has been reported in literature for improving image quality of PAT system with particular emphasis to the commercial LED-PAUS imaging system was provided. This chapter also presented a recently developed sub-pitch translation approach that demonstrated improved resolution, contrast and depth of penetration with fewer number of frames averaged when employed on a low energy PLD based PAT system. Studies using K-wave simulation and experimental validation using phantoms demonstrating the benefits of the sub pitch translation scheme were described. Further, presence of external noises such as EMI, jitter, electronic noise etc., has a pronounced detrimental effect on PAT image quality particularly when low energy sources such as LED/PLD are used. In this chapter a thresholding-based digital notch filtering approach to remove effect of EMI from the frequency spectrum of noisy data is also presented.

Acknowledgements The photoacoustic work from the authors were partially supported by IIT Madras through Exploratory Research Project funding.

References

1. M. Xu, L.V. Wang, Photoacoustic imaging in biomedicine. Rev. Sci. Instrum. **77**(4), 1–22 (2006)
2. P. Beard, Biomedical photoacoustic imaging. Interface Focus **1**(4), 602–631 (2011)
3. C. Kim, T.N. Erpelding, L. Jankovic, M.D. Pashley, L.V Wang, Deeply penetrating in vivo photoacoustic imaging using a clinical ultrasound array system. **1**(1), 335–340 (2010)
4. G. Ku, X. Wang, G. Stoica, L.V. Wang, Multiple-bandwidth photoacoustic tomography. Phys. Med. Biol. **49**(7), 1329–1338 (2004)
5. Y. Wang, T.N. Erpelding, J. Robert, L.V. Wang, In vivo three-dimensional photoacoustic imaging based on a clinical matrix array ultrasound probe. J. Biomed. Optics **17**(6), 1–5 (2012)
6. S.K. Biswas, P. Van Es, W. Steenbergen, S. Manohar, A method for delineation of bone surfaces in photoacoustic computed tomography of the finger. Ultrason. Imaging **38**(1), 63–76 (2016)
7. H.-P. Brecht, R. Su, M. Fronheiser, S.A. Ermilov, A. Conjusteau, A.A. Oraevsky, Whole-body three-dimensional optoacoustic tomography system for small animals. J. Biomed. Opt. **14**(6), 064007 (2009)
8. V.G. Andreev, A.A. Oraevsky, A.A. Karabutov, Breast cancer imaging by means of optoacoustic technique: initial experience. Hydroacoustics **4**, 1–4 (2001)
9. R.A. Kruger, W.L. Kiser, D.R. Reinecke, G.A. Kruger, Thermoacoustic computed tomography using a conventional linear transducer array. Med. Phys. **30**(5), 856 (2003)
10. T.J. Allen, P.C. Beard, Pulsed near-infrared laser diode excitation system for biomedical photoacoustic imaging. Opt. Lett. **31**(23), 3462 (2006)
11. R.G.M. Kolkman, W. Steenbergen, T.G. van Leeuwen, In vivo photoacoustic imaging of blood vessels with a pulsed laser diode. Lasers Med. Sci. **21**(3), 134–139 (2006)
12. K. Daoudi et al., Handheld probe integrating laser diode and ultrasound transducer array for ultrasound/photoacoustic dual modality imaging. Opt. Express **22**(21), 26365 (2014)
13. P.K. Upputuri, M. Pramanik, Pulsed laser diode based optoacoustic imaging of biological tissues. Biomed. Phys. Eng. Express **1**(4), 045010 (2015)
14. L. Zeng, G. Liu, D. Yang, X. Ji, 3D-visual laser-diode-based photoacoustic imaging. Opt. Express **20**(2), 1237 (2012)
15. W. Xia et al., Handheld real-time LED-based photoacoustic and ultrasound imaging system for accurate visualization of clinical metal needles and superficial vasculature to guide minimally invasive procedures. Sensors **18**(5), 1394 (2018)
16. T.J. Allen, P.C. Beard, *Light emitting diodes as an excitation source for biomedical photoacoustics, presented at the SPIE BiOS* (California, USA, San Francisco, 2013), p. 85811F

17. A. Hariri, M. Hosseinzadeh, S. Noei, M.R. Nasiriavanaki, Photoacoustic signal enhancement: Towards utilization of very low-cost laser diodes in photoacoustic imaging. in *Photons Plus Ultrasound: Imaging and Sensing,* vol. 10064 (2007), p. 100645L
18. M. Sun, N. Feng, Y. Shen, X. Shen, J. Li, Photoacoustic signals denoising based on empirical mode decomposition and energy-window method. Adv. Adapt. Data Anal. **04**(01, 02), 1250004 (2012)
19. S.A. Ermilov et al., Laser optoacoustic imaging system for detection of breast cancer. J. Biomed. Opt. **14**(2), 024007 (2009)
20. C. Zhang, K. Maslov, J. Yao, L.V. Wang, In vivo photoacoustic microscopy with 7.6-μm axial resolution using a commercial 125-MHz ultrasonic transducer. J. Biomed. Opt. **17**(11) (2012)
21. E.M.A. Anas, H.K. Zhang, J. Kang, E. Boctor, Enabling fast and high quality LED photoacoustic imaging: a recurrent neural networks based approach. Biomed Opt Express **9**(8), 3852–3866 (2018)
22. A. Alshaya, S. Harput, A.M. Moubark, D.M.J. Cowell, J. McLaughlan, S. Freear, Spatial resolution and contrast enhancement in photoacoustic imaging with filter delay multiply and sum beamforming techniquen. IEEE Int. Ultrasonics Symp. (IUS) **2016**, 1–4 (2016)
23. M. Mozaffarzadeh, A. Hariri, C. Moore, J.V. Jokerst, The double-stage delay-multiply-and-sum image reconstruction method improves imaging quality in a LED-based photoacoustic array scanner. Photoacoustics **12**, 22–29 (2018)
24. M.A.L. Bell, X. Guo, H.J. Kang, E. Boctor, Improved contrast in laser-diode-based photoacoustic images with short-lag spatial coherence beamforming. in *2014 IEEE International Ultrasonics Symposium* (Chicago, IL, USA, 2014), pp. 37–40
25. X.L. Den-Ben, D. Razansky, V. Ntziachristos, The effects of acoustic attenuation in optoacoustic signals. Phys. Med. Biol. **56**(18), 6129–6148 (2011)
26. X.L. Deán-Ben, V. Ntziachristos, D. Razansky, Effects of small variations of speed of sound in optoacoustic tomographic imaging. Med. Phys. **41**(7) (2014)
27. A. Buehler, A. Rosenthal, T. Jetzfellner, A. Dima, D. Razansky, V. Ntziachristos, Model-based optoacoustic inversions with incomplete projection data. Med. Phys. **38**(3), 1694–1704 (2011)
28. X.L. Deán-Ben, R. Ma, D. Razansky, V. Ntziachristos, Statistical approach for optoacoustic image reconstruction in the presence of strong acoustic heterogeneities. IEEE Trans. Med. Imaging **30**(2), 401–408 (2011)
29. X.L. Deán-Ben, R. Ma, A. Rosenthal, V. Ntziachristos, D. Razansky, Weighted model-based optoacoustic reconstruction in acoustic scattering media. Phys. Med. Biol. **58**(16), 5555–5566 (2013)
30. H. He, S. Mandal, A. Buehler, X.L. Dean-Ben, D. Razansky, V. Ntziachristos, Improving optoacoustic image quality via geometric pixel super-resolution approach. IEEE Trans. Med. Imaging **35**(3), 812–818 (2016)
31. A. Dima, N.C. Burton, V. Ntziachristos, Multispectral optoacoustic tomography at 64, 128, and 256 channels. J. Biomed. Optics **19**(3), 036021 (2014)
32. S.K. Kalva, M. Pramanik, Experimental validation of tangential resolution improvement in photoacoustic tomography using modified delay-and-sum reconstruction algorithm. JBO **21**(8), 086011 (2016)
33. M. Pramanik, G. Ku, L.V. Wang, Tangential resolution improvement in thermoacoustic and photoacoustic tomography using a negative acoustic lens. JBO **14**(2), 024028 (2009)
34. S. Selladurai, A.K. Thittai, Strategies to obtain subpitch precision in lateral motion estimation in ultrasound elastography. IEEE Trans. Ultrason. Ferroelectr. Freq. Control **65**(3), 448–456 (2018)
35. S. Chandramoorthi, A.K. Thittai, Enhancing image quality of photoacoustic tomography using sub-pitch array translation approach: Simulation and experimental validation. IEEE Trans. Biomed. Eng. **66**(12), 3543–3552 (2019)
36. E. Maneas et al., Human placental vasculature imaging using an LED-based photoacoustic/ultrasound imaging system. in *Photons Plus Ultrasound: Imaging and Sensing*, vol. 10494 (2018) p. 104940Y

37. Y. Zhu et al., Light Emitting Diodes based Photoacoustic Imaging and Potential Clinical Applications. Sci Rep **8**(1), 1–12 (2018)
38. J. Jo et al., Detecting joint inflammation by an LED-based photoacoustic imaging system: A feasibility study. JBO **23**(11), 110501 (2018)
39. A. Hariri, J. Lemaster, J. Wang, A.S. Jeevarathinam, D.L. Chao, J.V. Jokerst, The characterization of an economic and portable LED-based photoacoustic imaging system to facilitate molecular imaging. Photoacoustics **9**, 10–20 (2018)
40. M. Mozaffarzadeh, Linear-array photoacoustic imaging using minimum variance-based delay multiply and sum adaptive beamforming algorithm. J. Biomed. Opt. **23**(02), 1 (2018)
41. B. Huang, J. Xia, K. Maslov, L.V. Wang, Improving limited-view photoacoustic tomography with an acoustic reflector. J. Biomed. Opt. **18**(11) (2013)
42. A. Dima, N.C. Burton, V. Ntziachristos, Multispectral optoacoustic tomography at 64, 128, and 256 channels. J. Biomed. Opt. **19**(3), 36021 (2014)
43. M. Pramanik, Improving tangential resolution with a modified delay-and-sum reconstruction algorithm in photoacoustic and thermoacoustic tomography. J. Opt. Soc. Am. A JOSAA **31**(3), 621–627 (2014)
44. A. J. Pappano, W. Gil Wier, A.J. Pappano, W. Gil Wier, The microcirculation and lymphatics. Cardiovasc. Physiol. 153–170 (2013)
45. P.C. Johnson Overview of the microcirculation. Microcirculation (2008)
46. S. Arridge et al., Accelerated high-resolution photoacoustic tomography via compressed sensing. Phys. Med. Biol. **61**(24), 8908–8940 (2016)
47. P.D. Wilcox, J. Zhang, Quantification of the effect of array element pitch on imaging performance. IEEE Trans. Ultrason. Ferroelectr. Freq. Control **65**(4), 600–616 (2018)
48. M. Xu, L.V. Wang, Analytic explanation of spatial resolution related to bandwidth and detector aperture size in thermoacoustic or photoacoustic reconstruction. Phys. Rev. E Stat. Phys. Plasmas Fluids Related Interdisc. Top. **67**(5), 15 (2003)
49. M. Xu, L.V. Wang, *Photoacoustic Imaging and Spectroscopy*, vol. 144 (CRC Press, 2009)
50. M. Xu, L.V. Wang, Spatial resolution in three-dimensional photo-acoustic reconstruction. in *Proceedings of SPIE Vol. 5320, Photons Plus Ultrasound: Imaging and Sensing* (2004), p. 264
51. B.E. Treeby, B.T. Cox, *k-Wave : MATLAB Toolbox for the Simulation and Reconstruction of Photoacoustic Wave Fields*, vol. 15 (2015), pp. 1–12
52. B.E. Treeby, Acoustic attenuation compensation in photoacoustic tomography using time-variant filtering. Journal of Biomedical Optics **18**(3), 036008 (2013)
53. L. Fieramonti, *Feasibility, Development and Characterization of a Photoacoustic Microscopy System for Biomedical Applications* (2010), p. 120
54. M. Mozaffarzadeh, A. Mahloojifar, M. Orooji, S. Adabi, M. Nasiriavanaki, Double-stage delay multiply and sum beamforming algorithm: Application to linear-array photoacoustic imaging. IEEE Trans. Biomed. Eng. **65**(1), 31–42 (2018)
55. G. Paltauf, R. Nuster, P. Burgholzer, Weight factors for limited angle photoacoustic tomography. Phys. Med. Biol. **54**(11), 3303–3314 (2009)
56. Y. Xu, L. V Wang, *Limited View Thermoacoustic Tomography* (2002), pp. 937–938
57. Y. Xu, L.V. Wang, G. Ambartsoumian, P. Kuchment, Reconstructions in limited-view thermoacoustic tomography. Med. Phys. **31**(4), 724 (2004)
58. J. Wang, Y. Wang, An efficient compensation method for limited-view photoacoustic imaging reconstruction based on Gerchberg-Papoulis extrapolation. Applied Sciences **7**(5), 505 (2017)
59. B. Huang, J. Xia, K. Maslov, L.V. Wang, Improving limited-view photoacoustic tomography with an acoustic reflector. Journal of Biomedical Optics **18**(11), 110505 (2013)
60. M. Zhou, H. Xia, H. Zhong, J. Zhang, F. Gao, A noise reduction method for photoacoustic imaging in vivo based on EMD and conditional mutual information. IEEE Photonics J. **11**(1), 1–10 (2019)
61. J. Sun, B. Zhang, Q. Feng, H. He, Y. Ding, Q. Liu, Photoacoustic wavefront shaping with high signal to noise ratio for light focusing through scattering media. Sci Rep **9**(1), 4328 (2019)
62. P. Raumonen, T. Tarvainen, Segmentation of vessel structures from photoacoustic images with reliability assessment. Biomed. Opt. Express **9**(7), 2887 (2018)

63. N. Awasthi, S.K. Kalva, M. Pramanik, P.K. Yalavarthy, Image-guided filtering for improving photoacoustic tomographic image reconstruction. J. Biomed. Opt. **23**(09), 1 (2018)
64. E.R. Hill, W. Xia, M.J. Clarkson, A.E. Desjardins, Identification and removal of laser-induced noise in photoacoustic imaging using singular value decomposition. Biomed. Opt. Express **8**(1), 68–77 (2017)
65. A. Hariri et al., Development of low-cost photoacoustic imaging systems using very low-energy pulsed laser diodes. J. Biomed. Opt. **22**(7), 075001 (2017)

Vascular Complexity Evaluation Using a Skeletonization Approach and 3D LED-Based Photoacoustic Images

Kristen M. Meiburger, Alberto Vallan, Silvia Seoni, and Filippo Molinari

Abstract Vasculature analysis is a fundamental aspect in the diagnosis, treatment, outcome evaluation and follow-up of several diseases. The quantitative characterization of the vascular network can be a powerful means for earlier pathologies revealing and for their monitoring. For this reason, non-invasive and quantitative methods for the evaluation of blood vessels complexity is a very important issue. Many imaging techniques can be used for visualizing blood vessels, but many modalities are limited by high costs, the need of exogenous contrast agents, the use of ionizing radiation, a very limited acquisition depth, and/or long acquisition times. Photoacoustic imaging has recently been the focus of much research and is now emerging in clinical applications. This imaging modality combines the qualities of good contrast and the spectral specificity of optical imaging and the high penetration depth and the spatial resolution of acoustic imaging. The optical absorption properties of blood also make it an endogenous contrast agent, allowing a completely non-invasive visualization of blood vessels. Moreover, more recent LED-based photoacoustic imaging systems are more affordable, safe and portable when compared to a laser-based systems. In this chapter we will confront the issue of vessel extraction techniques and how quantitative vascular parameters can be computed on 3D LED-based photoacoustic images using an in vitro vessel phantom model.

K. M. Meiburger (✉) · A. Vallan · S. Seoni · F. Molinari
PoliToBIOMed Lab, Department of Electronics and Telecommunications, Politecnico di Torino, Turin, Italy
e-mail: kristen.meiburger@polito.it

A. Vallan
e-mail: alberto.vallan@polito.it

S. Seoni
e-mail: silvia.seoni@polito.it

F. Molinari
e-mail: filippo.molinari@polito.it

© Springer Nature Singapore Pte Ltd. 2020
M. Kuniyil Ajith Singh (ed.), *LED-Based Photoacoustic Imaging*,
Progress in Optical Science and Photonics 7,
https://doi.org/10.1007/978-981-15-3984-8_5

1 Introduction

Blood vessels play a fundamental role in the well-being of tissues, organs and organ systems, by providing them with oxygen and nutrients and subsequently eliminating waste products. Many diseases affect blood vessels and their attributes, such as their number, size, or pattern [1]. For example, tumors typically induce the growth of many vessel clusters with an abnormal tortuosity and smaller diameter, while chronic inflammations induce neoangiogenesis [1, 2]. It is therefore evident how the possibility of a non-invasive and quantitative evaluation of 3D vessel attributes is essential for early diagnosis and the staging of various diseases [1].

Many imaging techniques can be used for visualizing vasculature structures. For example, computed tomographic angiography (CTA) has an excellent spatial resolution and it is very common in clinics. As a downside, however, it uses ionizing radiations and iodinated contrast agents. Magnetic resonance angiography (MRA) in spite of very good contrast and temporal resolutions and lack of ionizing radiation, suffers from rapid extravasation of the contrast agent that affects the accuracy, and is a very expensive imaging modality. Doppler ultrasound imaging (DU) has much lower costs, large availability and it doesn't use nephrotoxic contrast agents, but it is operator-dependent, contrast agents typically have short duration and this imaging technique is typically sensitive only to larger vessels and is not able to highlight microvasculature. Also, a more recent technique, acoustic angiography, that uses dual-frequencies ultrasound transducers for the minimization of background [3] needs exogenous contrast agents and custom-made probes, while optical coherence tomography angiography (OCTA) has a limited penetration depth and a longer acquisition time [4, 5].

Photoacoustic imaging is an imaging modality that has seen an exponential growth over the last couple of decades. Using this technique, ultrasound signals are generated from the interaction between a pulsed light source at a given wavelength and the biological tissues that are irradiated. So, it is non-invasive and non-ionizing and it combines the high spatial resolution and the penetration depth of ultrasound with the high contrast and the spectral specificity of optical imaging [6, 7]. In particular, the visualization of blood vessels is a main application of photoacoustic imaging, as oxygenated and deoxygenated haemoglobin give forth a strong photoacoustic signal at various wavelengths and therefore present an endogenous contrast agent for this imaging modality [8, 9].

Typically, laser light sources are used for photoacoustic imaging, but these optical systems are typically cumbersome, expensive, and they usually have fluctuations of wavelength and power per pulse. Moreover, safety glasses or a shield is necessary to protect the operator and/or patient from the irradiation of the light source. Much recent research has focused on the use of different light sources, and in particular on the use of pulse laser diodes. In fact, light emitting diodes (LEDs) are inexpensive, compact, multi-wavelength and more stable. LED-based systems are therefore more portable and an enclosure or protective glasses aren't necessary [7]. However, due to the reduced energy the LED light source is able to emit compared to laser light sources, these systems typically are limited to more superficial imaging applications.

In this chapter, we will present a proof of concept and feasibility study of using 3D LED-based photoacoustic images for the quantitative evaluation of the vascular complexity network using a skeletonization approach and an in vitro phantom model. First of all, the numerous techniques for vessel extraction from images are presented and summarized. Then, quantitative vascular parameters that are used to describe vascular networks and that have been used in numerous studies are defined and explained. Finally, we then present our approach for the phantom model definition, image acquisition and processing steps, and validation results.

2   Blood Vessel Extraction Techniques

Many various methods have been introduced to automatically extract the vascular network from medical images. The main differences between techniques are due to pre-processing steps, computational time, accuracy, and the visual quality of the obtained results [10].

Four main categories of blood vessel extraction techniques can be defined: pattern recognition approaches, model-based approaches, vessel tracking approaches, and machine learning approaches. It is also possible to combine the use of different techniques together to improve the final results [11]. In this section, we will briefly explore the four main categories of blood vessel extraction methods and the numerous methods that are included in each main category.

2.1   Pattern Recognition Techniques

Pattern recognition techniques are methods that are used for the automatic detection and classification of various objects. In the specific application of vessel extraction, they detect vessel-like structures and features, and there are many different approaches that can be classified within this main category, such as multi-scale, skeleton-based, and ridge-based [12].

2.1.1   Multi-scale

Multi-scale approaches are based on extracting the vasculature at different levels of resolution. The vessels with a larger diameter are extracted using images with a lower resolution, since less detail is needed to correctly extract the vessel, whereas the smaller vessels and microvasculature are extracted using images with a higher resolution [12]. Instead of using images with an actual different resolution, multi-scale methods found in literature can also be based on using kernels with different scales that enhance vessels with diameters of a certain dimension, such as the well-known and applied Frangi filter [13].

2.1.2 Skeleton-Based

Skeleton-based vessel extraction techniques are employed to extract the blood vessel centerlines and the entire vessel structure is created by connecting the vessel centerlines. These kinds of techniques are based on first segmenting the vessels using various approaches (such as thresholding), and the segmentation is then thinned using a specific algorithm, such as the medial axis thinning algorithm [14]. The skeletonization process is used to reduce the segmentation to a minimal representation that keeps the morphology without redundancy.

Figure 1 shows some examples of skeletons obtained using various imaging modalities.

2.1.3 Ridge-Based

Ridge-based vessel extraction techniques are based on the idea that grayscale image can be seen as a 3D elevation map where intensity ridges approximate the skeleton of objects that adopt a tubular shape [12]. In this way, ridge points are simply local peaks in the direction of the maximal surface gradient and are invariant to affine transformations.

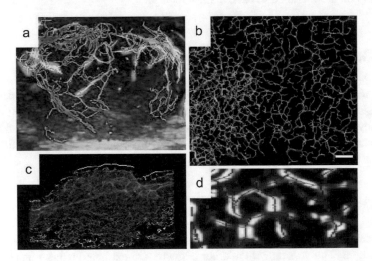

Fig. 1 Examples of skeletons obtained with different imaging modalities. **a** Doppler ultrasound imaging, skeleton in red. **b** Optical coherence tomography angiography, skeleton in green. **c** Contrast-enhance ultrasound imaging. **d** Photoacoustic imaging, skeleton in blue

2.1.4   Region Growing

Region growing approaches are those based on segmenting the vessel network through a region growing technique that segments images by analyzing neighboring pixels and assigning them to specific objects based on their pixel value similarity and spatial proximity [12]. A downfall to this kind of approach is that it is necessary to provide some form of seed point from which to start the region growing analysis, and these typically must be supplied by the user.

2.1.5   Differential Geometry-Based

Differential geometry-based vessel extraction methods consider the acquired images as hypersurfaces and therefore extracts features, thanks to the crest lines and curvature of the surface. The center lines of the vessels are therefore found as the crest points of the hypersurface. In this way, a 3D surface can be described by two principal curvatures (i.e., the eigen values of the Weingarten matrix) and their principal directions (i.e., the eigenvectors), which are their corresponding orthogonal directions [12].

2.1.6   Matching Filters

Vessel extraction techniques based on matching filters are used to find objects of interest by convolving the image with multiple matched filters. The design of different filters in order to detect vessels with different orientation and size plays a fundamental role with this type of approach, and the convolutional kernel size directly affects the computational load of the method.

2.1.7   Mathematical Morphology

Methods based on mathematical morphology schemes rely on the use of morphological operators to enhance vessel structures from the image. Morphological operators are defined by applying specific structuring elements to the image, which define the operator locality and can take on various geometries, such as a line, circle, square, diamond, etc. The two main morphological operators are dilation and erosion, which expands or shrinks objects, respectively. These operators can therefore be exploited to enhance vessel structures and/or remove areas of the image that are not vessels.

2.2 Model-Based Techniques

As their name implies, model-based techniques for vessel extraction apply explicit models to extract the vasculature from the images. These methods can be divided into four main different categories, which are briefly explored more in detail below.

2.2.1 Parametric Deformable Models

Parametric deformable models, often also known as snakes, are techniques that aim to find object contours using parametric curves that deform under the influence of internal and external forces. Internal forces are important for the smoothness of the surface, while external ones attract it to the vessel boundary. The smoothness constraint is the elasticity energy and makes the model more robust to the noise. A downside of these models is that in order to start the process, the surface has to be initialised and the model evolution depends on initial parameters that must be fine-tuned by the user. Moreover, it is fundamental that the final model is robust to its initialization. With recent implementations, it's also possible to insert constraints or a priori knowledge about geometry [12, 15]. These approaches are suitable for complex architecture or variable vessels, but they are very time consuming.

2.2.2 Geometric Deformable Models

Geometric deformable models are based on the theory of curve evolution, and are commonly known as level sets [12]. Level sets are based on the main concept that propagating curves are represented as the zero-level set of a higher dimensional function, which is typically given in the Eulerian coordinate system. This type of approach has the following advantages: (1) it can handle complex interfaces that present sharp corners and change its topology during the level set evolution; (2) the curvature and normal to the curve, which are intrinsic properties of the propagating front, can be easily extracted from the level set function; (3) it is easily extendable to problems of higher dimensions, and is therefore not limited to 2D images.

2.2.3 Parametric Models

Parametric models (PM), not to be confused with parametric deformable models, define parametrically the object of interest. In particular, for tubular objects, they are described as a set of overlapping ellipsoids. In some applications, the model of the vessel is circular. The estimation of parameters is done from the image, but the elliptic PM approximates healthy vessels well but not pathological shapes and bifurcations [12].

2.2.4 Template Matching

This method attempts to recognize a structure, an a priori model or template, in the image. This is a contextual top-down method. For the application of arterial extraction, the template is a set of nodes connected in segments, that then is deformed to fit the real structure. For the deformation, a stochastic process can be used [12].

2.3 Vessel Tracking Techniques

Vessel tracking approaches apply local operators on a focus known to be a vessel and track it. They differ from pattern recognition approaches in that they do not apply local operators to the entire image. So, starting from an initial point, these methods detect vessel centerlines or boundaries by analyzing the pixels orthogonal to the tracking direction [12].

2.4 Machine Learning

Machine learning is a subfield of artificial intelligence in which computers learn how to solve a specific problem from experimental data.

These approaches can be divided in unsupervised and supervised:

- Unsupervised approaches try to find a model that describes input images no having prior knowledge about them. This technique doesn't need the comparison with a gold standard.
- Supervised approaches learn the model from a training set of labelled images and then applies it to the input images. This technique has shown better performances, and testing the trained network is typically very fast. On the other hand, training the network typically requires a huge computational cost [11].

Recently, there has been a huge growth of the application of supervised machine learning approaches under the form of neural networks and specifically convolutional neural networks (CNNs) in the application of image processing. CNNs are characterized by the presence of convolutional layers for feature extraction, pooling layers for feature reduction and fully connected layers for classification [11].

3 Vessel Architecture Quantification

As discussed in the previous section, there are numerous methods that can be exploited to extract the vessel network from images acquired using various imaging modalities. All of these methods aim to extract the vessel network from the images, so that the vessels can further be classified and/or analyzed to gain important information about the tissue or organ health status.

Many studies in literature are based on qualitative or semi-quantitative analyses of the extracted vessel network, by either visually observing the enhanced or segmented network or by manually selecting specific vessels to analyze with more quantitative methods [1, 3, 16, 17].

In this chapter, and specifically in this section, we will go more into details about how a quantitative analysis of the vessel network can be obtained and what quantitative parameters can be computed from the skeleton of the vessel network.

As described previously, the skeleton of a vessel is a minimal representation of the segmented vessels, which can be independent of the imaging modality used to acquire the images. In fact, the main goal is to segment the vessels from the images and once the segmentation is obtained using the desired technique, the skeleton of the vessels can be obtained by applying, for example, the medial axis thinning algorithm [14]. Many techniques based on skeletonization have been used in literature to extract the vessel network and then used to calculate quantitative parameters that can help distinguish healthy from diseased tissue in numerous imaging modalities, such as in CT images of the lung [18], ultrasound contrast-enhanced clinical images of the thyroid to characterize thyroid nodules [19, 20], ultrasound contrast-enhanced images of tumors in murine models [21], photoacoustic images of burn wounds in rats to differentiate from healthy tissue [4], and optical coherence tomography angiography (OCTA) images of clinical dermatological lesions for the automatic segmentation of the lesion [22].

An important step before quantitative parameter calculation is the placement of a specific region-of-interest (ROI) within which to calculate the parameters. This is to help reduce the computational load, and is due to the fact that typically vasculature is present not only in the area that is of interest (for example, outside of the tumor or diseased tissue), and more importantly, due to the fact that these quantitative parameters should not be considered using their absolute values, but in comparison with the same parameters either at a different location or at a different time. So, the relative comparison between the parameters gives a better evaluation rather than the actual value by itself.

In all of the studies mentioned previously, the ROI is manually placed on the desired areas, except for in the most recent study by Meiburger et al. [22] in which the entire OCTA volume was analyzed by a sliding ROI. The quantitative vascular parameters computed inside each ROI were then employed to automatically define the lesion area. Subsequently, the ROI for the diseased zone was automatically placed in correspondence of the centroid of the defined lesion area and the healthy zone was automatically placed in correspondence of the ROI that was found to be furthest away from the considered diseased ROI.

In the next section we will go into more detail about what specific quantitative parameters can be computed on the skeleton of the vessel network within the defined ROI, which can be classified as either morphological or tortuosity parameters.

3.1 Morphological Parameters

As the name implies, morphological parameters give an idea of the morphology of the considered vessel network, taking into consideration their size, how many vessels are present, and how they are distributed between each other. The principal quantitative morphological parameters that have been used in previous studies are:

- Number of trees (NT): defined as the number of vessel trees in which the skeleton is decomposed
- Vascular density (VD): defined as the ratio between the number of skeleton voxels and the total number of voxels of the considered ROI
- Number of branching nodes (NB): defines as the number of branching nodes that are found in the vessel structure
- Mean radius (MR): mean radius of the segmented vessels of the structure.

While the first three parameters are consistently used in various studies, the mean radius is a quantitative parameter that is sometimes excluded, due to the fact that it the one that is most highly dependent on an accurate segmentation of the actual borders of the vessels. Thanks to the skeletonization process, a slightly oversegmented or undersegmented vessels do not influence the first three quantitative parameters (i.e., NT, VD, and NB). On the other hand, the mean radius is highly influenced by an inaccurate segmentation, which is the reason why this parameter is sometimes omitted in various studies.

3.2 Tortuosity Parameters

Tortuosity parameters are those parameters that analyze the path of the vessels and how curved, tortuous or tightly coiled the vessel path may be. In order to calculate these parameters, it is fundamental to first "isolate" a specific vessel to analyze and then begin from one end point and arriving at the other end point, various quantitative parameters can be calculated along the path, either by measuring angles, inflection points, or simply path length.

Specifically, three main quantitative tortuosity parameters are typically calculated to give an idea of the tortuosity of the considered vascular network:

- 2D distance metric (DM): defined as the ratio between the actual path length of the considered vessel and the linear distance between the first and last endpoint of the vessel
- Inflection count metric (ICM): defined as the 2D distance metric multiplied by the number of inflection points found along the vessel path
- 3D sum of angles metric (SOAM): defined as the sum of all the angles that the vessel has in space.

Fig. 2 Graphical
representation of 2D distance
metric computation

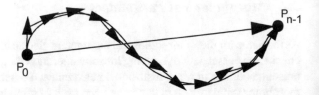

The mathematical descriptions of these tortuosity parameters can be found in previously published studies [1, 23].

Briefly, the DM gives a measure of the bidimensional tortuosity of the considered vessel, since a straight line would give forth a value of 1, and as the vessel potentially becomes more and more curved, the DM value will increase. Figure 2 shows a graphical representation of how the DM is computed. The ICM adds to the DM as it considers not only the overall curvature of the considered vessel, but also the number of times the vessel changes direction in its path. Finally, the SOAM parameters are helpful mostly in the case of tightly coiled vessels, which are not well-represented by either the DM or ICM.

4 Phantom Design

In this section of the chapter, we will describe how a possible vascular phantom can be designed to show the feasibility of evaluation the vascular complexity using a skeletonization approach and 3D LED-based photoacoustic images.

In medical imaging, phantoms are samples with known geometry and composition that mimic biological tissues with their physical and chemical properties for providing a realistic environment for clinical imaging applications. Stable and well characterized phantoms are very useful for routine quality controls, training, calibration and for evaluating the performance of systems and algorithms. They can be also used for the development of new applications before in vivo preclinical or clinical studies. Moreover, phantoms allow to understand reproducibility in time and among laboratories, to optimize signal to noise ratio, to compare detection limits and accuracies of different systems and to examine maximum possible depth [24–27].

4.1 Model Design

In order to correctly evaluate vascular complexity, it is first necessary to design a model that can represent in a simplified manner at least a section of a vascular network. An example of a method that can be used to mimic a vascular network is the creation of a 3D model which can then be printed using various materials.

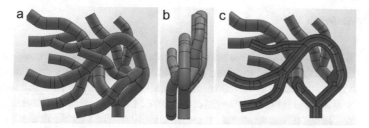

Fig. 3 3D model designed for vascular complexity analysis. **a** Front view. **b** Lateral view. **c** Section view

As a proof of concept, we designed a model using a computer aided design software that had the following dimensions: 39.23 mm × 34.37 mm × 12.78 mm with a wall thickness of 1 mm. The internal diameter of the designed vessels was equal to 1.5 mm. Figure 3 shows the designed model from a front view (a), lateral view (b) and section view (c).

4.1.1 3D Printing

Once the model was designed, we then proceeded to use a 3D printer to print the model. In this preliminary proof of concept study, we used the ProJet MJP 2500 Plus with the VisiJet R Armor (M2G-CL) material, a tough, ABS-like clear plastic that combines tensile strength and flexibility [28].

The ProJet MJP 2500 Plus is a 3D MultiJet printer that uses the inkjet printing process. In particular, a piezo printhead deposits a plastic resin and a casting wax material through the layer by layer technique.

Then the MJP EasyClean System is used to remove in a little time, the support material from plastic parts using steam and EZ Rinse-C. It is composed of two warmer units, one for bulk wax removal and one for fine wax removal. The support material is separated by melting or dissolving. This is a non-contact method, so there are less substrates or mask damages and contamination. Moreover, it permits a high resolution and is inexpensive. Figure 4a shows an image acquired during the 3D printing process and the final obtained model (Fig. 4b).

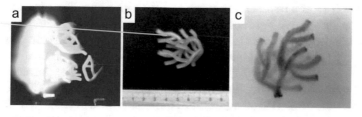

Fig. 4 Phantom manufacturing. **a** 3D printing process. **b** Final model. **c** Final phantom in agar

4.2 Phantoms Realization

Once the 3D model is printed and all wax is removed, the vascular network phantom must be filled with a liquid that can mimic blood, or at the very least absorb and respond to the photoacoustic light impulse. Ideally, real blood or a biocompatible contrast agent should be used. As what is reported here is a proof of concept idea to show the feasibility of the approach, here we simply used a liquid ink that gave forth a strong photoacoustic signal.

The final phantom was then realized using agar, which is a jellying polysaccharide, obtained from red algae and it is used to prepare transparent and neutral gels. Agar powder dissolves at around 90–100 °C and it solidifies at 45 °C. The dose for 1 kg of solution, is 7–10 g of powder.

The desired quantity of agar powder was weighed with a digital scale and then it was put in a small pot with the corresponding quantity of water stirring at the same time. The obtained solution was brought slowly to a boil with a burner continuing to stir and at this point, the warm solution was poured in a container with the vessel model. The phantom was left to cool down and after it solidified, it was pulled out from the container. Figure 4c shows the final phantom filled with dye and inserted in the solidified agar.

4.3 Acquisition Setup

In order to accurately assess vascular complexity, it is clear that a 3D volume of the network must be acquired. It is therefore of fundamental importance to have the phantom model fixed in the same spot and acquire 2D images at a given step size. Some ultrasound systems have a mechanical motor and a corresponding software that permits quick 3D acquisitions at a defined step size. On the other hand, if this is not an option for the system that is used, it is still possible to use a specific setup that guarantees the same position for the ultrasound probe as it runs along the phantom and 2D images are manually acquired at each step.

In our first tests that are presented here, we used the second solution along with a commercial LED-based photoacoustic and ultrasound imaging system (AcousticX, Cyberdyne, INC, Tsukuba, Japan). So, for the image acquisition, the phantoms were fixed to the base of a transparent container filled with water. The ultrasound probe and photoacoustic LED light source arrays were secured to a metallic angle beam, which in turn was fixed to a mobile support that could be moved along a binary in response to a knob rotation. Figure 5 shows the imaging setup used.

The ultrasound probe and LEDs were put underwater near the phantom and linear scans were made moving the system with a defined step size. The step size is what defines the resolution along the third dimension, so a smaller step size would give forth a more accurate volume reconstruction of the vascular network and is fundamental when considering microvasculature.

Fig. 5 Imaging setup. **a** Entire imaging setup with metallic angle beam. **b** Zoom on ultrasound probe and LED light sources

Due to the fact that here it was important mainly to show the feasibility of the approach of using 3D LED-based photoacoustic images to evaluate vascular network complexity, and that the phantom vessels had a large diameter compared to microvasculature, we chose to optimize processing time and used a large step size, equal to 1 mm. Considering the model that was designed, this gave forth a final volume that consisted of 65 2D frames.

4.4 Device Settings

The device used for this feasibility study is the AcousticX, a LED-based photoacoustic imaging system (PLED-PAI) that is commercially available [29, 30].

The excitation source are light emitting diodes characterized by high density and high power. Specifically, there are two LED arrays on either side of an ultrasound probe and each array is composed of 4 rows of 36 single embedded LEDs. The excitation wavelength is 850 nm. The dimensions of each array are 12.4 mm (height), 86.5 mm (length) and 10.2 mm (width). The pulse width is variable and can be set from 50 to 150 ns with steps of 5 ns. The pulse repetition rate can be 1, 2, 3 kHz or 4 kHz and it defines consequently the temporal resolution.

In order to reduce noise, it is also possible to control the frame averaging which then influences the frame rate and temporal resolution. The possible frame rates are 30, 15, 10, 6, 3, 1.5, 0.6, 0.3, and 0.15 Hz [7].

For the acoustic part, there is a 128 channels ultrasound linear array transducer with central frequency that can be set between 7 and 10 MHz that can pulse and receive.

For the volume acquisition of the model, only the PA mode was used. The depth was set to 3 cm and the frame rate was 6 Hz. The pulse repetition frequency was set to 4 kHz with 640 frames averaging.

5   Image Processing and Results

After image acquisition, it is then necessary to proceed to segment the images and extract the skeleton of the vascular network in order to compute quantitative vascular parameters that can give an idea of the complexity of the network. In this section of the chapter we will present an example workflow that can be used to extract the quantitative vascular parameters from the acquired images.

5.1   Segmentation and Skeletonization

Before the actual image segmentation, a few preprocessing steps often help in preparing the images and allowing a more accurate segmentation of the objects of interest, which are, in our case, the phantom vessels containing the contrast dye.

Firstly, a 3D median filter was applied to the entire volume, using a $3 \times 3 \times 3$ kernel and padding the volume by repeating border elements in a mirrored way. Then, a closing morphological operation was done using a disk-shaped structuring element with a radius equal to 5 pixels. This step helped fill the vessels where mainly the walls of the phantom were visible.

For the actual segmentation, the Otsu method [31] was used to find the global threshold of each slice and then, the maximum among these was chosen to define a unique threshold for all of the slices of the volume. The images were then segmented using the found threshold, which in our case was equal to 0.43.

Then, a brief cleaning process was used to refine the obtained segmentation. Specifically, each mask was processed by removing all the objects with area smaller than 2% of the biggest object found in the mask. Subsequently, dilatation with a disk-shaped structuring element with radius 3 and erosion with a disk-shaped structuring element with radius 1 were then applied. Finally, any remaining holes in the objects of the mask were then filled. Figure 6 shows a 3D representation of the original photoacoustic images and the obtained segmentation.

For the skeletonization, an algorithm based on the medial axis extraction algorithm by Lee et al. [14] that is implemented preserving the topology and the Euler

Fig. 6   3D representations. **a** Original photoacoustic image volume. **b** Volume after median filtering. **c** Segmented volume

number was used. This procedure is done to specifically reduce the segmented binary volume into a minimal representation of the vascular network while still preserving morphology.

An algorithm was then implemented with the aim to correct the defects of the skeletonization and to refine the final structure by removing the smallest branches. In some areas of the obtained skeleton, there can be an accumulation of skeleton voxels. In order to remove them, the branchpoints are identified and when, among them, there are connected objects with a value bigger than 10 pixels, they are removed. Thereafter, the branches with a length smaller than a defined threshold are removed.

5.2 Parameter Calculation and Validation

As discussed in a previous section of this chapter, quantitative parameters that give an idea of the morphology and tortuosity of the vascular network can be extracted from the skeleton of the segmented vessels.

In the feasibility study presented here, a 3D computer-aided design (CAD) model was specifically designed and was then printed. This allowed for not only real LED photoacoustic image acquisition once the phantom was correctly filled with a dye, but also the direct importation of the CAD model in the same processing environment (in our case, Matlab).

For validation purposes, the acquired images were also manually segmented so as to give an idea if the automatic segmentation (and therefore the subsequent skeleton) could be considered reliable or not. Then, the recall, precision, and Jaccard index were calculated. These parameters are defined as follows:

$$Recall = \frac{TP}{TP + FN} \tag{1}$$

$$Precision = \frac{TP}{TP + FP} \tag{2}$$

$$Jaccard Index = \frac{TP}{TP + FP + FN} \tag{3}$$

where TP is a true positive, a pixel that was segmented in both the automatic and manual masks; FN is a false negative, a pixel that was segmented only in the manual mask; FP is a false positive, a pixel that was segmented only in the automatic mask.

Furthermore, thanks to the 3D model the quantitative vascular parameters were able to be calculated using the experimental data with the 3D printed phantom and LED photoacoustic image acquisition and also on the imported model using the same skeletonization and vascular parameter computation processes. This type of approach also allows a direct comparison of the quantitative vascular parameters obtained using the various methods. Figure 7 shows different views of the 3D model skeleton together with the automatic skeleton obtained using the acquired images.

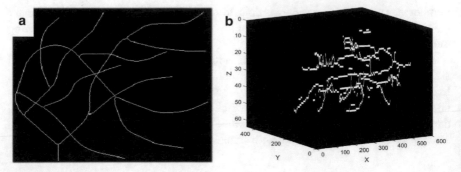

Fig. 7 **a** Top view of skeleton of imported 3D model. **b** 3D skeleton view of automatically segmented volume

6 Feasibility Study Results

Table 1 shows the results of the comparison between the manual and the automatic segmentation of the entire volume of the model. As can be seen, the recall parameter is quite high, showing that when compared to a manual segmentation, the automatic segmentation did not produce many false negatives. This means that the thresholding technique was capable of accurately capturing the photoacoustic signal when it was present within the image. On the other hand, however, the precision is only equal to approximately 72%, meaning that there is a reasonably high number of false positives, so the automatic algorithm was quite sensitive to noise and tended to oversegment the acquired images.

As can be seen in Table 2, the quantitative vascular parameters that were calculated corresponded quite well. In this table, the first column corresponds to the parameters

Table 1 Automatic segmentation validation results

Recall	Precision	Jaccard index
0.94±0.11	0.72±0.18	0.68±0.18

Table 2 Automatic quantitative vascular parameters validation results

Vascular parameter	3D model	Automatic segmentation	Manual segmentation
NT	1	6	6
VD	5.16×10^{-5}	13.69×10^{-5}	15.39×10^{-5}
NB	9	57	77
DM	2.164	2.229	2.289
ICM	67.935	70.197	89.244
SOAM	0.041	0.241	0.545
MR (mm)	0.688±0.174	0.732±0.352	0.591±0.311

computed using the directly imported 3D model, so it can be considered the ground truth. The middle column shows the values computed using the automatic algorithm and segmentation, whereas the last column displays the values obtained when using the manual segmentation.

The biggest discrepancies can be seen within the SOAM tortuosity parameter and the number of trees and number of branch nodes of the vascular network. It is important to point out here how not only the automatic segmentation but also the manual segmentation provided an overestimation of these parameters. This is most likely due to the fact that, during the phantom manufacturing process, it was seen that some parts of the phantom were not properly filled with the ink due to the presence of remaining wax, resulting in no or less photoacoustic signals from those points. At the same time, it is also important to underline how the 3D model was imported into MATLAB with a very good spatial resolution, providing a perfectly clean and rounded vessel mask. So, the acquired images were limited by a number of various issues. Specifically, the obtained results were limited by (a) the high step size and therefore low resolution between slices, (b) any small air bubble or imperfect filling of the model with the dye, and (c) photoacoustic imaging artefacts which are common especially when employing linear ultrasound probes for the photoacoustic signal reception.

7 Conclusion

While the feasibility study presented here showed some limitations, mainly due to phantom manufacturing and an imperfect wax removal technique, the results are promising and merit a further investigation using even more complex vascular phantoms at first and then using in vivo images considering micro-vasculature to evaluate the resolution limits of this approach. Overall, the proof of concept study shown here in this chapter demonstrates the potential of evaluating vascular complexity using 3D LED-based photoacoustic images.

References

1. E. Bullitt, K.E. Muller, I. Jung, W. Lin, S. Aylward, Analyzing attributes of vessel populations. Med. Image Anal. **9**(1), 39–49 (2005)
2. E. Bullitt, S.R. Aylward, T. Van Dyke, W. Lin, Computer-assisted measurement of vessel shape from 3T magnetic resonance angiography of mouse brain. Methods **43**(1), 29–34 (2007)
3. S.E. Shelton et al., Quantification of microvascular tortuosity during tumor evolution using acoustic angiography. Ultrasound Med. Biol. **41**(7), 1896–1904 (2015)
4. K.M. Meiburger, S.Y. Nam, E. Chung, L.J. Suggs, S.Y. Emelianov, F. Molinari, Skeletonization algorithm-based blood vessel quantification using in vivo 3D photoacoustic imaging. Phys. Med. Biol. **61**(22) (2016)
5. J.T. Perry, J.D. Statler, Advances in vascular imaging. Surg. Clin. North Am. **87**(5), 975–993 (2007)

6. Y. Zhang, H. Hong, W. Cai, *Photoacoustic Imaging* (Spring, Berlin, 2011)
7. A. Hariri, J. Lemaster, J. Wang, A.K.S. Jeevarathinam, D.L. Chao, J.V. Jokerst, The characterization of an economic and portable LED-based photoacoustic imaging system to facilitate molecular imaging. Photoacoustics **9**, 10–20 (2018)
8. P. Beard, Biomedical photoacoustic imaging. Interface Focus **1**(4), 602–631 (2011)
9. E.Z. Zhang, J.G. Laufer, R.B. Pedley, P.C. Beard, In vivo high-resolution 3D photoacoustic imaging of superficial vascular anatomy. Phys. Med. Biol. **54**(4), 1035–1046 (2009)
10. B. Preim, S. Oeltze, 3D visualization of vasculature: an overview (2008), pp. 39–59
11. S. Moccia, E. De Momi, S. El Hadji, L. S. Mattos, Blood vessel segmentation algorithms— review of methods, datasets and evaluation metrics. Comput. Methods Programs Biomed. **158**, 71–91 (2018) Elsevier Ireland Ltd
12. C. Kirbas, F.K.H. Quek, Vessel extraction techniques and algorithms : a survey, in *Proceedings—3rd IEEE Symposium on BioInformatics and BioEngineering, BIBE 2003*, (2003), pp. 238–245
13. A.F. Frangi, W.J. Niessen, K.L. Vincken, M.A. Viergever, Multiscale vessel enhancement filtering, in *International Conference on Medical Image Computing and Computer-Assisted Intervention*, (1998), pp. 130–137
14. T.C. Lee, R.L. Kashyap, C.N. Chu, Building skeleton models via 3-D medial surface axis thinning algorithms. CVGIP Graph. Model. Image Process. **56**(6), 462–478 (1994)
15. T. Chan, L. Vese, Active contours without edges. IEEE Trans. Image Process. **10**(2), 266–277 (2001)
16. Y.S. Zhang, J. Yao, C. Zhang, L. Li, L.V. Wang, Y. Xia, Optical-resolution photoacoustic microscopy for volumetric and spectral analysis of histological and immunochemical samples. Angew. Chem. Int. Ed. Engl. **53**(31), 8099–8103 (2014)
17. Z. Yang et al., Multi-parametric quantitative microvascular imaging with optical-resolution photoacoustic microscopy in vivo. Opt. Express **22**(2), 1500–1511 (2014)
18. T. Tozaki, Y. Kawata, N. Niki, H. Ohmatsu, N. Moriyama, 3-D visualization of blood vessels and tumor using thin slice CT images, in *IEEE Nuclear Science Symposium and Medical Imaging Conference*, vol. 3, (1995), pp. 1470–1474
19. U.R. Acharya, O. Faust, S.V. Sree, F. Molinari, J.S. Suri, ThyroScreen system: high resolution ultrasound thyroid image characterization into benign and malignant classes using novel combination of texture and discrete wavelet transform. Comput. Methods Programs Biomed. **107**(2), 233–241 (2012)
20. F. Molinari, A. Mantovani, M. Deandrea, P. Limone, R. Garberoglio, J.S. Suri, Characterization of single thyroid nodules by contrast-enhanced 3-D ultrasound. Ultrasound Med. Biol. **36**(10), 1616–1625 (2010)
21. F. Molinari et al., Quantitative assessment of cancer vascular architecture by skeletonization of high-resolution 3-D contrast-enhanced ultrasound images: role of liposomes and microbubbles. Technol. Cancer Res. Treat. **13**(6), 541–550 (2014)
22. K.M. Meiburger et al., Automatic skin lesion area determination of basal cell carcinoma using optical coherence tomography angiography and a skeletonization approach : Preliminary results. J. Biophotonics, no. April, 1–11 (2019)
23. E. Bullitt, G. Gerig, S.M. Pizer, W. Lin, S.R. Aylward, Measuring tortuosity of the intracerebral vasculature from MRA images. IEEE Trans. Med. Imaging **22**(9), 1163–1171 (2008)
24. W.C. Vogt, C. Jia, K.A. Wear, B.S. Garra, T.J. Pfefer, Biologically relevant photoacoustic imaging phantoms with tunable optical and acoustic properties. J. Biomed. Opt. **21**(10), 101405 (2016)
25. M. Fonseca, B. Zeqiri, P. Beard, B. Cox, Characterisation of a PVCP based tissue-mimicking phantom for Quantitative Photoacoustic Imaging, in *Opto-Acoustic Methods and Applications in Biophotonics II*, (2015), p. 953911
26. S.E. Bohndiek, S. Bodapati, D. Van De Sompel, S.-R. Kothapalli, S.S. Gambhir, Development and application of stable phantoms for the evaluation of photoacoustic imaging instruments. PLoS ONE **8**(9), e75533 (2013)

27. J.R. Cook, R.R. Bouchard, S.Y. Emelianov, Tissue-mimicking phantoms for photoacoustic and ultrasonic imaging. Biomed. Opt. Express **2**(11), 3193 (2011)
28. Y. Guo, H.S. Patanwala, B. Bognet, A.W.K. Ma, Inkjet and inkjet-based 3D printing: Connecting fluid properties and printing performance. Rapid Prototyp. J. **23**(3), 562–576 (2017)
29. N. Sato, M. Kuniyil Ajith Singh, Y. Shigeta, T. Hanaoka, T. Agano, High-speed photoacoustic imaging using an LED-based photoacoustic imaging system, in *Photons Plus Ultrasound: Imaging and Sensing 2018*, (2018), p. 128
30. J. Joseph et al., Characterization and technical validation of a multi-wavelength LED-based photoacoustic/ultrasound imaging system (Conference Presentation), in *Photons Plus Ultrasound: Imaging and Sensing 2018*, 2018, p. 34
31. N. Otsu, A threshold selection method from gray-level histograms. IEEE Trans. Syst. Man. Cybern. **9**(1), 62–66 (1979)

Multiscale Signal Processing Methods for Improving Image Reconstruction and Visual Quality in LED-Based Photoacoustic Systems

Kausik Basak and Subhamoy Mandal

Abstract Light-emitting diodes (LED) based photoacoustic imaging (PAI) systems have drastically reduced the installation and operational cost of the modality. However, the LED-based PAI systems not only inherit the problems of optical and acoustic attenuations encountered by PAI but also suffers from low signal-to-noise ratio (SNR) and relatively lower imaging depths. This necessitates the use of computational signal and image analysis methodologies which can alleviate the associated problems. In this chapter, we outline different classes of signal domain and image domain processing algorithms aimed at improving SNR and enhancing visual image quality in LED-based PAI. The image processing approaches discussed herein encompass pre-processing and noise reduction techniques, morphological and scale-space based image segmentation, and deformable (active contour) models. Finally, we provide a preview into a state-of-the-art multimodal ultrasound-photoacoustic image quality improvement framework, which can effectively enhance the quantitative imaging performance of PAI systems. The authors firmly believe that innovative signal processing methods will accelerate the adoption of LED-based PAI systems for radiological applications in the near future.

1 Introduction

Photoacoustic imaging (PAI) emerged in the early 2000s as a novel non-invasive and non-ionizing imaging method, harnessing the advantages of optical and ultrasound imaging modalities to provide high-contrast characteristic responses of functional and molecular attributes without sacrificing resolution (for depths of millimeters to

K. Basak
Centre for Health Science and Technology, JIS Institute of Advanced Studies and Research Kolkata, JIS University, Kolkata, West Bengal, India
e-mail: kausikbasak@ieee.org

S. Mandal (✉)
Technical University of Munich (TUM), Munich, Germany
e-mail: s.mandal@ieee.org

© Springer Nature Singapore Pte Ltd. 2020
M. Kuniyil Ajith Singh (ed.), *LED-Based Photoacoustic Imaging*,
Progress in Optical Science and Photonics 7,
https://doi.org/10.1007/978-981-15-3984-8_6

centimeters) in highly optically scattering biological tissues. In PAI, acoustic waves are generated by the absorption of short-pulsed electromagnetic waves, followed by detection of generated acoustic waves using sensors, e.g., piezoelectric detectors, hydrophones, micro-machined detectors. The term photoacoustic (also optoacoustic) imaging is synonymous with the modalities where one uses visible or near-infrared light pulses for illumination while using electromagnetic waves in the radiofrequency or microwave range is referred to as thermo-acoustic imaging. The research efforts in PAI have been directed towards the development of new hardware components and inversion methodologies allowing an increase in imaging speed, depth, and resolution, as well as on investigating potential biomedical applications. Further, the unique capabilities of the recently developed small animal imaging systems and volumetric scanners have opened up the unexplored domain of post-reconstruction image analysis. Despite these advantages and massive growth in PAI modalities, it is still operational mostly in research fields and for preclinical studies due to high-cost associated with the instrumentation and so-called limited-view effects, offering sub-optimal imaging performance and limited quantification capabilities. Significant limitations yet remain in terms of inadequate penetration depth and lack of high-resolution anatomical layout of whole cross-sectional areas, thereby encumbering its application in the clinical domain.

With the emergence of the light-emitting diode (LED)-based PAI modalities, the operational cost of the imaging system drastically reduced, and the instrumentation becomes compact and portable while maintaining the imaging depth of nearly 40 mm with significant improvisation in resolution as well. Being cost-effective and more stable compared to the standard optical parametric oscillator (OPO)-based systems, the LED-based systems have made PAI technology more accessible and open to new application domains. Recently several early clinical studies using PAI have been reported, e.g., gastrointestinal imaging [1], brain resection guidance [2], rheumatoid arthritis imaging [3]. Further, it is the capability of real-time monitoring of disease biomarkers that makes it an impeccable tool for longitudinal supervision of circulating tumour cells, heparin, lithium [4]. However, this probing modality, especially LED-PAI is still characterized by low signal to noise ratio (SNR) and lower image saliency when compared to several other clinically adopted imaging modalities. Therefore, enhancing the SNR in both signal and image domain, as well as the use of image enhancement techniques and pre-processing of PA images is of significant interest to obtain clinically relevant information and characterize different tissue types based on their morphological and functional attributes. In this context, this chapter aims to provide the use of signal and image analysis in conjunction with imaging and post-processing techniques to improve the quality of PA images and enable optimized workflows for biological, pre-clinical and clinical imaging.

This chapter will illustrate relevant signal analysis techniques and is organized with the following sections. Section 2 introduces a generalized PAI system where different aspects of the imaging instrumentation are highlighted with a precise description, followed by discussion of different signal processing techniques, e.g., ensemble empirical mode decomposition, wavelet-based denoising, Wiener deconvolution using the channel impulse response and principal component analysis to increase the

SNR of the acquired acoustic signals prior to reconstruction, in Sect. 3. Section 4 entails image analytics, intends to implement at different stages in post-processing, ranging from noise removal to segmentation of different biological features. Various techniques, to significantly increase the SNR of PA images, are discussed while maintaining a great deal in both spatial domain and frequency domain processing techniques. Besides, segmentation of different biological structures, based on their structural and functional properties, can be achieved through numerous approaches e.g. morphological processing, feature-based segmentation, cluster techniques, deformable objects. This entire section describes various image analysis methods to comprehend PA image analysis further and helps to ascertain problem-specific processing methodologies in the application domain. Additionally, Sect. 5 covers advanced solutions to improve image quality by rectifying various PAI parameters such as optical and acoustic inaccuracies, generated sue to practical limitations and approximations in PAI modality. In this context, different experimental and algorithmic approaches are discussed with the help of recent findings in PA research. In summary, this chapter provides a holistic approach of performing LED-based PA signal and image processing at various stages starting from acoustic signal acquisition to post-reconstruction of PA images through the different computational algorithms with prospective dimensions of probable research areas to improve the efficacy of PA imaging system in clinical settings.

2 Block Diagram of Imaging and Signal Acquisition

A generalized schematic of a PAI system is shown in Fig. 1. Generally, a nano-second pulse duration light source (in Fig. 1, we show a laser diode-based PAI system) with a repetition rate from 10 Hz to several kHz and wavelength in the range of visible to

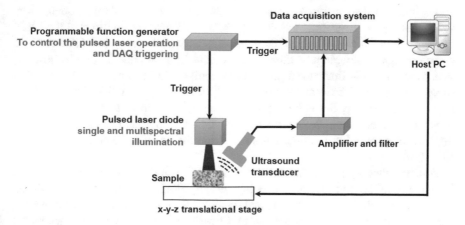

Fig. 1 Schematic block diagram of a generalized laser-diode based PAI system

NIR is used to irradiate the sample under observation. Importantly, it is also empirical to keep the laser energy exposure within the maximum permissible limit following the guideline by ANSI standard.[1] Due to the thermoelastic effect, the absorbing tissue compartments will produce characteristic acoustic responses which are further acquired using an ultrasound transducer unit. An ultrasound transducer unit, working in the range of several MHz, can be placed adjacent to the sample body to capture these PA signals and converts them into their corresponding electrical signal levels to transfer it to the data acquisition unit which is directly connected and controlled with the host PC for post-processing, reconstruction and storage of the signal and image data for further offline processing if required.

Prior to acquisition by data acquisition system, these PA signals are amplified due to their low order of amplitude and filtered to reduce the effect of noises, usually combines electronic noise, system thermal noise and most importantly the measurement noise that arises due to the highly scattering tissue media for which the acoustic waves undergo multiple attenuation event before acquisition using ultrasound transducers [5]. These noises are capable of deteriorating the signal strength and eventually the quality of PA images. Therefore, a significant amount of signal processing both at the hardware level and software platform needs to be carried out to mitigate this challenge. These techniques are discussed in the following sections.

3 Signal Domain Processing of PA Acquisitions

In most cost-effective PA imaging systems, researchers are using low energy PLD or LED which in turn significantly reduces signal strength, hence affecting SNR and quality of reconstructed images. Such imaging set-ups need significant improvement in signal processing to enhance the SNR so that it would eventually produce considerably good quality PA images after reconstruction. With recent scientific deductions and technological advancement, several researchers have targeted this problem from a different perspective. Zhu et al. introduced a low-noise preamplifier in the LED-PAI signal acquisition path to increase the sensitivity of PA reception, followed by a two-steps signal averaging: 64 times by data acquisition unit and 6 times by host PC, thus combining 384 times averaging which significantly improves the SNR with a square root factor of the total averaging times [6]. They also established such an SNR improvement strategy through the phantom model experiment and in vivo imaging of vasculature on a human finger. However, such a technique can also lead to losing high-frequency information that stems from the small and subtle structures in LED-PAI.

Among other signal enhancement techniques, several conventional approaches include ensemble empirical mode decomposition (EEMD), wavelet-based denoising, Wiener deconvolution using the channel impulse response, and principle component analysis (PCA) of received PA signals [5]. EEMD is a time-space analysis technique

---

[1] ANSI-American National Standards Institute

that relies on shifting an ensemble noisy (white noise) data, followed by averaging out the noisy counterpart with a sufficient number of trials [7]. In this mechanism, the added white noise provides a uniform reference frame in time-frequency space. The advantage over the classical empirical mode decomposition technique is that EEMD scales can be easily separated without any a priori knowledge. However, challenges arise while specifying an appropriate criterion for PA image reconstruction based on intrinsic mode function.

In the case of wavelet-based denoising, although the acoustic signal can be tracked-out from the background noise with significant accuracy, it is empirical to optimize the thresholding parameter to suppress undesired noise and preserve signal details optimally. Moreover, wavelet-based denoising requires prior knowledge about the signal and noisy environment as the choice of wavelet function and threshold necessitate the characteristics knowledge of the signal and noise. One way to overcome such difficulty is to make the process parametric and adaptive [8, 9]. They introduced polynomial thresholding operators which are parameterized according to the signal environment to obtain both soft and hard thresholding operators. Such methodology not only enables increased degrees of freedom to preserve signal details optimally in a noisy environment but also adaptively approach towards the optimal parameter value with least-square based optimization of polynomial coefficients. However, such a heuristic analogy for optimally finding the threshold values is cumbersome in LED-PAI imaging modality, thereby increasing the computational burden of the overall denoising process.

Another category of methods that follow a deconvolution based strategies to restore signal content and suppress noisy counterpart. Wiener deconvolution plays a vital role in reducing noise by equalizing phase and amplitude characteristics of the transducer response function [10]. Such a technique can greatly diminish both the noisy and signal degradation part with an accurate assumption of the transducer impulse response, failing to which it may bring additional signal artifacts and interpretation of signal becomes difficult in those scenarios. The algorithm is hugely influenced by the accurate estimation of correlation function between signal and noise which firmly controls the SNR of the output. In case, where the prior knowledge about the transducer response function and noisy power are unknown, researchers undergo a probabilistic measure of the response function using Bayesian or maximum a posteriori estimation-based approach, which on the other hand, increases the computational cost of the signal recovery mechanism. In the PCA mechanism, although the algorithm searches for principal components distributed along the perpendicular directions, often, it shows insignificant results due to its baseline assumption that the ratio of PA energy to the total energy of detected signals is more than 75%, which is not always the case [5].

Recently, researchers are exploring adaptive filtering mechanism, which does not require any prior knowledge of the signal and noise parameters, which could yield significant noise reduction. The ground assumption of such methodology stems from the fact that signal and noise are uncorrelated in consecutive time points, which can be satisfied with the general physics of the LED-PAI signal generation [11, 12]. Moreover, such techniques also attract the eye corner due to its fewer computations and reduced sensitivity to tuning parameters. One such technique where the SNR can be significantly increased is an adaptive noise canceller (ANC). Although, in ANC, there is a specific need to define a reference signal that significantly correlates with the noise which is hard to deduce in a real-time environment. This challenge can be adjusted in another form of ANC—adaptive line enhancer (ALE), in which the reference signal is prepared by providing a de-correlation delay to the noisy signal [13]. The reference signal consists of a delayed version of the primary (input) signal, instead of being derived separately. The delay is provided to de-correlate the noise signal so that the adaptive filter (used as a linear prediction filter) cannot predict the noise signal while easily predicting the signal of interest. Thus, the output contains only the signal of interests which is again subtracted from the desired signal, and the error signal is thereafter used to adapt the filter weights to minimize the error. It adaptively filters the delayed version of the input signal in accordance with the least mean square (LMS) adaptation algorithm. The time-domain computation of the ALE can be summarized as follows.

$$x(n) = pa(n) + noi(n) \tag{1}$$

$$r(n) = x(n - d) \tag{2}$$

$$y(n) = \sum_{k=0}^{L-1} w_k(n) r(n - k) \tag{3}$$

$$e(n) = x(n) - y(n) \tag{4}$$

$$w_k(n + 1) = w_k(n) + \mu \, e(n) x(n - k - d) \tag{5}$$

where, $x(n)$ is the primary input signal corresponding to the individual sensor element of ultrasound (US) transducer array, consists of PA signal component [$pa(n)$] and wideband noise component [$noi(n)$]. The reference input signal $r(n)$ is the delayed version of the primary input signal by a delaying factor d. The output $y(n)$ of the adaptive filter represents the best estimate of the desired response and $e(n)$ is the error signal at each iteration. $w_k(n)$ represents the adaptive filter weights, and L represents the adaptive filter length. The filter is selected as a linear combination of the past values of the reference input. Three parameters determine the performance of the LMS-ALE algorithm for a given application [14]. These parameters are ALE adaptive filter length (L), the de-correlation delay (d), and the LMS convergence parameter (μ). The performance of the LMS-ALE includes: adaptation rate, excess mean squared error (EMSE) and frequency resolution.

The convergence of the mean square error (MSE) towards its minimum value is commonly used performance measurement in adaptive systems. The MSE of the LMS-ALE converges geometrically with a time constant τ_{mse} as:

$$\tau_{mse} \approx \frac{1}{4\mu\lambda_{\min}} \tag{6}$$

where, $\lambda_{\min}$ is the minimum eigenvalue of the input vector autocorrelation matrix. Because τ_{mse} is inversely proportional to μ, a large τ_{mse} (slow convergence) corresponds to small μ. The EMSE ξ_{mse} resulting from the LMS algorithm noisy estimate of the MSE gradient is approximately given by:

$$\xi_{mse} \approx \frac{\mu L \lambda_{av}}{2} \tag{7}$$

where, λ_{av} is the average eigenvalue of the input vector autocorrelation matrix. EMSE can be calibrated by choosing the values of μ and L. Smaller values of μ and L reduce the EMSE while larger values increase the EMSE. The frequency resolution of the ALE is given by:

$$f_{res} = \frac{f_s}{L} \tag{8}$$

where, f_s is the sampling frequency. Clearly, f_{res} can be controlled by L. However, there is a design trade-off between the EMSE and the speed of convergence. Larger values of μ results in faster convergence at the cost of steady-state performance. Further, improper selection of μ might lead to the convergence speed unnecessary slower, introducing more EMSE in steady-state. In practice, one can choose larger μ at the beginning for faster convergence and then change to smaller μ for a better steady-state response. Again, there is an optimum filter length L for each case, because larger L results in higher algorithm noise while smaller L implies the poor filter characteristics. As the noise component of the delayed signal is rejected and the phase difference of the desired signal is readjusted, they cancel each other at the summing point and produce a minimum error signal that is mainly composed of the noise component of the input signal.

Moreover, researchers have proposed signal domain analysis to retrieve the acoustic properties of the object to be reconstructed from characteristic features of the detected PA signal prior to image reconstruction. In the proposed method, the signals are transformed into a Hilbert domain to facilitate analysis while retaining the critical signal features that originate from absorption at the boundary. The spatial and the acoustic propagation properties are strongly correlated with the PA signal alteration, and the size of an object governs the time of flight of the PA signal from the object to the detections. The relationship between object shape and signal acquisition delay exists partly because the smaller speed of sound (SoS) within the object will delay the arrival of the signal and vice versa. A simplistic low dimensional model as predicted by Lutzweiler et al. can forecast the corresponding time of arrival given the known phantom shape or the SoS (Fig. 2) [15]. Based on a similar assumption, the inverse problem of obtaining the unknown acoustic parameters can be solved from the extracted signal features. Lutzweiler et al. implemented the signal domain approach for the segmentation of PA images by addressing the heterogeneous optical

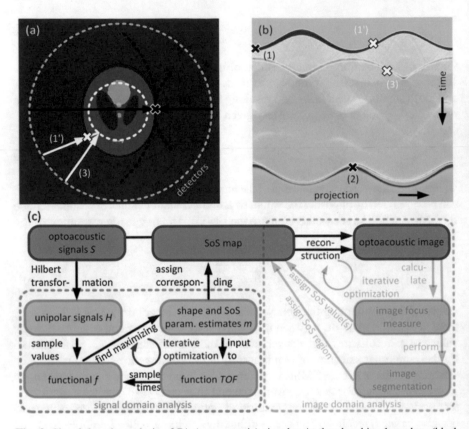

Fig. 2 Signal domain analysis of PA (optoacoustic) signal **a** At the absorbing boundary (black cross) of the numerical phantom huge signals will be detected at detector locations [(1) and (2)] with a tangential integration arc (dashed black line). Opposite detectors provide partially redundant information and, consequently information on the SoS. Accordingly, boundary signals (white cross) with direct (1′) and indirect (3) propagation provide information on the location of a reflecting boundary (white dashed line). **b** The corresponding sinogram with signal features corresponding to those in the image domain in (**a**). **c** The workflow of the proposed algorithm: Instead of performing reconstructions (red) with a heuristically assumed SoS map, signal domain analysis (green) is performed prior to reconstruction. Unipolar signals H are generated from the measured bipolar signals S by applying a Hilbert transformation with respect to the time variable. The optimized SoS parameters are obtained by retrieving characteristic features in the signals via maximizing the low dimensional functional f depending on acoustic parameters m through TOF and on the signals H. Subsequently, only a single reconstruction process with an optimized SoS map has to be performed. Conversely, for image domain methods (pale blue) the computationally expensive reconstruction procedure has to be performed multiple times as part of the optimization process. Adapted with permission from [15]

and acoustic properties of the tomographic reconstruction. Later in this chapter we will discuss about the use of image analytics in improving the visual quality.

4 Image Processing Applications

Post image reconstruction, the PA images need to be further processed in both spatial and transform domain for better visual perception of subtle features within the object and in advance level to classify/cluster different regions based on the morphological and functional attributes of the image. Several approaches that need to be performed in this domain starting with pre-processing, object recognition and segmentation based on morphological attributes and feature-based methodologies, including clustering of various regions within the object to image super-resolution techniques which are detailed in the following sub-sections.

4.1 Pre-processing and Noise Removal

Pre-processing in image domain majorly targets intensity enhancement of LED-PAI images and filtering of noises through spatio-frequency domain techniques. Although the implementation of pre-processing steps is subjective, indeed it is essential to readjust the dynamic scale of intensity and contrast for better understanding and perception of PA images, further helping in figuring out the significant regions or structures within the sample of interest. Generally, the dynamic range of reconstructed grayscale PA images is of low contrast, the histogram of which is concentrated within a narrow range of gray intensities. Therefore, a substantial normalization in the grayscale range needs to performed to increase the dynamic range of intensities and eventually enhance the contrast at both global and local scale. Although the intensity transformations—gamma, logarithmic, exponential functions play a crucial role in the intensity rescaling process, it is quite evident that the exact transformation that would possibly provide better-enhanced result is modality dependent.

Let $f(x, y)$ denotes a reconstructed PA image, while $f(x, y, t)$ corresponds to the successive time frames of PA images and $g(x, y)$ is the intensity enhanced PA image. Following the gamma transformation, the rule, $s = cr^\gamma$ maps the input intensity value r into output intensity s with the power-law factor γ. This law works well in general sense because most of the digital devices obey power-law distribution. However, the exact selection of γ is instrument-specific and depends on the image reconstruction methodology as well. Another frequently used technique is the logarithmic transformation function, $s = c\log(1 + r)$ that expands the intensity range of dark pixels of the input image while narrowing down the intensity range of brighter pixels of the input PA images. The opposite is true for the exponential transformation function. It is quite apparent that these two intensity transformations are experiment specific and are used to highlight the significant area/region in a case specific manner. More subjectively, the intensity transformations are quite limited in use and can facilitate image details at a very crude level. Histogram equalization, on the other hand, works at both global and local scale, stretching out the dynamic range of the intensity gray scales. The transformation function for the histogram equalization, of particular interest in image processing, at a global scale and can be written as,

$$s = T(r) = (L - 1) \int_0^r p_r(\omega) d\omega \tag{9}$$

where, $p_r(r)$ denotes the probability density function of input intensities, L is the maximum gray level value and ω is a dummy integration variable. In the discrete domain, the above expression is reduced to,

$$s = T(r) = \frac{(L - 1)}{MN} \sum_{j=0}^k n_j \quad for \, k = 0, 1, 2, \dots (L - 1) \tag{10}$$

where, n_k is the number of pixels having gray level r_k and MN stands for the total number of pixels in the input PA image. Although histogram processing at a global scale increases the contrast level significantly, often, it turns out that several subtle features in the imaging medium cannot be adequately distinguished from its neighborhood background due to proximity in gray levels values between these two. To mitigate this effect, researchers chose to implement local histogram processing for contrast enhancement which works on relatively smaller regions (sub-image) to implement histogram equalization technique. Similar to the global scale, local histogram analysis stretch-out the intensity levels within the sub-image part, thereby enhancing the subtle structural features at those locations.

For noise removal, PA image $f(x, y)$ undergoes filtering operations based on the PA imaging instrumentation and type of noises that hamper the image quality. The filtering operations can be performed either in spatial or in frequency domain. Depending on the nature of the associativity of noise (additive or convolutive), the filtering domains are finalized. In general, for additive noises, one can go forward with the spatial domain filtering, whereas for a convolutive type of noise it is advised to carry out the filtering process in frequency or transform domain. Spatial filtering operations are performed using the convolution operation using a filter kernel function $h(x, y)$, a generalized form of which is presented below,

$$g(x, y) = f(x, y) \otimes h(x, y) = \sum_{s=-a}^a \sum_{t=-b}^b h(s, t) f(x - s, y - t) \tag{11}$$

where $\otimes$ denotes the convolution operation which is linear spatial filtering of an image of size $M \times N$ with a kernel of size $m \times n$ and (x, y) are varied so that the origin of the kernel $h(x, y)$ visits every pixel in $f(x, y)$. In case of filtering based on the correlation, only the negative sign in the above equation will be replaced by a positive sign. Now, depending on the type of noise that corrupt the image content or produce an artifact in the PA images, the filter kernel can be any of the types— Low-pass filters: Gaussian, simple averaging, weighted averaging, median and High-pass filters: first-order derivative, Sobel, Laplacian, Laplacian of Gaussian functions. Details of these filtering kernels and related operations are described in [16, 17]. In general, Gaussian smoothing operation can reduce the noisy effect which follows a Gaussian distribution pattern, whereas simple averaging can reduce the blurry noise globally, and median filtering reduces the effect of salt and pepper noise from input reconstructed PA images. While the low-pass filters are working on the images to reduce the effect of high frequency noises, high-pass filtering is performed to sustain the edge and boundary information as well as to keep the subtle high-frequency structures in PA images. A special category of filtering operation which reduces the high frequency noises as well as restores the high-frequency edge information is unsharp masking and high-boost filtering, expression of which is presented below,

$$g_{mask}(x, y) = f(x, y) - \bar{f}(x, y) \tag{12}$$

$$g(x, y) = f(x, y) + k.g_{mask}(x, y) \tag{13}$$

where, $\bar{f}(x, y)$ is a blurred version of the input image $f(x, y)$ and for unsharp masking k is kept at 1 whereas, $k > 1$ signifies high-boost filtering. However, the above filters work globally irrespective of the changes in local statistical patterns, and there is a class of filtering techniques through an adaptive approach which includes adaptive noise removal filter, adaptive median filtering, etc. [16]. Apart from these generalized filtering approaches, there is a special class of techniques that controls the intensity values using fuzzy statistics, enabling the technique to regulate the inexactness of gray levels with improved performance [18]. The fuzzy histogram computation is based on associating the fuzziness with a frequency of occurrence of gray levels $h(i)$ by,

$$h(i) \leftarrow h(i) + \sum_x \sum_y \mu_{\tilde{I}(x,y)i} \quad \text{for} \quad k \in [a, b] \tag{14}$$

where, $\mu_{\tilde{I}(x,y)}$ is the fuzzy membership function. This is followed by the partitioning of the histogram based on the local maxima and dynamic histogram equalization of these sub-histograms. The modified intensity level corresponding to j-th intensity level on the original image is given by,

$$y(j) = start_i + range_i \sum_{k=strat_i}^{j} \frac{h(k)}{M_i} \tag{15}$$

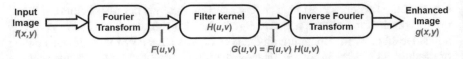

Fig. 3 Generalized block diagram for implementation of the frequency domain filtering process

where $h(k)$ denotes histogram value at k-th intensity level of the fuzzy histogram and M_i specifies the whole population count within i-th partition of a fuzzy histogram. In the last stage, the final image is obtained by normalization of image brightness level to compensate for the difference in mean brightness level between input and output images. Such a technique not only provides better contrast enhancement but also efficiently preserves the mean image brightness level with reduced computational cost and better efficiency.

On the other hand, frequency domain techniques are also essential in the scope of denoising PA images as it can efficiently and significantly reduce the effect of convolutive type of noises. Periodic noises or noises arrived due to specific frequency bands can be reduced through the frequency domain filtering approach as well. Moreover, such transform domain filtering can also be used to reduce the effect of degradation sources that hinders the image details after reconstruction. The general block diagram of the frequency domain filtering technique is depicted below in Fig. 3.

The input PA images are transformed into frequency domain counterpart, followed by the implementation of filtering kernel and again bringing back the images to a spatial domain at the end. Through this frequency domain approach, one can become aware of the noise frequencies and their strength, which further enables frequency selective filtering of the PA images. Several filtering kernels can significantly reduce the noisy part, like Butterworth and Gaussian low- and high-pass filters, bandpass, notch filtering kernels, homomorphic filtering etc. While transforming the input image into its frequency domain counterpart, the noisy part in $f(x, y)$ becomes in multiplicative form, which can be further reduced by homomorphic filtering technique [16]. A more generalized form of homomorphic filtering which works on a Gaussian high-pass filtering approach is given below,

$$H(u, v) = (\gamma_H - \gamma_L)\left[1 - exp\left\{-c\left(\frac{D(u, v)}{D_0}\right)^2\right\}\right] + \gamma_L \qquad (16)$$

where, D_0 is the cut-off frequency, $D(u, v)$ is the distance between coordinates (u, v) and the center frequency at $(0, 0)$. c advocates the steepness of the slope of the filter function. γ_H and γ_L are the high and low-frequency gains. This transfer function simultaneously compresses the dynamic range of intensities and enhances the contrast, thereby preserving the high-frequency edge information while reducing the noisy components. Another form of such filter is given in [17] which has the transfer function,

$$H(u, v) = \frac{1}{1 + exp\{-a(D(u, v) - D_0)\}} + A \qquad (17)$$

where, the high frequency and low-frequency gains are manipulated by following rules,

$$\gamma_H = 1 + A \quad \text{and} \quad \gamma_L = \frac{1}{1 + exp[aD_0]} + A \qquad (18)$$

Although preprocessing of PA images in spatial and frequency domains significantly improve the image quality and enhances the contrast level, the selection of proper filter function is purely subjective, and the parametrization of filtering attributes is PAI model specific. Therefore, it is of more significant importance while choosing the filter function and optimizing its parameters either through heuristic approaches or iterative solutions which can further help in reducing the artifacts and noisy components in PA reconstructed images.

4.2 Segmentation of Objects in PA Images

Post-reconstruction image analysis is an integral part of PAI as it aids in understanding different sub-regions through the processing of different morphological features. In case of functional imaging, it also helps to reduce artefacts and noise that unnecessary hampers functional parameters. Being a challenging task due to relatively low intrinsic contrast of background structures and increased complexity due to limited view problems of LED-PAI, various researchers have worked on the segmentation of objects in PA images through the implementation of different algorithms that are detailed in the following sections.

4.2.1 Morphological Processing

Classical approaches in image processing for edge detection and segmentation of objects with different shapes and sizes can be implemented to segment out any object of interest in PA images. In this context, classical Sobel operators (works on approximating the gradient of image intensity function), Canny edge detector (implements a feature synthesis step from fine to coarse-scale) and even combination of morphological opening and closing operators (through a specific size of structuring element) can help in identifying the object edge/boundary. However, the parameterization of these kernel operators is specific to the LED-PAI system as the contrast resolution between object and background varies significantly, and intra-object intensity distribution is modality dependent. Another mechanism driven by anisotropic diffusion can estimate the object boundary in PA images through the successful formulation of a scale-space adaptive smoothing function [19]. The operation works on successive smoothing of the original PA image $I_0(x, y)$ with a Gaussian kernel $G(x, y : t)$ of variance t (scale-space parameter), thereby producing a successive number of more and more blurred images [20]. Such anisotropic diffusion can be modelled as,

$$I(x, y, t) = I_0(x, y) * G(x, y; t) \tag{19}$$

with the initial condition $I(x, y, 0) = I_0(x, y)$, the original image. Mathematically, the AD equation can be written as,

$$I_t = div(c(x, y, t)\nabla I) = c(x, y, t)\Delta I + \nabla c \cdot \nabla I \tag{20}$$

For a constant $c(x, y, t)$, the above diffusion equation becomes isotropic as given by,

$$I_t = c\Delta I \tag{21}$$

Perona and Malik have shown that the simplest estimate of the edge positions providing excellent results are given by the gradient of the brightness function [19]. So, the conduction coefficient can be written as,

$$c(x, y, t) = g(\|\nabla I(x, y, t)\|) \tag{22}$$

The Gaussian kernel, used in smoothing operation, blurs the intra-region details while the edge information remains intact. A 2D network structure of 8 neighboring nodes is considered for diffusion conduction. Due to low intrinsic contrast stemming from the background structures on PA modality, often this anisotropic diffusion filtering can be used to create a rough prediction of the object boundary which can be further utilized as seed contour for active contour technique (described later in chapter) for actual localization and segmentation of the object boundary under observation.

In another work, researchers show that low-level structures in images can be segmented through a multi-scale approach, facilitating integrated detection of edges and regions without restrictive models of geometry or homogeneity [20]. Here, a vector field is created from the neighborhood of a pixel while heuristically determining its size and spatial scale by a homogeneity parameter, followed by integrating the scale into a nonlinear transform which makes structure explicit in transformed domain. The overall computation of scale-space parameters is made adaptive from pixel to pixel basis. While such methodology can identify structures at low-resolution and intensity levels without any smoothing at even coarse scales, another technique that serves as a boundary segmentation through the utilization of color and textural information of images to track changes in directions, creating a vector flow [21]. Such an edge-flow method detects boundaries when there are two opposite directions of flow at a given location in a stable state. However, it depends strongly on color information and requires a user-defined scale to be input as a control parameter. However, reconstructed PA images lack color information and edge-flow map has to be extracted solely from edge and textural information.

Recently, Mandal et al. have developed a new segmentation method by integrating multiscale edge-flow, scale-space diffusion and morphological image processing [22]. As mentioned earlier, this method draws inspiration from edge-flow methods but circumvents the lacking color information by using a modified subspace sampling method for edge detection and iteratively strengthens edge strengths across scales. This methodology reduces the parameters that need to be defined to achieve a segmented boundary between imaged biological tissue and acoustical coupling medium by integrating anisotropic diffusion and scale-space dependent morphological processing, followed by a curve fitting to link the detected boundary points. The edge flow algorithm defines a vector field, such that the vector flow is always directed towards the boundary on both its sides, in which the relative directional differences are considered for computing gradient vector. The gradient vector strengthens the edge locations and tracks the direction of the flow along x and y directions. The search function looks for sharp changes from positive to negative signs of flow directions and whenever it encounters such changes, the pixel is labelled as an edge point. The primary deciding factor behind the edge strength is the magnitude of change of direction for the flow vector, which is reflected as edge intensity in the final edge map. The vector field is generated explicitly from fine to coarse scales, whereas the multiscale vector conduction is implicitly from coarse to finer scales. The algorithm essentially localizes the edges in the finer scales. The method achieves it by preserving only the edges that exist in several scales and suppressing features that disappear rapidly with an increment of scales.

Often in LED-PAI imaging, noisy background is present in reconstructed images due to low SNR, limited view, and shortcomings of inversion methodologies. Additionally, signals originate from the impurities or inhomogeneities within the coupling medium. Such noises are often strong enough to be detected by edge detection algorithm as true edges. Thus, the use of an anisotropic diffusion process is useful to further clean up the image, where it smoothens the image without suppressing the edges. Thereafter non-linear morphological processing is done on the binary (diffused) edge mask. Mandal et al. took a sub-pixel sampling approach (0.5 px), rendering the operation is redundant beyond the second scale level [22]. Further in PA images, the formation of smaller edge clusters and open contours is quite apparent. Thus, getting an ideal segmentation using edge linker seems to perform poorly. The proposed method first generates the centroids for edge clusters and then tries to fit on a geometric pattern (deformable ellipse) iteratively through a set of parametric operations. The method is self-deterministic and requires minimal human intervention. Thus, the algorithm is expected to help automate LED-based PA image segmentation, with important significance towards enabling quantitative imaging applications.

4.2.2 Clustering Through Statistical Learning

Pattern recognition and statistical learning procedures play an essential role in segmentation through clustering of different tissue structures in LED-PAI based on their intensity profiles which are directly associated with the wavelength-specific absorption of optical radiation, followed by characteristic emission of acoustic waves. Depending on the constituents at vascular and cellular levels, various structures and regions can be segmented through structural and functional attributes in PA images using machine learning-based approaches. Guzmán-Cabrera et al. shown a segmentation technique, performed using an entropy-based analysis, for identification and localization of the tumor area based on different textural features [23]. The local entropy within a window $M_k \times N_k$ can be computed as,

$$E(\Omega_k) = - \sum_{i=0}^{L-1} P_i \log(P_i) \quad \text{where} \quad P_i = \frac{n_i}{M_k \times N_k} \tag{23}$$

where, Ω_k is the local region within which the probability of grayscale i is P_i with number of pixels having the grayscale i is n_i. The whole contrast image is then converted to a texture-based image, in which the bottom texture represents a background mask. This is used as the contrast mask to create the top-level textures, thus obtaining the segmentation of different classes of objects with region-based quantification of tumor areas.

In another research, Raumonen and Tarveinen worked on developing a vessel segmentation algorithm following a probabilistic framework in which a new image voxel is classified as a vessel if the classification parameters are monotonically decreased [24]. The procedure follows an iterative approach by uniformly sweeping over the parameter space, resulting in an image where the intensity is replaced with confidence or reliability value of how likely the voxel is from a vessel. The framework is initiated with the smoothing of PA images, followed by clustering and vessel segmentation of clusters and finally filling gaps in the segmented image. A small ball-supported kernel is convolved with the reconstructed PA images to smooth-out the noisy parts, followed by a threshold filtering. Clustering is approached using a region growing procedure in which the vessel structures are labelled as connected components. A large starting intensity and a large neighbor intensity leads the voxels to be classified as a vessel with high reliability, and decreasing these values increases the number of voxels classified as vessel but with less reliability. Post-clustering, each vessel network is segmented into smaller segments without bifurcations and finally filling the gaps in vessel-segmented data and potential breakpoints of vessels are identified and filled based on a threshold length of the gap and threshold angle between the tip directions.

Furthermore, statistical learning procedures have shown to perform significantly well in this context [25], showing a new dimension in LED-PAI research towards automatic segmentation and characterization of pathological structures. Different learning techniques, comprising supervised, unsupervised and even deep neural networks, can be implemented for segmentation and classification of breast cancer, which substantially improves the segmentation accuracy at the cost of computational expenses. In a nutshell, Bayesian decision theory quantifies the tradeoffs between various classification decisions using probability theory. Considering a two-class problem with n features having feature space, $X = [X_1, X_2 . . . . . X_n]^T$, the Bayes' theorem forms the relationship,

$$p(\omega_i/x) = \frac{P(\omega_i)p(x/\omega_i)}{P(x)} \tag{24}$$

where, $p(\omega_i/x)$ is termed as posterior, $P(\omega_i)$ is prior, $p(x/\omega_i)$ likelihood, $P(x)$ is evidence. Based on the various statistical and morphological features, the decision can be made as,

$$P(\omega_1)p(x/\omega_1) > P(\omega_2)p(x/\omega_2) \text{ for class1 else class2} \tag{25}$$

At a bit higher level, support vector machine (SVM), which is a highly non-linear statistical learning network, works on maximizing the distance between the classes and separating hyper-plane. Considering a two-class problem in which the region of interest belongs to a particular class and all other areas are comprising another class in PA images, let $\{x_1, x_2 . . . . x_n\}$ be our data set and let y_i be the class label of x_i.
Now,

(a) The decision boundary should be as far away from the data of both classes as possible. Distance between the origin and the line $W^T X = k$ is,

$$Distance = \frac{k}{\|W\|} \tag{26}$$

And we have to maximize m where $m = \frac{2}{\|W\|}$.
(b) For this the linear Lagrangian objective function is

$$J(w, \alpha) = \frac{1}{2}w^T w - \sum \alpha_i \{y_i (w_0 + w^T X) - 1\} \tag{27}$$

(c) Differentiating this with respect to w and α, w can be recovered as $w = \sum \alpha_i y_i x_i$.
(d) Now for testing a new data z, compute $(w^T z + b)$, and classify z as class 1 if the sum is positive, and class 2 otherwise.

Although the algorithm is well capable of segmenting the region of interest, the training procedure requires prior knowledge and annotation of the region of interest to work in a supervised manner. In contrast to this, K-means clustering approach is purely unsupervised and works relatively fast in clustering various regions as per their different statistical and image-based feature sets. For a 2-class clustering problem, initially two points are taken randomly as cluster centers. The main advantages of

K-mean clustering are its simplicity and computational speed. The disadvantage of this technique is that it does not give the same result due to random initialization. It minimizes the intracluster variances but does not ensure that the result has a global minimum of variance. After the initialization of the center for each class, each sample is assigned to its nearest cluster. To find nearest cluster one can use different distance measures e.g. Euclidian, city-block, Mahalanobis distances etc. the simplest one to use the Euclidian distance with the following form,

$$d(x_i, y_i) = \sqrt{\sum_{i=1}^{n}(x_i - y_i)^2} \tag{28}$$

where x_i and y_i are the coordinates of 'i'th sample, and n is the total number of samples. The new cluster center is obtained by,

$$c_i = \frac{1}{p_k}\sum_{x_i \in c_k} x_i \tag{29}$$

where, p_k is the number of the points in kth cluster and c_k is the kth cluster. In order to include the degree of belongingness, fuzzy c-means (FCM) based approach is more accurate over the K-means clustering process. FCM contrasts K-means clustering with the fact that in FCM each data point in the feature set has a degree of belonging to a cluster rather than belonging entirely to a cluster. Let us define a sample set of n data samples that we wish to classify into c classes as $X = \{x_1, x_2, x_3, \ldots, x_n\}$ where each x_i is an m-dimensional vector of m elements or features. The membership value of kth data point belonging to ith class is denoted as μ_{ik} with the constraint that,

$$\sum_{I=1}^{c}\mu_{ik} = 1 \ \forall k = 1, 2, \ldots, n \quad \text{and} \quad 0 < \sum_{k=1}^{n}\mu_{ik} < n \tag{30}$$

The objective function is,

$$J(\mu, v) = \sum_{k=1}^{n}\sum_{i=1}^{c}\mu_{ik}^{b}(d_{ik})^2 \tag{31}$$

where b is the index of fuzziness and d_{ik} is the Euclidean distance measure between the kth sample x_k and ith cluster center v_i. Hence, d_{ik} is given by (23),

$$d_{ik} = \|\mathbf{x}_k - \mathbf{v}_i\| = \left[\sum_{j=1}^{m}\left(x_{kj} - v_{ij}\right)^2\right]^{1/2} \tag{32}$$

Minimization of the objective function with respect to μ and v leads to the following equations. The ith cluster center is calculated by,

$$v_i = \frac{\sum_{k=1}^{n} \mu_{ik}^b \cdot x_k}{\sum_{k=1}^{n} \mu_{ik}^b} \tag{33}$$

And membership values are, $\mu_{ik} = \frac{(1/d_{ik})^{2/(b-1)}}{\sum_{r=1}^{c}(1/d_{rj})^{2/(b-1)}}$

Now to find the optimum partition matrix μ, an iterative optimization algorithm is used. The step-by-step procedure is given below,

(a) Initialization of the partition matrix $\mu(0)$ randomly.
(b) Then do $r = 0, 1, 2, …..$
(c) Calculation of c cluster centre vectors $v_i^{(r)}$ using $\mu^{(r)}$
(d) Updating the partition matrix $\mu^{(r)}$ using the cluster center values, if $\left\| \mu^{(r+1)} - \mu^{(r)} \right\|_F \leq \varepsilon$ where ε is the tolerance level and $\|\cdot\|_F$ is Frobenius distance, stop; otherwise set $r = r + 1$ and return to step 2.

After obtaining the optimized partition matrix, depending upon the highest membership value, the data points are assigned to that particular class.

In more recent work, Zhang et al. have demonstrated the deep-learning procedures to segment out tumor area in breast PA images [25]. The area of deep learning is becoming very broad with the recent advancement in artificial intelligence-based approaches through mathematical formulations which is beyond the scope of this chapter. In short, deep learning is a powerful technique that not only reduces the labor in manually computing various features and curse of feature dimensionality but also provides a powerful and robust mechanism of creating any decision. Zhang et al. have shown different deep learning networks like AlexNet and GoogleNet for PA images and established their efficacy in segmenting the breast tumor area [25]. Furthermore, the final contour selection was implemented using a dynamic programming architecture: active contour model which is elaborated in the next section.

4.2.3 Deformable Segmentation

Segmentation of the region of interest through deformable objects plays a significant role in PAI as it is indeed necessary to locate a region / area from the background in-homogeneous reflection model. An example of such deformable object formation is through designing an active contour (AC) algorithm which can regulate the boundary based on various parameters such as energy, entropy, class levels etc. AC can be modeled using geodesic and level set methods. Here, we focus on such an algorithm designed using an improved snake-based AC method which works on a greedy approach [13]. The idea is to fit an energy-minimizing spline along the boundary, characterized by different internal and external image forces. The goal is to reach for a curve where the weighted sum of internal and external energy will be minimum.

The basic equation can be formulated as,

$$E_{snake} = \int_0^1 \{E_{\text{int}}(v(s)) + E_{ext}(v(s))\}ds \tag{34}$$

where, the position of snake is represented by a planar curve $v(s) = (x(s), y(s))$, E_{int} is the internal energy force, used to smooth the boundary during deformation. E_{ext} represents the external energies, pushing the snake towards the desired object. Seed contour for the initial labeling can be identified through various segmentation techniques discussed earlier in this chapter. Coordinates of seed contour is transformed into polar form (ρ, θ). The contour is now represented with a set of such discrete polar coordinates $v_i = (\rho_i, \theta_i)$ for $i = 0, 1, 2, \ldots, (n-1)$; where $\theta_i = i \times \theta_s$. For example, quantization step size for angel θ is $\theta_z = 1°$ ($n = 360$) and for ρ is $\rho_s = 1$ pixel. The energy function of this model is given by,

$$E = \sum_{i=0}^{n-1} \left(a E_{cont}(v_i) + b E_{curv}(v_i) + c E_{image}(v_i) + d E_{grow}(v_i)\right) \tag{35}$$

According to Fig. 4, for each point v_i for $i = 0, 1, 2, \ldots, (n-1)$, the energies at the points $\Omega_i = \{v_i^-, v_i, v_i^+\}$ are calculated and v_i are moved to the point with the minimum energy among these three where v_i^- and v_i^+ are the two discrete points adjacent to v_i at the radial direction. This operation is performed iteratively until the number of moved contour points is sufficiently small or the iteration time exceeds a predefined threshold. The energy functions are: E_{cont} is the internal continuity spline energy that helps to maintain the contour to be continuous, E_{curv} is the internal curvature energy for smoothing the periphery, E_{image} is external image force that depends on the image intensity points and E_{grow} represents the external grow energy that helps to expand the contour from the center towards the boundary. Mathematically they can be represented as [26, 27],

Fig. 4 Representation of active contour (snake) in a polar coordinate system

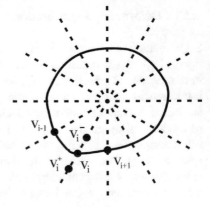

$$E_{cont}(v_j) = |\bar{d} - |v_j - v_{i-1}|| + |\bar{\rho} - |\rho_j - \rho_{i-1}|| \quad (v_j \in \Omega_i) \quad (36)$$

where, $\bar{d} = \sum \frac{|v_t - v_{t-1}|}{n}$ and $\bar{\rho} = \sum \frac{|\rho_t - \rho_{t-1}|}{n}$

$$E_{curv}(v_j) = |v_{i+1} - 2v_j + v_{i-1}|^2 + |\rho_{i+1} - 2\rho_j + \rho_{i-1}|^2 \quad (v_j \in \Omega_i) \quad (37)$$

$$E_{image}(v_j) = \frac{1}{R} \sum_{r=1}^{R} I(\rho_j + r \times \rho_s, \theta_j) - \frac{1}{R} \sum_{r=1}^{R} I(\rho_j - r \times \rho_s, \theta_j) \quad (v_j \in \Omega_i)$$

$$(38)$$

$$E_{grow}(v_j) = \begin{cases} e & \text{if } v_j = v_i^+ \text{ and } |\bar{I}_{v_j} - \bar{I}_{origin}| < T \\ 0 & \text{else} \end{cases} \quad (39)$$

where, $\bar{I}_{v_j} = \frac{1}{k \times k} \sum_{v_i \in \Psi_{v_j}} I(v_i)$ and $\bar{I}_{origin} = \frac{1}{k \times k} \sum_{v_i \in \Psi_0} I(v_i)$

Ψ_{v_j} and Ψ_0 are two $k \times k$ (e.g., $k = 3$) sub-blocks with center points at v_i and the centroid of the contour respectively. The energy will decrease at v_i^+ if both the sub-blocks are of the same intensity approximately, resulting in an outward movement of the contour. This movement stops while the sub-blocks have different intensities. Threshold T determines the range up to which the change in intensity is allowed. e is a negative constant, small value of which will limit the algorithm for more shape restrictions where large value of e also nullifies the effect of image energy for which the contour can exceed the actual boundary.

5 Reconstructed PA Image Quality Improvement Using a Multimodal Framework

Biological tissues show significant depth-dependent optical fluence loss and acoustic attenuations. Correcting for the optical and acoustic variations are critical for delivering a quantitative imaging performance [28, 29]. Several techniques have been proposed for alleviating this problem including the use of exogenous contrast agents and computing differences in the spatial characteristics of the absorption coefficient over length scales [30, 31], using multiple optical sources together with non-iterative reconstruction and use of context encoding within a machine learning framework [32]. Most of the applied methods use a model of light transport equation considering a homogeneous medium [33]. Furthermore, for characterizing heterogeneous medium, the use of intrinsic (segmented) priors [34], and extrinsic priors obtained by combining PA with diffused optical tomography [35], acousto-optical imaging [36] and other imaging modalities have been investigated. However, these methods require additional computational resources and often hardware support for multimodal imaging. On the other hand, most current state-of-the-art PA imaging

Fig. 5 The algorithmic
workflow for multimodal
prior based image correction
for small animal PA imaging.
Reprinted with permission
from [38]

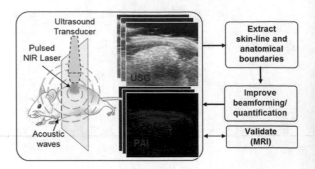

systems come equipped with hybrid ultrasound imaging capabilities. Thus, there is an increased interest in employing the co-registered US image to improve the performance of quantitative PA imaging [37, 38]. Additionally, the use of integrated PA-US imaging can further be utilized for the correction of small SoS changes. Mandal et al. aimed to correct for the optical and acoustic inaccuracies in PA imaging using extrinsic imaging priors obtained through segmentation of concurrently acquired high-frequency US images [38]. The US prior is used to create a localized fluence map and apply the correct SoS during advanced beamforming. Figure 5 depicts the process diagram. The method outlined by [38] shows that the use of multimodal priors can significantly improve the quantification of PA signals, and further computer vision methods can be employed to obtain the performance enhancement. Related publications by Naser et al. [39] have further shown that combining finite-element-based local fluence correction (LFC) with SNR regularization can estimate oxygen saturation (SO_2) in tissue accurately. Though a detailed discussion on tissue oxygenation measurement is beyond the scope of the chapter, the readers should reconcile to the fact that a quantitative measurement is only possible by producing an accurate estimate of tissue absorption profile. The B-mode ultrasound images provided a mean for surface segmentation and an initiation point for building the FEM mesh, which was employed by both research groups.

The PA signal received from a high-frequency linear array system is often not suitable for proper segmentation of anatomical structures. Therefore, the co-registered B-mode US signal is used as a reference frame to segment skin boundaries and delineated organ structure as well as tumor masses. The segmented prior information from the US is then used for iteratively correcting the PA images. A two-step approach is used to generate the US priors: (1) the skin line is detected using graph cuts [40, 41], and (2) internal structures were detected using active contour models (Fig. 5) [27, 42]. The lazy snapping method based on graph cuts separates coarse and fine-scale processing and enhances object specification and boundary detection even in low contrast conditions. The satisfactory low SNR performance of the method with suitable convergence speed makes it an ideal choice for skin line detection in PA-US images. Earlier in this chapter, we have described active contour methods in sufficient detail. Modified AC segmentation methods have been used extensively for visual quality enhancement in PA images. The methods performed efficiently for whole-body tomographic images, as well as for 2D linear array geometries. The majority of commercial LED-PAI systems utilize linear array geometry for signal acquisition; thus, the outlined methods are translatable to such instrumentations without many changes.

The algorithmic workflow consists of acquiring the PA signal and beamforming using the delay and sum algorithm. An automatic SoS estimation is implemented based on prior temperature information [42]. The images are spectrally unmixed using 10 optical wavelengths for finding out the tissue oxygenation profile. The US images are individually segmented and superimposed on the PA image. A deformable active contour segmentation (snakes) model is used for the segmentation of US images. The segmented tissue boundary is considered as the starting point for the model, followed by an iterative segmentation of the tumor region using multiscale edge detectors.

The (segmented) prior information from the US image is used to delineate the tumor mass and improve the fidelity of the optical fluence and multiparametric SoS fitting. Based on the segmented US mask (Fig. 6a), the process can accurately model the decay of light fluence used. The fluence field, thus created, is used to compensate for the depth-dependent decrease in the PA signal (Fig. 6b–d). Additionally, given the prior information about the tissue/coupling medium background, a two-compartment model for SoS calibration can be implemented and fit two different SoS for the object and the background. In summary, the multimodal segmentation framework is helpful in addressing both the optical attenuation as well as the acoustic attenuation, providing an improved visual image quality and a more quantitative

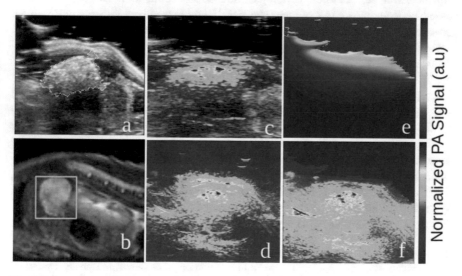

Fig. 6 Fluence correction improves CNR performance and quantitative information of PA images, **a** segmented ultrasound image, **b** reference MRI image for validation, **c** co-registered PA-US image, **d** fluence field map generated FEM method with US prior information, **e** PA image without correction, and **f** PA image after correction. Adapted in parts from [38]

imaging performance in vivo. The advanced multimodal methodologies can be integrated with crucial image processing techniques and imaging physics to achieve better LED-PAI imaging performance. Interestingly, the small form factor, as well as the ease of handling LED-based illumination arrays, can make it a modality of choice for exploring for such multimodal imaging, especially as we enter the realms of radiological imaging.

6 Summary

The last couple of decades have seen rapid developments in the field of biomedical PA imaging with the evolution of state-of-the-art small animal imaging scanners and experimental clinical hand-held platforms. The technology has graduated from the engineering laboratories to commercial products for pre-clinical imaging, and further into biomedical/translational imaging platforms. So far, the focus of development in PA imaging was primarily focused on hardware improvements and solving complex inverse problems. More recently, researchers have shown the applicability of image analysis to the current state-of-the-art PA imaging instrumentation. Post reconstruction signal and image processing methods are increasingly becoming practical tools for improving the visual image quality of PA imaging. The imaging physics—image analysis corroboration, as illustrated in this chapter, has led to the development of new methods for quantitative inversion and parameter self-calibration, resolution enhancement, and accurate mapping of fluence and acoustic heterogeneities. LED-based PA systems are in a nascent state itself, and these developments in PA signal processing will accelerate the growth and clinical adoption of LED-PAI. However, several additional challenges (e.g., low SNR, reduced imaging depths, errors in multimodal image registration due to high signal averaging requirement) exist in the application of intelligent image processing techniques in LED-PAI images. In the future, these advancements will be helpful in enabling quantitative molecular and oncological imaging using multispectral LED-PAI imaging [43, 44]. This opens up the possibility of a plethora of new developments, including the development of machine learning (ML) based algorithms for parameter estimation and image enhancement. ML-based algorithms can vastly be useful for improved reconstruction, identification, and segmentation of organs and vascular structures [45]. Finally, the relatively lower cost, accessibility, and low-profile form factor of LED-based PA system is bound to accelerate its use in computational PA imaging and encourage further development in signal and image processing methodologies.

Acknowledgements The authors acknowledge the support of several students and research scholars (PS Viswanath, XL Dean Ben, Jayaprakash, V Periyasamy, HT Garud) and faculty mentors (D Razansky, PK Dutta, D Komljenovic, M Pramanik) whose work and/or comments have contributed directly or indirectly to this chapter.

References

1. Y. Zhang, M. Jeon, L.J. Rich, H. Hong, J. Geng, Y. Zhang, S. Shi, T.E. Barnhart, P. Alexandridis, J.D. Huizinga, Non-invasive multimodal functional imaging of the intestine with frozen micellar naphthalocyanines. Nat. Nanotechnol. **9**(8), 631–638 (2014)
2. M.F. Kircher, A. De La Zerda, J.V. Jokerst, C.L. Zavaleta, P.J. Kempen, E. Mittra, K. Pitter, R. Huang, C. Campos, F. Habte, A brain tumor molecular imaging strategy using a new triple-modality MRI-photoacoustic-Raman nanoparticle. Nat. Med. **18**(5), 829–834 (2012)
3. J. Jo, G. Xu, Y. Zhu, M. Burton, J. Sarazin, E. Schiopu, G. Gandikota, X. Wang, Detecting joint inflammation by an LED-based photoacoustic imaging system: a feasibility study. J. Biomed. Opt. **23**(11), 110501 (2018)
4. C.M. O'Brien, K. Rood, S. Sengupta, S.K. Gupta, T. DeSouza, A. Cook, J.A. Viator, Detection and isolation of circulating melanoma cells using photoacoustic flowmetry. J. Vis. Exp. 57 (2011)
5. M. Zafar, R. Manwar, K. Kratkiewicz, M. Hosseinzadeh, A. Hariri, S. Noei, M. Avanaki, Photoacoustic signal enhancement using a novel adaptive filtering algorithm. In: Proceeding of SPIE BiOS. vol. 10878 (2019)
6. Y. Zhu et al., Light emitting diodes based photoacoustic imaging and potential clinical applications. Sci. Rep. **8**(1), 9885 (2018)
7. Z. Wu, N.E. Huang, Ensemble empirical mode decomposition: a noise-assisted data analysis method. Adv. Adapt. Data Anal. **1**(01), 1–41 (2009)
8. X. Jing et al., Adaptive wavelet threshold denoising method for machinery sound based on improved fruit fly optimization algorithm. Appl. Sci. **6**(7), 199 (2016)
9. C.B. Smith, A. Sos, A. David, A wavelet-denoising approach using polynomial threshold operators. IEEE Signal Process. Lett. **15**, 906–909 (2008)
10. C.V. Sindelar, N. Grigorieff, An adaptation of the Wiener filter suitable for analyzing images of isolated single particles. J. Struct. Biol. **176**(1), 6074 (2011)
11. J. Xia, J. Yao, L.V. Wang, Photoacoustic tomography: principles and advances. Electromagn. Waves (Cambridge, Mass.), **147**, 1 (2014)
12. Z. Yu, H. Li, P. Lai, Wavefront shaping and its application to enhance photoacoustic imaging. Appl. Sci. **7**(12), 1320 (2017)
13. K. Basak, X.L. Deán-Ben, S. Gottschalk, M. Reiss, D. Razansky, Non-invasive determination of murine placental and foetal functional parameters with multispectral optoacoustic tomography. Nat. Light. Sci. App. **8**(1), 1–10 (2019)
14. J.R. Zeidler, Performance analysis of LMS adaptive prediction filter. Proc. IEEE **78**(12) (1990)
15. C. Lutzweiler, R. Meier, D. Razansky, Optoacoustic image segmentation based on signal domain analysis. Photoacoustics **3**, 151–158 (2015)
16. R.C. Gonzalez, Woods RE. *Digital Image Processing*. 4th edition. (Pearson, 2018)
17. M. Petrou, C. Petrou, *Image Processing: The Fundamentals* (Wiley, 2010)
18. D. Sheet, H. Garud, A. Suveer, M. Mahadevappa, J. Chatterjee, Brightness preserving dynamic fuzzy histogram equalization. IEEE Trans. Consumer Electro. **56**(4), 2475–2480 (2010)
19. P. Perona, J. Malik, Scale-space and edge detection using anisotropic diffusion. IEEE Trans. Patt. Ana. Mach. Intelli. **12**(7), 629–639 (1990)
20. M. Tabb, N. Ahuja, Multiscale image segmentation by integrated edge and region detection. IEEE Trans. Image Process. **6**(5), 642–655 (1997)
21. W.Y. Ma, B.S. Manjunath, Edgeflow: a technique for boundary detection and image segmentation. IEEE Trans. Image Process. **9**(8), 1375–1388 (2000)
22. S. Mandal, V.P. Sudarshan, Y. Nagaraj, X.L. De´an-Ben, D. Razansky, Multiscale edge detection and parametric shape modeling for boundary delineation in optoacoustic images. in *Proceeding of 37th Annual International Conference of the IEEE Engineering in Medicine and Biology Society (EMBC)*. (2015). pp. 707–710
23. R. Guzmán-Cabrera, J.R. Guzmán-Sepúlveda, A. González-Parada, M. Torres-Cisneros, A system for medical Photoacoustic image processing. Pensee J. **75**(12), 374–381 (2013)

24. P. Raumonen, T. Tarveinen, Segmentation of vessel structures from photoacoustic images with reliability assessment. Biomed. Opt. Express. **9**(7), 328251 (2018)
25. J. Zhang, B. Chen, M. Zhou, H. Lan, F. Gao, Photoacoustic image classification and segmentation of breast cancer: a feasibility study. IEEE Access **7**, 5457–5466 (2019)
26. K. Basak, R. Patra, M. Manjunatha, P.K. Dutta, Automated detection of air embolism in OCT contrast imaging: anisotropic diffusion and active contour-based approach. in *Proceeding of 3rd International Conference on Emerging Applications of Information Technology*. (2012), pp. 110–115. https://doi.org/10.1109/eait.2012.6407874
27. M. Kass, A. Witkin, D. Terzopoulos, Snakes: active contour models. Int. J. Comp. Vis. **1**(4), 321–331 (1988)
28. S.L. Jacques, Coupling 3D monte carlo light transport in optically heterogeneous tissues to photoacoustic signal generation. Photoacoustics **2**(4), 137–142 (2014)
29. B. Cox, J.G. Laufer, S.R. Arridge et al., Quantitative spectroscopic photoacoustic imaging: a review. J. Biomed. Optics. **17**, 061202 (2012)
30. A. Rosenthal, D. Razansky, V. Ntziachristos, Quantitative optoacoustic signal extraction using sparse signal representation. IEEE Trans. Med. Imaging **28**(12), 1997–2006 (2009)
31. T. Jetzfellner, A. Rosenthal, A. Buehler et al., Optoacoustic tomography with varying illumination and non-uniform detection patterns. J. Opt. Soc. Am.. A, Optics, image science, and vision. **27**(11), 2488–2495 (2010)
32. T. Kirchner, J. Grohl, L. Maier-Hein, Context encoding enables machine learning-based quantitative photoacoustics. J. Biomed. Opt. **23**(5), 056008 (2018)
33. J. Laufer, B. Cox, E. Zhang et al., Quantitative determination of chromophore concentrations from 2d photoacoustic images using a nonlinear model-based inversion scheme. Appl. Opt. **49**(8), 1219–1233 (2010)
34. S. Mandal, X.L. Dean-Ben, D. Razansky, Visual quality enhancement in optoacoustic tomography using active contour segmentation priors. IEEE Trans. Med. Imaging **35**, 2209–2217 (2016)
35. A.Q. Bauer, R.E. Nothdurft, J.P. Culver et al., Quantitative photoacoustic imaging: correcting for heterogeneous light fluence distributions using diffuse optical tomography. J. Biomed. Optics. **16**(9), 096016 (2011)
36. K. Daoudi, A. Hussain, E. Hondebrink et al., Correcting photoacoustic signals for fluence variations using acousto-optic modulation. Opt. Express **20**(13), 14117–14129 (2012)
37. M.A. Naser, D.R. Sampaio, N.M. Munoz et al., Improved photoacoustic-based oxygen saturation estimation with snr-regularized local fluence correction. IEEE Trans. Med. Imaging **38**(2), 561–571 (2018)
38. S. Mandal, M. Mueller, D. Komljenovic, Multimodal priors reduce acoustic and optical inaccuracies in photoacoustic imaging. in *Photons Plus Ultrasound: Imaging and Sensing. 2019*, vol. 10878. International Society for Optics and Photonics, p. 108781M
39. M.A. Naser et al., Improved photoacoustic-based oxygen saturation estimation with SNR-regularized local fluence correction. IEEE Trans. Med. Imaging **38**(2), 561–571 (2018)
40. Y. Li, J. Sun, C.K. Tang et al., Lazy snapping. ACM Trans. Graph. (ToG) **23**(3), 303–308 (2004)
41. J. Shi, J. Malik, Normalized cuts and image segmentation. Dep. Pap. (CIS). 107 (2000)
42. S. Mandal, E. Nasonova, X.L. Dean-Ben et al., Fast calibration of speed-of-sound using temperature prior in whole-body small animal optoacoustic imaging. in *Photonics West—Biomedical Optics*. (2015), p. 93232Q
43. K.S. Valluru, K.E. Wilson, J.K. Willmann, Photoacoustic imaging in oncology: translational preclinical and early clinical experience. Radiology **280**(2), 332–349 (2016)
44. V. Ermolayev, X.L. Dean-Ben, S. Mandal et al., Simultaneous visualization of tumour oxygenation, neovascularization and contrast agent perfusion by real-time three-dimensional optoacoustic tomography. Eur. Radiol. **26**(6), 1843–1851 (2016)
45. S. Mandal, A.B. Greenblatt, J. An, Imaging intelligence: AI is transforming medical imaging across the imaging spectrum. IEEE Pulse (2018)

Data Structure Assisted Accelerated Reconstruction Strategy for Handheld Photoacoustic Imaging

Samir Kumar Biswas and Nitin Burman

Abstract In photoacoustic computed tomography (PACT), advanced model based iterative image reconstruction (MoBIIR) offers several advantages over analytical methods such as back-projection, time reversal, Fourier transform, delay and sum algorithms. However, MoBIIR also shows some disadvantages such as requirement of large storage memory, higher matrix computation time and necessity of selecting optimum parameters for the right solution. When using model based reconstruction methods for high resolution photoacoustic and ultrasound tomography, large matrix computation time is an important concern. In this chapter, we will discuss about filtered back-projection, time reversal methods, F-K migration and a specific model based iterative photoacoustic image reconstruction scheme where the direct non-symmetric photoacoustic system matrix of form $Hx = z$ (where H is m by n matrix and $m > n$) has been analyzed in detail using Least Squared Conjugate Gradient (LSCG) method where the computation of the $H^T H$ and thereafter regularization are explicitly avoided. Apart from this, a unique pseudo-dynamical systems approach based iterative algorithm is also discussed to demonstrate the insensitivity of tikhonov type physical regularization (λ), which is used frequently in normal equation of form $H^T Hx = H^T z$. However, to implement the algorithms, the photoacoustic equation is usually discretized over the spatial and temporal domain to form spatial-temporal interpolated model photoacoustic system matrix (H), where the data structure for sparsity is considered for accelerating the computation and hence the reconstruction. Finally, the applications of algorithms in photoacoustic imaging modality are shown. The computational requirements of different reconstruction strategies suitable for handheld photoacoustic imaging are also analyzed and discussed in detail.

Keywords Photoacoustic tomography · Iterative · Reconstruction · Regularization · Low noise

S. . K. Biswas (✉) · N. Burman
Bio-NanoPhotonics Group, Department of Physical Sciences, IISER Mohali,
P.O. Box 160062, Mohali, India
e-mail: skbiswas@iisermohali.ac.in

© Springer Nature Singapore Pte Ltd. 2020
M. Kuniyil Ajith Singh (ed.), *LED-Based Photoacoustic Imaging*,
Progress in Optical Science and Photonics 7,
https://doi.org/10.1007/978-981-15-3984-8_7

1   Introduction

Parameter recovery based on information carried by radiation has been the essential
methodology employed in the development of many valuable tools for medical diagnostic imaging. Examples are the well known X-ray computer assisted tomography
(CAT), magnetic resonance imaging (MRI), ultrasonic imaging (USI), positron emission tomography (PET), etc. Each of the above modalities possess certain advantages
and some unavoidable disadvantages. For example, the ultrasound based imaging
is affordable and uses non-ionizing radiation, but provides images with low soft-
tissue contrast and probe-dependent spatial resolution, which does not give useful
functional information (for example, the metabolic state of an organ being imaged),
especially when the required resolution is below 200–500 mu. Our observation shows
that contrast plays a vital role in enhancing the resolution. The MRI provides good
quality images which can also give functional information with the administration of
external contrast agents, but is prohibitively expensive. To this collection of imaging
techniques, photoacoustic computed tomography (PACT) is a recent addition which
uses the physics of near infrared (NIR) light for enhancing contrast and principle of
ultrasound for improving high resolution in tissue. In PACT, a pulsed light is used for
excitation of certain tissue substances and as a result of light absorption, ultrasound
signals generation occur in the tissue substances through thermo-elastic phenomena.
The process of generating ultrasound with light-matter interaction and use of those
light induced ultrasound signals around the object for reconstructing the absorption
image of tissue substances altogether is known as photoacoustic computed tomography (PACT). Near-infrared and ultrasound radiation are non-ionizing and therefore
both can be repeatedly employed without harm to the patient.

Photoacoustic phenomena generates a pressure gradient locally within tissue in the
ultrasound frequency regime, by the processes of optical absorption and thermoelastic
expansion [2, 3]. The physics of photoacoustic wave propagation can be used to map
the spatial distribution of light absorption in tissue substances (Fig. 1). Photoacoustic
imaging shows clinical level potential in providing deep-tissue high resolution images
with optical spectroscopic contrast (Fig. 1). This new imaging modality has several
potential clinical applications in cancers [2–5], inflammatory arthritis [6], diabetes,
metabolic rate estimation in healthy as well in disease affected patients. This is also
proven clinically (Fig. 1) through the efforts of a number of researchers around the
globe [1–4, 7]. Usually, light with nanosecond (*ns*) pulse widths illuminates the
tissue sample. Ultrasound is produced by the PA effect following absorption of light
by tissue substances such as hemoglobin. The pressure waves propagates from high
gradient location to low gradient area and the propagated wave can be detected at
the tissue surface using ultrasound detectors.

The partial differential equation that models the photoacoustic wave propagation through the acoustically homogeneous medium due to nanosecond pulsed laser
irradiance can be described [8, 9] as

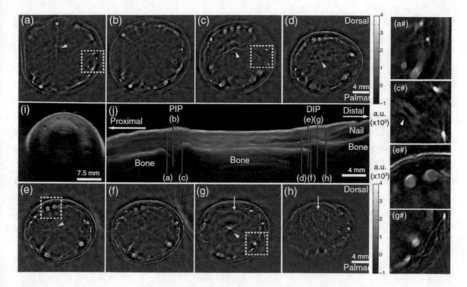

Fig. 1 Photoacoustic computed tomographic image and ultrasound image in human subject. Comparing the resolution and what wee see in photoacoustic and ultrasound image under rheumatoid arthritis disease in human finger joints. Reproduced with permission from [1]

$$\nabla^2 p(r, t) - \frac{1}{c^2}\frac{\partial^2 p(r, t)}{\partial^2 t} = -\frac{\beta}{C_p}\frac{\partial I(t)}{\partial t}A(r) \tag{1}$$

where $A(r)$ is the absorbed thermal energy density generated at position r and time t due to nanosecond pulsed laser irradiance. $I(t)$ is temporal pulsed laser profile. C_p is the isobaric specific heat of the medium, β is the isobaric volume expansion coefficient, and c is the acoustic speed. Now if the pulsewidth of the nanosecond laser is much shorter than the stress relaxation time of the tissue like medium then we can write the temporal pulse laser profile $I(t)$ with stress confinement condition [8, 9] as $\delta(t)$. The forward solution of above pressure wave equation can be obtained by the use of Green's function approach [8, 9] with absorbed energy ($A(r)$) distribution as;

$$p(r, t) = \frac{\beta}{C_p}\frac{\partial}{\partial t}\left[\frac{1}{ct}\int\limits_{R=ct} A(r')dr'\,\delta(t - |r' - r|/c)\right] \tag{2}$$

The above equation is the pressure propagation equation where the propagation is estimated by spatial-temporal correlated impulse response function $\delta(t - |r - r'|/c)$. The pressure $p(r, t)$ is an integrated pressure over a circle (in 2D) of radius $R = ct$ with a spatial sample width dr'. The equation shows that the contribution of the pressure at time t at detector location (r) is only from a circular strep of width dr' at a radial distance $ct = \| r' - r \|$. Here r' is an arbitrary point in the space where pressure build up occurs due to the photoacoustic effect. Both analytical and model based iterative image reconstruction (MoBIIR) have been developed for solving and modeling Eq. 2 for knowing the initial energy deposition at the site of the tissue substances. To reconstruct the map of total absorbed energy, we will discuss frequently used filtered back-projection and time reversal methods. In addition to the above mentioned algorithms, we will discuss in details about Fourier transform assisted F-K Migration based reconstruction, variants of model based iterative methods such as a specific model based iterative photoacoustic (PA) image reconstruction scheme where the direct non-symmetric PA system matrix of form $Hx = z$ (where H is m by n matrix and $m > n$) has been analysed in detail using Least Squared Conjugate Gradient (LSCG) method where the computation of the $H^T H$ and thereafter regularization are explicitly avoided. Apart from this, an unique pseudo-dynamical systems approach based iterative algorithm is also discussed to demonstrate the insensitivity of tikhonov type physical regularization (λ), frequently used in normal equation of form $H^T Hx = H^T z$. The following sections describe various methods in more details.

2   Analytic Equation Based Algorithms

2.1   Filtered BackProjection Based PACT Imaging

Notable among the algorithms used for solving Eq. 2 are the analytic algorithms based on filtered backprojection (BP) [3, 10] in the time-domain, which assume that the measured data is a Radon Transform of the object function. The algorithms are easy to apply for planar, cylindrical and spherical geometries [11]. A drawback of this method is that it requires a full-view of object with a high number of projections, and does not provide quantitative solution.

Following are the steps to reconstruct the source image using the back projection algorithm in photoacoustic tomography. The original photoacoustic source, recorded signal and their amplitude,time graph and reconstruction with backprojection are presented conceptually in Fig. 2a–c

1. Start recording photoacoustic signals when t = 0.0 by using ultrasound detector, as shown in Fig. 2a
2. Filter the signal as per the interest and then extract the time of flight details from the signal, as shown in Fig. 2b
3. Drawing a circle by taking a detector position as its center and its time of flight data as the radius(calculated using c = speed of sound), as shown in Fig. 2c
4. Repeating this for each detectors, as shown in Fig. 2c.

Backprojection or filtered backprojection is widely used in qualitative photoacoustic image reconstruction due to its simplistic approach, ease of implementation and speed suitable for quasi real time PA imaging. Downside of this method is that it does not provide quantitative information.

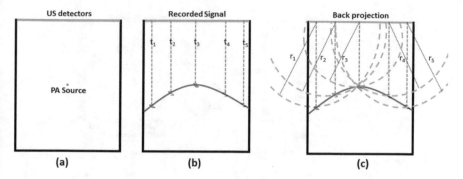

Fig. 2 Backprojection's conceptual presentation through visual graphs

2.2 Time Reversal Based PACT Imaging

In Time reversal (TR) approach, image reconstruction is performed by numerically propagating the recorded data in reversed temporal order back into the domain [12, 13]. When $t = 0$ is arrived, the initial pressure distribution is recovered. The advantages of the algorithm are that it can be applied to arbitrarily shaped measurement surfaces, and has generally been described to be the least sensitive PA algorithm to restrictions [12]. The method is also gaining its popularity due to the availability of a free third party MATLAB toolbox, which performs the time reversal image reconstruction using k-space methods [12]. The drawback of the TR approach lies in the requirement for the time-reversed waves to traverse the entire domain from detector coordinates which may entail unnecessary computations in regions which hold little interest. In cases when the propagation model assumes acoustic homogeneity while the measurement domain has unknown variations in density and speed-of-sound, image artifacts can result from the phenomenon of artefact trapping [14]. TR method needs large number projections to obtain high resolution images [12]. Time reversal conceptual presentation is shown in Fig. 3.

Figure 3a represents source and the array detector location, Fig. 3b shows recorded signal's amplitude at linear detector array as per the arrival time at their correspond detectors' spatial locations. Figure 3c–e shows reversed temporal order back into the spatial domain at three different time samples.

2.3 F-K Migration Based PACT Imaging

Fourier transform can be used to migrate the wave field as per the amplitude and phase. Originally, migration technique is developed based on the reflecting source model which assumes that all the field scatterers generate secondary acoustic sources. The main aim of migration is to reconstruct the secondary source position. Under the plane wave model, the scattering source estimation problem can be made suitable

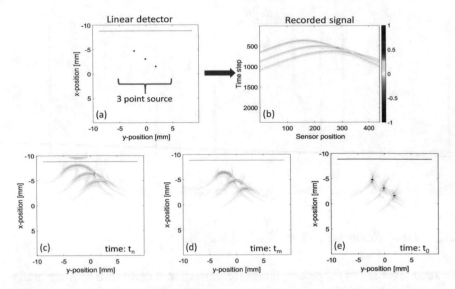

Fig. 3 Time reversal's conceptual presentation. **a** Represent source and the array detector location in space, **b** recorded signal's amplitude at linear detector array as per the arrival time at their correspond detector locations, **c–e** shows reversed temporal order back into the spatial domain at three different times

for plane wave imaging by a spatial transformation (F-K) of the hyperbolic traces present in the raw data. To produce an image of the scatterers, all the hyperbolas must be migrated back to their apexes. However, the advantage of migration technique is that it improves focusing by use of amplitude and phase rectifications where the correction is done for the effects of spreading of ray paths as the waves propagate. This technique has been used as a basic tool in geophysics since the 1950s [15].

F-K migration takes back the recorded US signal to that time at which the wave emerges out of the secondary source. It was first developed by Stolt in 1978 for B scan seismic imaging [15]. Later, it was developed for plane wave ultrasound imaging by Garcia in 2014 [16]. This algorithm is limited by the assumption of constant wave velocity [16]. However, its fastest computation time makes it suitable for real-time ultrasound imaging and same is true for photoacoustic imaging because both imaging modality uses raw ultrasound data. The assumption to neglect the downward going waves exactly matches with the PACT. Whereas, in plane wave ultrasound imaging we have to fit the travel time with the exploding reflector model as shown in Fig. 4.

In plane wave imaging, all the transducer elements emit the ultrasound at the same time to generate a plane wave. The plane wave proceeds towards the transducer and interacts with the reflector surface. After the interaction, the reflector(at $S(s_x, s_z)$, see Fig. 4) becomes the secondary source and starts to emit radially outwards. usually the reflected signal is recorded by the linear transducer. The travel time of the wave, varying with the detector position(x), is given below:

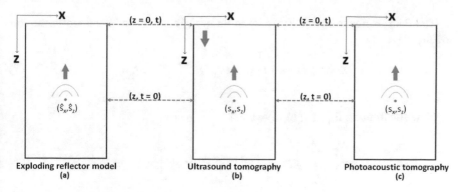

Fig. 4 x-z plane, where linear transducer is placed on the x axis and reflector in x-z plane. The arrows represents the direction of propagation

- For exploding reflector model:

$$t(\hat{x}) = \frac{1}{\hat{c}}\sqrt{(\hat{s}_x - x)^2 + (\hat{s}_z - z)^2} \tag{3}$$

- For plane wave ultrasound imaging:

$$t(x) = \frac{1}{c}\left(s_z + \sqrt{(s_x - x)^2 + (s_z - z)^2}\right) \tag{4}$$

- For plane wave photoacoustic imaging:

$$t(x) = \frac{1}{c}\sqrt{(s_x - x)^2 + (s_z - z)^2} \tag{5}$$

In order to use F-K migration in PWI we need to fit its travel time equation with the exploding reflector model. However, doing so is an unachievable task. Due to the dependency of wave amplitude with distance, most of the its energy is located at the apex of the hyperbola. By equating the 0th–2nd order derivative of the Eqs. 3, 4 and 5 we can find out approximate fitting parameters. It yields $\hat{c} = \frac{c}{\sqrt{2}}$ and $\hat{s}_z = \sqrt{2}s_z$ for plane wave ultrasound imaging [15] and $\hat{c} = c$ and $\hat{s}_z = s_z$ for photoacoustic imaging.

Lets assume that $\Psi(x, z, t)$ is the scalar wavefield that is a solution to

$$\nabla^2\Psi - \frac{1}{c}\frac{\partial^2}{\partial t^2}\Psi = 0 \tag{6}$$

We know the scalar wavefield at $z = 0$, time t. We need to know the scalar wavefield at distance z at time $t = 0$ i.e. $\Psi(x, z, t = 0)$ (see Fig. 4).

The Fourier transform of $\Psi(x, z, t)$ in the (k_x, f) spectrum is defined in the following way:

$$\Psi(x, z, t) = \iint\limits_{-\infty}^{\infty} \phi(k_x, z, f) e^{2\pi\iota(k_x x - ft)} dk_x df \qquad (7)$$

Now substituting Eqs. (7) in (6) we get

$$\nabla^2 \left[\iint\limits_{-\infty}^{\infty} \phi(k_x, z, f) e^{2\pi\iota(k_x x - ft)} dk_x df \right]$$

$$- \frac{1}{c} \frac{\partial^2}{\partial t^2} \left[\iint\limits_{-\infty}^{\infty} \phi(k_x, z, f) e^{2\pi\iota(k_x x - ft)} dk_x df \right] = 0 \qquad (8)$$

These derivatives can easily be taken inside the integral and can be evaluated to get

$$\iint\limits_{-\infty}^{\infty} \left[\frac{\partial^2 \phi(k_x, z, f)}{\partial z^2} + 4\pi^2 \left[\frac{f}{c^2} - k_x^2 \right] \phi(k_x, z, f) \right] e^{2\pi\iota(k_x - ft)} dk_x df = 0 \qquad (9)$$

The left hand side of Eq. (9) is the Fourier transform of the terms in the square bracket in Eq. (9). Now since its right hand side is equal to zero, the function will also be equal to zero.

$$\frac{\partial^2}{\partial z^2} \phi(z) + 4\pi^2 k_z^2 \phi(z) = 0 \qquad (10)$$

where,

$$k_z^2 = \frac{f^2}{v} - k_x^2 \qquad (11)$$

Now we have formulated the problem in the (k_x, f) domain i.e. is a Fourier domain of (x, t). The boundary condition is now the Fourier transform of $\Psi(x, z = 0, t)$ over (x, t) i.e. $\phi(k_x, z = 0, f)$. Since, Eq. (10) is a second order differential equation, its unique general solution can be written as

$$\phi(k_x, z, f) = A(k_x, f) e^{2\pi\iota k_z z} + B(k_x, f) e^{-2\pi\iota k_z z} \qquad (12)$$

where $A(k_x, f), B(k_x, f)$ are to be determined from the boundary condition. It is important to note that in Eq. (12) the two terms can be interpreted as the upgoing $(B(k_x, f) e^{-2\pi\iota k_z z})$ and downgoing $(A(k_x, f) e^{2\pi\iota k_z z})$ wavefield (Fig. 5).

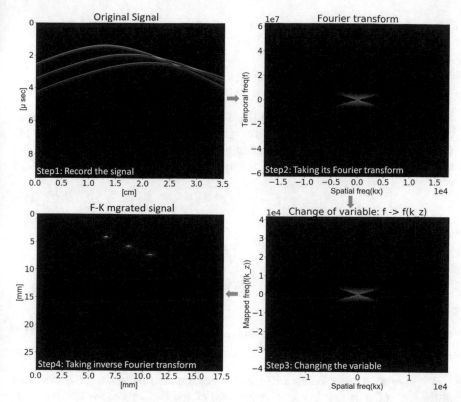

Fig. 5 Various steps for implementing F-K migration and photoacoustic image reconstruction strategies

Since, we only have one boundary condition, i.e. $\phi(k_x, z = 0, f)$, in order to solve the problem we have to assume a limited model which assumes waves propagating in one direction only. This means that

$$A(k_x, f) = 0, \quad B(k_x, f) = \phi(k_x, z = 0, f) \tag{13}$$

Substituting (13) in (12) we get

$$\phi(k_x, z, f) = \phi(k_x, z = 0, f)e^{-2\pi \iota k_z z} \tag{14}$$

Substituting (13) solution in (7) we get

$$\Psi(x, z, t) = \int\!\!\!\int_{-\infty}^{\infty} \phi(k_x, z = 0, f)e^{2\pi \iota (k_x x - k_z z - ft)} dk_x df \tag{15}$$

Now migrating (15) from time t to $t = 0$ we get our migrated solution

$$\Psi(x, z, t = 0) = \iint\limits_{-\infty}^{\infty} \phi(k_x, z = 0, f)e^{2\pi\iota(k_x x - k_z z)}dk_x df \tag{16}$$

This solution has a disadvantage that it is not an inverse Fourier transform of function $\phi(k_z, z = 0, f)$. Stolt in 1978 suggested a change of variable from (k_x, f) to $(k_x, f(k_z))$ to make the migrated solution an inverse fourier transform of $\phi(k_x, z = 0, f(k_z))$. The variable change is defined by Eq. (11) which then can be solved for f as:

$$f = c \times \sqrt{k_x^2 + k_z^2} \tag{17}$$

$$\implies df = \frac{ck_z}{\sqrt{k_x^2 + k_z^2}}dk_z \tag{18}$$

Now substituting (17) and (18) in (16) we get

$$\Psi(x, z, t = 0) = \iint\limits_{-\infty}^{\infty} \frac{ck_z}{\sqrt{k_x^2 + k_z^2}}\phi(k_x, z = 0, f(k_z))e^{2\pi\iota(k_x x - k_z z)}dk_x dk_z \tag{19}$$

We have seen that the new FFT-based F-K migration determines the wavefield at the time of start that is $t = 0$. Advantage of the F-K migration is that it uses few mapping and FFT techniques which makes it faster for real time imaging. The reconstructed images with backprojection (Fig. 6a), time reversal (Fig. 6b) and F-K migration (Fig. 6c) are compared visually and also based on the computation time. For computation we have used a PC with Intel(R) Core(TM) i7-6700 CPU @ 3.40 GHz, DDR4 RAM: 32 GB. The computation time for reconstructing images using BP is 2.6 s, TR is 99.3 s and for F-K migration it is 1.14 s.

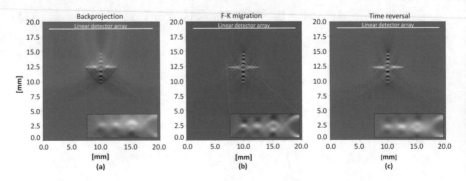

Fig. 6 Comparing the reconstruction methods through visual perception and computed time

3 Model Based Iterative Image Reconstruction Algorithms

Due to constant demand for quantitative and high resolution photoacoustic imaging, model based algorithms are gaining importance. In model based iterative image reconstruction (MoBIIR) methods, algebraic and matrix methods within an iterative framework are used to minimize the residue between the model-generated data and measured data. To implement the MoBIIR algorithms, first we shall discretize Eq. 2 and formulate forward model in such a way that it serves to model PA wave propagation [17–22] and relate the spatially discretized initial pressure distribution to the acquired signals. Such an approach lends itself to application of algebraic and matrix methods within an iterative framework, for image reconstruction. Now by shifting r', in Eq. 2, we find a new t and then corresponding integrated pressure at a new radial distance. So a spatial and temporal (spacetime) matrix H can be formed by shifting the position $r'(i, j)$ over the discretized space for a series of sample time t_k and then estimate the boundary pressure (z_{r_d, t_k}) at detector location r_d. The deposited energy $A(r)$ can be expressed over a discretized spatial domain ($\Omega \in R^2, r, r' \in \Omega$) as $A(i, j)$. Spatially correlated temporal impulse term for a sampled time, t_k over space $r_{i,j}$ can be written as $h(t_k - |r' - r_{i,j}|/c_{i,j}) \simeq \delta(t - |r' - r|/c)$. The boundary pressure estimation forward problem can be formulated (assumed $\eta = \frac{\beta}{C_p} \frac{\partial}{\partial t}$) from Eq. 2 over discretized spatial-temporal domain as,

$$z_p(r_d, t_k) = \eta \sum_{i,j} \left\{ \int_{R=ct} \frac{1}{c_{i,j} t_k} \times h(t_k - |r_d - r'_{i,j}|/c_{i,j}) dr' \right\} A(i, j) \tag{20}$$

Algorithm 1 : Algorithm for estimating forward model matrix H and boundary pressure (z_p)

---
for
   all detector positions r_d
   for
      all pixels (i, j) over the discretized image domain x_A^l
      calculate the time of flight $t_k = \| (r_{i,j} - r_d)/c_{i,j} \|$
      Calculate interpolation coefficient p_k at neighboring time points of t_k
      Estimate $H(t_k, l)$= coefficient of $p_k/(t_k c^l) \forall l$;
      Integrate the pressure over constant sampled time points t_k with corresponding coefficient
p_k as;
        $z_{d,t_k} = z_{d,t_k} + \frac{p_k}{(t_k c^l)} \times x_A^l$
   end for
end for

A series of constant time samples ($t_k = k/f_s$, where $k = 1 \ldots m$) are considered and $h(t)$ is evaluated over space ($r_{i,j} \leftrightarrow r'$) to form the propagation system matrix H. The simplified form of pressure propagation equation can be written with system matrix H and initial pressure vector x_A^l (with $\beta = 1$, $C_p = 1$) as;

$$z_p(r_d, t_k) = \eta \sum_{l,k} H(t_k, r^l) x_A^l \tag{21}$$

where the initial pressure vector (x_A^l) is formed from energy absorption matrix $A(i, j)$. The propagation system matrix H is formed with $h(t)$ and its interpolated value around the sampled time t_k. The forward photoacoustic projection integral is computed as shown in Algorithm 1.

The PA forward problem (Eq. 21) with system matrix H can be written in matrix multiplication form as;

$$z_p = H x_A \tag{22}$$

where z_p is measurement vector $(z_p \in \mathbb{R}^m)$, H is the system matrix $(H \in M_{m \times n}(\mathbb{R})$ where $m > n)$ which model the propagation of PA signal and the vector x_A $(x_A \in \mathbb{R}^n)$ represents initial pressure rise. A photoacoustic reconstruction algorithm is used to solve the PA inversion problem, that is to recover an image of the initial pressure rise distribution x_A inside the tissue from z_p, the noisy PA signals measured at tissue boundary. The photoacoustic pressure z_p at boundary is obtained by integrating pressure over a constant sampled time points with linear interpolated coefficients.

The solution is obtained with minimizing the residue between the model-generated $(H x_A)$ data and measured data (z_p) by iterations. The minimization function [17–22] for iterative method can be written as;

$$\chi = \arg \min_{\chi} \| z_p - H x_A \|^2 + \lambda \| x_A \|^2 \tag{23}$$

where λ is a regularization parameter to stabilize the ill-condition of the system matrix in normal equation. The minimization equation (Eq. 23) with Gauss-Newton scheme lands to the normal equation [21, 22] required to be solved is then of the form;

$$[H^T H + I \lambda] x_A = H^T z_p \tag{24}$$

The normal equation for photoacoustic inverse problem can be solved with variant of regularization schemes. The simplest regularization selection method is L-Curve method where a series of regularization set is formed and the best one which minimizes the residue is chosen. However, Dean-Ben et al. [21] has used a least squares QR (LSQR) based regularized reconstruction method where the direct solution vector from full view data is used as regularization and shown an added advantage of being highly efficient. Inversion of limited-view data is stabilized using Tikhonov or Total Variation regularization [17, 19–23] which require explicit selection of an optimal regularization parameter. Recently, Shaw et al. [22] presented a regularization optimization scheme based on LS-QR decomposition which shows good performance and computational efficiency with reconstruction time of 444.9 s.

In order to achieve high accuracy with flexibility in using limited-view data, H should be a highly discretized large system matrix. Further, the matrix H is transposed to form a square matrix $[H^T H + I\lambda]$ and the inverse is computationally expensive, requiring large memory storage. $H^T H$ is often dense even when H is sparse [24]. Further, if H is ill-conditioned then the $H^T H$ is more ill-conditioned since its condition number is the square of the condition number $-\{\kappa(H)\}^2$ [24, 25]. One of the challenging issues of iterative PA imaging is to balance the trade-off between the computational efficiency of the reconstruction algorithms and the resolution of reconstructed images.

To avoid the regularization selection and explicit formulation of $H^T H$, non-symmetric system matrix equation (Eq. 22) can be solved by least squared CGS method. The steps for solving the non-symmetric PA system matrix are shown in Algorithm 2.

Algorithm 2: Solving non-symmetric matrix equation $Hx_A = z_p$ using regularization free LSCGS scheme.

Input: H and z_p

Initialize:

$$x_A^0 \Leftarrow 0$$
$$s^0 \Leftarrow z_p - Hx_A^0$$
$$r^0 \Leftarrow f^0 = H^T s^0$$
$$q^0 \Leftarrow Hf^0$$

Compute for output: x_A^k

For each iteration "k" the Least Squared Conjugate Gradient (LSCG) algorithm becomes as;;

$$\alpha = \frac{\|r^{k-1}\|_2^2}{\|q^{k-1}\|_2^2}$$
$$x_A^k \Leftarrow x_A^{k-1} + \alpha f^{k-1}$$
$$s^k \Leftarrow s^{k-1} - \alpha q^{k-1}$$
$$r^k \Leftarrow H^T s^{k-1}$$
$$\beta \Leftarrow \frac{\|r^k\|_2^2}{\|r^{k-1}\|_2^2}$$
$$f^k \Leftarrow r^k + \beta f^{k-1}$$
$$q^k \Leftarrow Hf^k$$

It can be noticed that the inverse Algorithm 2 explicitly avoids calculation of $H^T H$. The main aim of the algorithm is to estimate the residual $z_p - Hx$ and then multiply it by H^T rather than subtracting $H^T Hx_A$ from $H^T z_p$. The algorithm 2 uses few simple vector multiplications and additions which helps to execute it faster due to use of the sparsity property of H and H^T.

3.1 Symmetry Conjugate Gradient Search (CGS) and Least Square Conjugate Gradient Search (LSCGS) Based Reconstruction

It is clear from several studies [17, 19–23] that symmetric normal equation is commonly used to model the photoacoustic reconstruction strategies and it is considered as a starting platform for adding various type of regularization and improving the reconstruction thereafter. Main goal of the inverse problem is to compute the eigenvalues of normal equation $H^T Hx = H^T z_p$ or in other suitable mathematical form. Symmetric CG algorithm can be applied to the normal equation either from $[H^T H + I\lambda]x_A = H^T z_p$ explicitly or simply extending it through applying vector multiplications on H^T and H in succession. By applying vector multiplication on H^T and H in succession, we can avoid formulating matrix $H^T H$ which will generally be dense even when H is sparse. Formulating the normal equation ($H^T H$) from data structure assisted formulated sparse rectangular matrix (H or H^T) does not solve the problem where the condition number of $H^T H$ is the square of the condition number of H and, it loses the sparsity which increase the required storage memory for $H^T H$. However, conjugate gradient algorithms use the matrix and matrix-vector multiplications only. So it is not mandatory to form the matrix $H^T H$ which leads to cancellation or loss of sparsity. Due to serious amplification of the spectral condition number (as it is squared), it introduces error in eigenvalues. The idea of least-squares CG was originally proposed by Hestenes and Stiefel [26] and, later it came to be known as CGLS which involves vector computing terms of the form $H^T(Hx - H^T z_p)$ instead of $H^T Hx - H^T z_p$. The difference between the two methods is entirely the rounding error, which is important in practical problems where sparsity of the large matrix need to be preserved for fast computing.

However, in the non-symmetric case, as it is shown in algorithm 2, all previous search directions are used in order to calculate a new approximation and the rate of convergence is determined by the Krylov sequence Hr^0, $H^2 r^0$ and for symmetric CGS method, the rate of convergence is determined by $(H^T H)Hr^0, (H^T H)^2 Hr^0, \ldots$ as it would have been the case if the normal equations had been used. Here r is the residue.

3.2 Pseudo-dynamical Systems Approach for PACT Imaging

A regularization-insensitive route to computing the parameter updates using the normal equations (Eq. 21) is to introduce an artificial time variable [23, 27–29]. Such pseudo-dynamical systems, typically in the form of ordinary differential equations (ODE-s), may then be integrated and the updated parameter vector is recovered once when either a pseudo steady-state is reached or a suitable stopping rule is applied to the evolving parameter profile (the latter being necessary if the measured data is limited and noisy). Indeed, it has been shown [23, 28–30] that the well known Landweber iterations correspond to an explicit Euler time discretization of the pseudo-time ODE-s for the Gauss-Newton's updates (Eq. 21) and appear to exhibit a self-regularization character depending upon the pseudo time step. This is also desirable in view of the fact that the addition of the regularization term alters the minimization problem which we are trying to solve. Moreover, if one adopts an explicit pseudo-time integration scheme for the ODE-s, an explicit inversion of the discretized (linear case) operator in the normal equation for PA can be avoided. This is the best feature of this method which has several advantages when dealing with singularity and rank deficiency issues.

Here, we will develop a concept of solving the optimized normal equation for photoacoustic problem as it was said in previous paragraph. The normal form of minimized photoacoustic equation can be further simplified with a notion of $A = H^T H$ and $b = H^T z_p$. The optimized system of linear or non-linear equation (Eq. 21) of many physical, biological, and economic processes can be expressed in a generalized form as,

$$Ax = b \tag{25}$$

where $A \in \mathbb{R}^{N \times N}$ is a state companion matrix (for PAT, $A = H^T H + \lambda I$), $b \in \mathbb{R}^{N \times m}$ ($b = H^T z_p$ in our case) is the constant force matrix and $x \in \mathbb{R}^{N \times 1}$ is the unknown solution vector (x). Since e^{tA} and $e^{tA}\mu_0$ are solutions of ordinary differential equations, it is natural to consider methods based on numerical integration. For an example, in a simple case, we solve a homogeneous matrix differential equation such as $Ax = 0$ with an introduced fictitious time at steady state as $\dot{x}(t) = Ax(t)$ describing the evolution of the system on pseudo time. With an initial condition $x(t = 0) = x_0$, the solution would be of form $e^{tA}x_0$. The solution of Eq. 25 can be obtained without inverting the square matrix A. In principle, the solution is obtained from $x(t) = e^{tA}x_0$ and can be formally defined by the convergent power series as,

$$e^{tA}x_0 = Ix_0 + tAx_0 + \frac{t^2 A^2 x_0}{2!} + \cdots + \cdots \tag{26}$$

Generally, A is large, dense and non-sparse (in some cases, partially sparse A is observed) due to formulation of normal equation from the sparse matrix with its own transpose form. In particular the system of ordinary differential equation arises from the spatial discretization of a partial differential equation. Typically e^A is dense even if A is sparse, we would like to compute this in an iterative way. The iterative methods are used for sparse matrix problems where only matrix vector products are needed. A powerful class of methods that are applicable to many problems are the Krylov space methods [31, 32], in which approximations to the solution are obtained from the Krylov spaces spanned by the vectors $\{x_0, Ax_0, A^2 x_0, \ldots, A^m x_0\}$ for some 'm' that is typically small compared to the dimension of A. The Lanczos method [33] for solving symmetric eigenvalue problems is of this form and for non-symmetric matrices the Arnoldi iteration [33] can be used. In this method the eigenvalues of a large matrix are approximated by the eigenvalues of a Hessenberg matrix of dimension 'm'.

Now, we will show how to form a pseudo-dynamic time integral equation of minimized system Eq. 26 of form $Ax + b = 0$. The steady state equation of $Ax + b = 0$ (obtained from Eq. 26) with a fictitious time can be written as time derivative form,

$$\dot{x}(t) = Ax(t) + b(t). \tag{27}$$

Multiplying by factor of e^{-At} throughout and integrating, we obtain

$$e^{-At}\dot{x}(t) - e^{-At}Ax(t) = e^{-At}b(t) \tag{28}$$
$$e^{-At}\dot{x}(t) - Ae^{-At}x(t) = e^{-At}b(t)$$
$$\frac{d(e^{-At}x)}{dt} = e^{-At}b(t) \tag{29}$$

Now integrating the above equations in $[t, t + \Delta t]$ with initial boundary condition $x(t) = x^*$, we obtained the pseudo-dynamic time integration equation for updating the solution vector with $f(s) = A(x^*)x^* - b$ as,

$$x(t + \Delta t) = e^{A\Delta t}x^* + \int\limits_{t}^{t+\Delta t} e^{A(t+\Delta t-s)}f(s)ds \tag{30}$$

which converges to $\hat{x}$ as $t \to \infty$. Note that we have assumed A is square and positive definite. Following the concept of local linearization [29], the linearization point $t = t^*$ (such that $x^* := x(t)^*$) could be chosen anywhere in the closed interval $[t, t + \Delta t]$ without affecting the formal error order. While choosing $t = t^*$ yields the explicit phase space linearization (PSL) [29], $t = t^*$ results in the implicit locally transversal linearization (LTL) [34]. Denoting $h_d = t_{k+1} - t_k$ to be the time step and $x_k := x(t_k)$, the explicit PSL map corresponding to the continuous update Eq. 30 is written as:

$$x_{k+1} = e^{A(t_{k+1}-t_k)}x_k + \int\limits_{t}^{t+\Delta t} e^{A(t_{k+1}-s)}f(s)ds \tag{31}$$

An explicit strategy for obtaining the parameter updates via a semi-analytical integration of the pseudo-dynamic linear equation is proposed in this chapter. Despite the ill-posedness of the inverse problem associated with photoacoustic computed tomography, adoption of the first derivative update scheme combined with the pseudo-time integration appears a muted sensitivity to the regularization parameter which includes the pseudo-time step size for integration (Fig. 7).

A digital numerical phantom with two holo circular shape photoacoustic emitting sources is considered to be embedded in a surrounding medium of size 20×20 mm and the medium is discretized with 100×100 square grids where the size of each pixel is $20.0\,\mu$m. The initial pressure rise is assumed to be 1 kPa to the both local shapes and 0 kPa pressure for the background medium. The speed of sound ($c = 1500$ m/s)

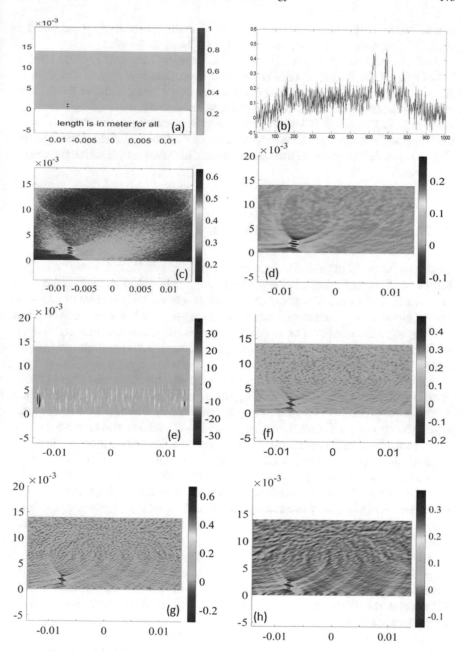

Fig. 7 Reconstructed with forward model generated data which were detected with a line detector of 128 sensors placed at y = 14.5 mm. Original image is shown in (**a**), noisy signal is shown in (**b**). The reconstruction are carried out by **c** BP method, **d** L-Curve method with regularization (0.001), **e** L-Curve method without (near zero, 0.000001) regularization, **f** LSCGS method, **g** pseudo dynamic approach with Tikhonov type physical regularization (0.001), **h** pseudo dynamic approach with negligible (near zero, 0.000001) Tikhonov type physical regularization

is taken to be uniform all over the simulated domain. Numerically generated data with 128 detectors and data are corrupted with 40% random noise.

For computation we have used a PC with Intel(R) Core(TM) i7-6700 CPU @ 3.40 GHz, DDR4 RAM: 32 GB. The computation time was found to be 0.3 s for BP, more than 8000.3 s for pseudo dynamic system approach, 51.8 s for L-Curve method with regularization, 128 s for L-Curve method without regularization, and 1.5 s for LSCGS method. Selection of time step in pseudo dynamic system is one of the drawback where computation time depends on the time step size. As the time step size reduces, the computation time increases due to increase of iteration for smooth solutions.

4   Numerical Phantom Experiment

Phantoms (physical or numerical) are of paramount importance to evaluate the performance of reconstructed photoacoustic tomographic image either in experiment or in simulation model. The test objects were developed with the aim of providing known ground truths with complexity approaching the level of the context in which the imaging and reconstruction is intended for. For all cases, a numerical phantom with rectangular shape and circular shape photoacoustic emitting source is designed and simulations are performed.

A digital numerical phantom with rectangular shape and circular shape photoacoustic emitting source were considered to be embedded in a surrounding medium of size 30×30 mm and the medium is discretized with 2048×2048 square grids where the size of each pixel is $14.6\,\mu$m. The initial pressure rise is assumed to be 1 kPa to the both local shapes and 0 kPa pressure for the background medium. The speed of sound ($c = 1500$ m/s) is taken to be uniform all over the simulated domain. Figure 8 shows simulated phantom, detector orientation and the pressure variation over the domain. We have considered almost discrete helical shape detector array orientation for getting very low number of linearly dependent algebraic equations in the measurement sets. Numerically generated data are corrupted with 40% random noise.

The image is reconstructed with full view where total number of projection is 12 and total number of detectors in the measurement is 384. Measurement data expands around $0°–360°$ of phantom surface. The images are reconstructed with (a) TR method, (b) BP method, (c) L-Curve method, (d) proposed LSCGS method and their corresponding images are shown in Fig. 9.

The image is reconstructed with half view where total number of projection is 6 and total number of detectors in the measurement is 192. Measurement data expands around $0°–180°$ of phantom surface. The images are reconstructed with (a) TR method, (b) BP method, (c) L-Curve method, (d) proposed LSCGS method and their corresponding images are shown in Fig. 10.

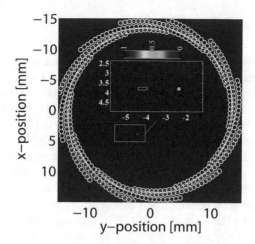

Fig. 8 Photoacoustic experimental set up for side illumination and the experimentation with ultrasound array detector

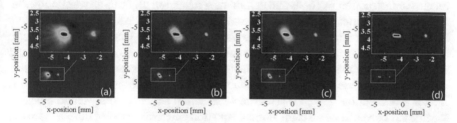

Fig. 9 Reconstructed with forward model generated 12 projections (half view, 384 detectors) data expands around 0°–180° of phantom surface. The reconstruction are carried out by **a** TR method, **b** BP method, **c** L-Curve method, **d** proposed LSCGS method

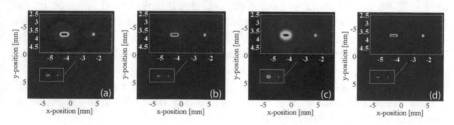

Fig. 10 Reconstructed with forward model generated 6 projections (half view, 192 detectors) data expands around 0°–180° of phantom surface. The reconstruction are carried out by **a** TR method, **b** BP method, **c** L-Curve method, **d** proposed LSCGS method

5 Discussion and Conclusions

Here we have shown several reconstruction schemes where the motivation was to develop or implement a suitable algorithm for real time handheld photoacoustic imaging. Analytic equation based reconstruction strategies are quite simple, fast and provide reasonably acceptable results though it faces challenging drawbacks due to quantification which can be addressed with well defined reference and standardization. Model based iterative reconstruction algorithm that permits to solve a non-symmetric matrix ($H \in M_{m \times n}(\mathbb{R})$ where $m > n$) without explicit formation of $H^T H$ and regularization can be used to obtain the photoacoustic solution of large system matrix. It is shown that this procedure is computationally simple and gives reasonably good results in terms of computation and resolution. It achieves low computation time by explicitly avoiding the computation of $H^T H$ and regularization. A major advantage of the proposed method is that it takes less memory compared to the normal equation and is fast in execution compared to the time reversal methods, but slower than backprojection. Computation time and memory requirement for conventional image reconstruction methods and certain new inversion algorithms were studied in detail using numerical phantoms. The computation details have been shown for both limited view data and full view data when a considerable Gaussian random noise is added to simulated boundary measurements. The resolution of non-symmetric system matrix inversion with LSCGS method can be further improved with suitable interpolation scheme which may introduce larger computation time and this needs further investigation. A new class of reconstruction strategy with pseudo-dynamic scheme has been discussed using normal equation where we showed the way to avoid direct inversion of the system matrix and makes it tikhonov type regularization free.

The total reconstruction computation time with 220×220 grid points, 80 MSPS sampling rate for back projection is 2.46 s and it used 3 GB memory, time reversal took 1405.4 s and used 2.8 GB memory, L-Curve based normal equation method took 6510.6 s and used 59.8 GB memory, non-symmetric matrix inversion took 7.4 s and used 4.5 GB memory, when half view data is considered. When full view data is considered the computation time with back projection is 4.22 s and it used 3 GB memory, time reversal took 1411.4 s and used 2.9 GB memory, L-Curve based normal equation method took 4588.6 s and used 59.9 GB memory, non-symmetric matrix inversion took 17.2 s and used 5.3 GB memory. The computer used has Processor: Xenon(R) CPU @2.67 MHz, 60 GB RAM).

The most formidable difficulty in crossing over to a full-blown 3D problem is the disproportionate increase in the parameter vector dimension (a typical tenfold increase) compared to the data dimension where one cannot expect an increase beyond two to three folds [35]. Thus, if 3D iterative image reconstruction algorithms are used, they would require implementation on highly parallelized processing architectures as in graphics processing units (GPUs) [36]. However, considering resolution and computation time, real time imaging may be possible with F-K migration based reconstruction both for 2D and 3D imaging, provided some calibration is performed for gathering quantitative information. Getting quantitative information from F-K migration method is still a debatable subject. However, combination of F-K migration and accelerated model based imaging may solve the purpose, where F-K migration will provide real time imaging capability and accelerated model method will quantify the parameters.

Acknowledgements This was work was financially supported by the IISER Mohali startup grant.

References

1. P. Van Es, S.K. Biswas, H. Moen, W. Steenbergen, S. Manohar, Initial results of finger imaging using photoacoustic computed tomography. J. Biomed. Opt. Lett. **19**(6), 060501 (2014)
2. L.H.V. Wang, S. Hu, Photoacoustic tomography: in vivo imaging from organelles to organs. Science **335**, 1458–1462 (2012)
3. C. Lutzweiler, D. Razansky, Optoacoustic imaging and tomography: reconstruction approaches and outstanding challenges in image performance and quantification. Sensors **13**, 7345–7384 (2013)
4. M. Heijblom, D. Piras, W. Xia, J.C.G. van Hespen, J.M. Klaase, F.M. van den Engh, T.G. van Leeuwen, W. Steenbergen, S. Manohar, Visualizing breast cancer using the twente photoacoustic mammoscope: what do we learn from twelve new patient measurements? Opt. Express **20**, 11582–11597 (2012)
5. K. Wang, M.A. Anastasio, A simple Fourier transform-based reconstruction formula for photoacoustic computed tomography with a circular or spherical measurement geometry. Phys. Med. Biol. **57**, N493–N499 (2012)
6. G. Xu, J.R. Rajian, G. Girish, M.J. Kaplan, J.B. Fowlkes, P.L. Carson, X. Wang, Photoacoustic and ultrasound dual-modality imaging of human peripheral joints. J. Biomed. Opt. **73**, 491–496 (2012)
7. S.K. Biswas, P.V. Es, W. Steenbergen, S. Manohar, A method for delineation of bone surfaces in photoacoustic computed tomography of the finger. Ultrason. Imaging **38**(1), 63–76 (2015)
8. L.V. Wang, H. Wu, *Biomedical Optics: Principles and Imaging* (Wiley, Hoboken, NJ, 2007)
9. L.V. Wang, Tutorial on photoacoustic microscopy and computed tomography. IEEE J. Sel. Top. Quant. Electron. **14**, 171–179 (2008)
10. M. Xu, L.V. Wang, Universal back-projection algorithm for photoacoustic computed tomography. Phys. Rev. E. **71**, 016706(1-7) (2005)
11. M. Xu, L.V. Wang, Analytic explanation of spatial resolution related to bandwidth and detector aperture size in thermoacoustic or photoacoustic reconstruction. Phys. Rev. E. **65**, 056605(1-7) (2003)
12. B.E. Treeby, B.T. Cox, k-Wave: MATLAB toolbox for the simulation and reconstruction of photoacoustic wave fields. J. Biomed. Opt. **15**, 021314–22 (2010)
13. P. Burgholzer, G.J. Matt, M. Haltmeier, G. Paltauf, Exact and approximative imaging methods for photoacoustic tomography using an arbitrary detection surface. Phys. Rev. E **75**, 046706(1-10) (2007)
14. B.E. Treeby, B.T. Cox, Artifact trapping during time reversal photoacoustic imaging for acoustically heterogeneous media. IEEE Trans. Med. Imag. **29**, 387–396 (2010)
15. R.H. Stolt, Migration by Fourier transform. Geophysics **63**, 23 (1978)
16. D. Garcia, L. Tarnec, S. Muth, E. Montagnon, J. Porée, G. Cloutier, Stolt's f-k migration for plane wave ultrasound imaging. IEEE Trans. Ultrason. Ferroelectr. Freq. Control **60**, 1853 (2014)
17. A. Buehler, A. Rosenthal, T. Jetzfellner, A. Dima, D. Razansky, V. Ntziachristos, Model-based optoacoustic inversions with incomplete projection data. Med. Phys. **38**, 1694–1704 (2011)

18. G. Paltauf, J.A. Viator, S.A. Prahl, S.L. Jacques, Iterative reconstruction algorithm for optoacoustic imaging. J. Acoust. Soc. Am. **112**, 1536–1544 (2002)
19. X.L. Dean-Ben, V. Ntziachristos, D. Razansky, Acceleration of optoacoustic model-based reconstruction using angular image discretization. IEEE Trans. Med. Imag. **5**, 1154–1162 (2012)
20. K. Wang, R. Su, A.A. Oraevsky, M.A. Anastasio, Investigation of iterative image reconstruction in three-dimensional optoacoustic tomography. Phys. Med. Biol. **57**, 5399–5424 (2012)
21. X.L. Dean-Ben, A. Buehler, V. Ntziachristos, D. Razansky, Accurate model-based reconstruction algorithm for three-dimensional optoacoustic tomography. IEEE Trans. Med. Imag. **31**, 1922–1928 (2012)
22. C.B. Shaw, J. Prakash, M. Pramanik, P.K. Yalavarthy, Least squares QR-based decomposition provides an efficient way of computing optimal regularization parameter in photoacoustic tomography. J. Biomed. Opt. **15**, 021314–22 (2013)
23. C. Vogel, *Computational Methods for Inverse Problems* (SIAM, 2002)
24. J.A. Scale, Tomographic inversion via the conjugate gradient method. Geophysics **52**, 179–185 (1987)
25. Y. Saad, *Iterative Methods for Sparse Linear Systems*, 2nd edn. (SIAM, 2003)
26. M.R. Hestenes, E. Stiefel, Methods of conjugate gradients for solving linear systems. J. Res. Natl. Bur. Standards **49**, 409–436 (1952)
27. S.K. Biswas, K. Rajan, R.M. Vasu, D. Roy, A pseudo-dynamical systems approach based on a quadratic approximation of update equations for diffuse optical tomography. J. Opt. Soc. Am. A. **28**, 1784 (2011)
28. U. Ascher, E. Haber, H. Huang, On effective model for implicit piecewise smooth surface recovery. SIAM J. Comput. **28**, 339 (2006)
29. D. Roy, A numeric-analytic principle for non-linear deterministic and stochastic dynamical systems. Proc. R. Soc. A **A457**, 539 (2001)
30. K. van den Doel, U. Ascher, On level set regularization for highly ill-posed distributed parameter estimation problem. J. Comput. Phys. **216**, 707 (2006)
31. Y. Saad, Krylov subspace methods on supercomputers. SIAM J. Sci. Stat. Comput. **10**, 1232 (1989)
32. C. Moler, C.V. Loan, Nineteen dubious ways to compute the exponential of a matrix, twenty-five years later. SIAM Rev. **45**, 49 (2003)
33. Y. Saad, *Numerical Methods for Large Eigenvalue Problems*, vol. 1232 (SIAM, 2011)
34. D. Roy, A family of lower and higher order transversal linearization techniques in non-linear stochastic engineering dynamics. Int. J. Numer. Methods Eng. **61**, 790 (2004)
35. S.K. Biswas, K. Rajan, R.M. Vasu, D. Roy, Practical fully three-dimensional reconstruction algorithms for diffuse optical tomography. J. Opt. Soc. Am. **29**, 1017–1026 (2012)
36. K. Wang, C. Huang, Y. Kao, C.Y. Chou, A.A. Oraevsky, M.A. Anastasio, Accelerating image reconstruction in three-dimensional optoacoustic tomography on graphics processing units. Med. Phys. **40**, 023301–15 (2013)
37. S. Gutta, S.K. Kalva, M. Pramanik, P.K. Yalavarthy, Accelerated image reconstruction using extrapolated Tikhonov filtering for photoacoustic tomography. Med. Phys. **45**, 3749 (2018)
38. L. Li, L. Zhu, C. Ma et al., Single-impulse panoramic photoacoustic computed tomography of small-animal whole body dynamics at high spatiotemporal resolution. Nat. Biomed. Eng. **1**, 0071 (2017)
39. A. Rosenthal, V. Ntziachristos, D. Razansky, Acoustic inversion in optoacoustic tomography: a review. Curr. Med. Imaging Rev. **9**, 318–336 (2013)
40. P.C. Hansen, D.P. O'Leary, The use of the L-curve in the regularization of discrete ill-posed problems. SIAM J. Sci. Comput. **14**, 1487–1503 (1993)
41. M. Bhatt, A. Acharya, P.K. Yalavarthy, Computationally efficient error estimate for evaluation of regularization in photoacoustic tomography. J. Biomed. Opt. **21**, 106002 (2016)
42. D. Vigoureux, J.L. Guyader, A simplified time reversal method used to localize vibrations sources in a complex structure. Appl. Acoust. **73**, 491–496 (2012)

43. S.K. Biswas, K. Rajan, R.M. Vasu, Accelerated gradient based diffuse optical tomographic image reconstruction. Med. Phys. **38**, 539–547 (2011)
44. C.C. Paige, M.A. Saunder, LSQR: an algorithm for sparse linear equations and sparse leasr squares. ACM Trans. Math. Software **8**, 195–209 (1982)
45. J. Jose, R.G.H. Willemink, W. Steenbergen, C.H. Slump, T.G. van Leeuwen, S. Manohar, Speed-of-sound compensated photoacoustic tomography for accurate imaging. Med. Phys. **39**, 7262–7271 (2012)

Democratizing LED-Based Photoacoustic Imaging with Adaptive Beamforming and Deep Convolutional Neural Network

Haichong K. Zhang

Abstract The current standard photoacoustic (PA) imaging technology includes two hardware requirements: high power pulsed laser for light illumination and multi-channel data acquisition device for PA signal recording. These requirements have been limiting factors to democratize PA imaging because a laser is heavy, expensive and includes hazardous risk, and most parallel data acquisition technology is available only in specialized research systems. The goal of this chapter is to provide an overview of technologies that will enable safer and more accessible PA imaging, as well as introduce the use of safe and fast light emitting diode (LED) light sources in combination with clinical ultrasound machines. There are two limiting factors that prevent achieving this. First, clinical ultrasound machines typically only provide post-beamformed data based on an ultrasound delay function, which is not suitable for PA reconstruction. Second, a PA image based on the LED light source suffers from low signal-to-noise-ratio due to limited LED-power and requires a large number of averaging. To resolve these challenges, an adaptive synthetic aperture beamforming algorithm is applied to treat defocused data as a set of pre-beamformed data for PA reconstruction. An approach based on deep convolutional neural network trains and optimizes the network to enhance the SNR of low SNR images by guiding its feature extraction at different layers of the architecture. We will review and discuss these technologies that could be the key to advancing LED based PA imaging to become more accessible and easier to translate into clinical applications.

H. K. Zhang (✉)
Department of Biomedical Engineering,
Worcester Polytechnic Institute, Worcester, MA 01609, USA
e-mail: hzhang10@wpi.edu

Robotics Engineering Program,
Worcester Polytechnic Institute, Worcester, MA 01609, USA

© Springer Nature Singapore Pte Ltd. 2020
M. Kuniyil Ajith Singh (ed.), *LED-Based Photoacoustic Imaging*,
Progress in Optical Science and Photonics 7,
https://doi.org/10.1007/978-981-15-3984-8_8

1 Introduction

Photoacoustic imaging is an emerging modality offering unique contrast of optical absorption and imaging depth of ultrasound for a wide range of biomedical applications [1]. The clinical accessibility of photoacoustic (PA) imaging is limited because of specific hardware requirements including high energy pulsed laser and channel data acquisition system [2–5]. Most laser systems used for PA imaging provide high pulse energy in the mJ scale with a low pulse repetition frequency (PRF) of 10–20 Hz. These laser systems are bulky, expensive, and unsafe, requiring eye protection, such as laser glasses. Installation at a hospital would require a special room that meets the laser safety requirements.

To democratize PA imaging toward broader clinical applications and its usage in research, a light source that is compact, low-cost, and safe to use is desired. Light emitting diode (LED) light sources have been considered as a viable alternative [6, 7]. Compared to high power laser systems, the LED-based light source has the advantage in terms of size, cost, and safety. Most importantly, the LED light source is not classified as a laser, so laser safety regulations such as the light shield and laser safety glasses are not required. The limitation of an LED light source is its low output power. Series of LEDs can generate energy only in the range of μJ, while common high-power pulsed laser used for PA imaging produce energy in the mJ range. Due to the low power output, the received PA signal of an LED-based system suffers from low signal-to-noise-ratio (SNR). Current technology aiming to improve the SNR is based on acquiring multiple frames of PA signals, and subsequently perform an averaging over them to minimize the noise. Though the pulse repetition frequency of a LED-based system is much higher (in range of kHz) than the high-power laser, an averaging over many frames, typically thousands, reduces the effective frame rate of PA images. Furthermore, a large number of averaging frames require longer scanning times, leading to potential motion artifacts in reconstructed PA images.

The other hardware challenge is the accessibility of data acquisition (DAQ) devices used for PA imaging [8, 9]. Pre-beamformed channel data from acquisition devices are required to collect the raw PA signals because PA reconstruction requires a delay function calculated based on the time-of-flight (TOF) from the light source to the receiving probe element, while US beamforming considers the round trip initiated from the transmitting and receiving probe element. Thus, the reconstructed PA image with an ultrasound beamformer would be defocused due to the incorrect delay function. Real-time channel data acquisition systems are only accessible from limited research platforms. Most of them are not FDA approved, which hinders the development of PA imaging in the clinical setting. Therefore, there is a demand to implement PA imaging on more widely used clinical machines.

To broaden the impact of clinical PA imaging, this paper presents a vendor-independent PA imaging system utilizing ultrasound post-beamformed data, which is readily accessible in some clinical scanners. While a LED light source with low energy output and high PRF is used to replace a conventional high energy laser, a deep neural networks-based approach is presented to improve the quality of PA

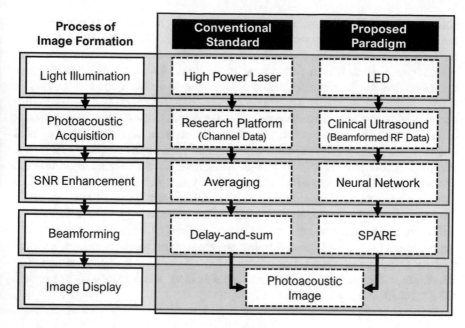

Fig. 1 The conventional photoacoustic imaging architecture and the new paradigm introduced in this chapter using LED light source and clinical ultrasound machine

images as well as reduce the number of averaging frames in image reconstruction. Figure 1 summarizes the process of PA image formation based on conventional architecture compared with the proposed paradigm incorporating a LED light source and a clinical ultrasound machine.

In this chapter, we review two enabling technologies for a LED-based and PA imaging system integrated with clinical ultrasound scanners; the image reconstruction approach using a post-beamformed RF data and the deep neural network-based SNR enhancer.

2 Image Reconstruction from Post-beamformed RF Data

2.1 Problem Statement

The acquisition of channel information is crucial to form a PA image, since typical clinical ultrasonic machines only provide access to beamformed data with delay-and-sum [2, 8]. Accessing pre-beamformed channel data needs customized hardware and parallel beamforming software and is available for dedicated research ultrasound platforms, such as the Ultrasonix DAQ system [9]. In general, these systems are costly with fixed data transfer rates that prohibit high frame rate, real-time imaging [10].

More importantly, PA beamforming is not supported by most clinical ultrasound systems. Harrison et al. has suggested changing the speed of sound parameter of clinical ultrasound systems [11]. Software access to alter the sound speed is not prevalent, however, and the range for this change is restricted when available, making this choice inadequate for reconstruction of PA images. In addition, the applicability of this technique is restricted to linear arrays, because angled beams (e.g. as in curvilinear arrays) change beamformer geometry and the speed of sound. Thus, compensation cannot be made by merely altering the sound velocity. In contrast, several clinical and research ultrasound systems have post-beamformed radio frequency (RF) data readily available. The objective in this section is to devise a PA image reconstruction approach based on ultrasound RF data that the system has already beamformed. A synthetic aperture-based beamforming algorithm, named $\underline{S}$ynthetic-aperture based $\underline{P}$hoto$\underline{A}$coustic $\underline{RE}$-beamforming (SPARE), utilizes ultrasound post-beamformed RF data as the pre-beamformed data for PA beamforming [12, 13]. When receive focusing is applied in ultrasound beamforming, the focal point can be regarded as a virtual element [14–16] to form a set of pre-beamformed data for PA beamforming. The SPARE beamformer takes the ultrasound data as input and outputs a PA image with the correct focal delay applied.

2.2 Technical Approach

2.2.1 Ultrasound Beamforming

The difference between ultrasound and PA beamforming is the acoustic time-of-flight (ToF) and related delay function. The delay function in delay-and-sum beamforming is calculated from the distance between the receivers and the target in ultrasound image reconstruction [17]. The acoustic wave is first transmitted from the ultrasound transducer via a medium with a specific velocity, reflected at boundaries, and the backscattered sound is received by the ultrasound transducer. The acoustic ToF during this process can be formulated as,

$$t_{US}(r_F) = \frac{1}{c}(|r_T| + |r_R|),\tag{1}$$

where r_F is the focus point originating from the ultrasound image coordinates, r_T is the vector from the transmit element to the focal point, r_R is the vector from the focal point to the receive element, and c is the speed of sound. Sequential beamforming with dynamic focus or fixed focus is applied as a delay-and-sum algorithm in clinical ultrasound systems. In dynamic focusing, the axial component, z_F, of the focusing point differs with depth, while a single fixed depth focus is used for the fixed focusing.

The acoustic TOF of PA signals is half of that of ultrasound, because the acoustic wave is produced at the target by absorbing light energy, and the time to travel from the

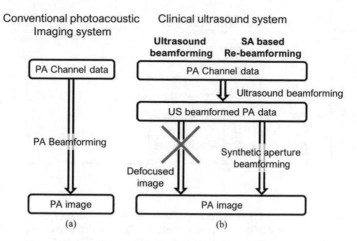

Fig. 2 Conventional PA imaging system (**a**) and proposed PA imaging system using clinical ultrasound scanners (**b**). Channel data is required for PA beamforming because ultrasound beamformed data is defocused with the incorrect delay function, where the introduced approach treats this information as pre-beamformed data for additional beamforming

optical transmission side negligible. Therefore, the acoustic TOF for photoacoustic imaging is

$$t_{PA}(r_F) = \frac{|r_R|}{c}. \tag{2}$$

Considering the differences between Eqs. (1) and (2), when beamforming is applied to the received PA signals based on Eq. (2), the beamformed RF signals are defocused (Fig. 2).

2.2.2 Synthetic Aperture-Based Re-beamforming

In the SPARE beamforming, the beamformed RF data from the ultrasound scanner is not considered as defocused useless data, but as pre-beamformed RF data for PA beamforming. The additional delay-and-sum step is applied on the beamformed RF data, and it is possible to reconstruct the new photoacoustically beamformed RF data. The focus point in the axial direction is constant with depth when fixed focusing is applied in the ultrasound beamforming process, suggesting that optimal focusing has been implemented at the particular focal depth with defocused signals appearing elsewhere. Initiating from the single focal depth, the defocused signals appear as if they were transmitted from the focal point (i.e. a virtual element as illustrated in Fig. 3b). In this sense, the ultrasound post-beamformed RF data is considered as PA pre-beamformed RF data. The TOF from the virtual element, when a fixed focus at z_F is applied, becomes

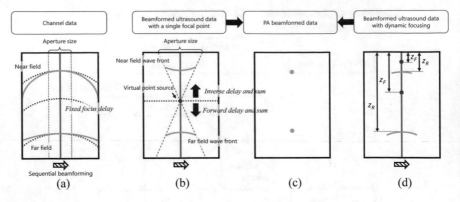

Fig. 3 Illustration of channel data and the SPARE-beamforming process [55]. **a** In channel data, the wave front of received RF signals expand corresponding to the depth (green line). The red lines indicate fixed focus delay function. **b** When fixed receive focusing is applied, the delay function is only optimized to the focus depth (red line). **c** As a result of fixed receive focusing, the focal point can be regarded as a virtual point source, so that inverse and forward delay and sum can be applied. **d** Similarly, dynamic focusing could be regarded as a specific case of that in which the virtual element depth z_F is the half distance of re-beamforming focal depth z_R

$$t\left(r'_F\right) = \frac{|r'_R|}{c},\tag{3}$$

where

$$|r'_R| = \sqrt{(x_R)^2 + (z_R - z_F)^2},\tag{4}$$

and $r'_F = r_F - z_F$. x_R and z_R is the lateral and axial components of r_R, respectively. The dynamic receive delay function is applied in the positive axial direction when $z_R \geq z_F$, and negative dynamic focusing delay is applied when $z_R < z_F$. The diagrams in Fig. 3b, c show the re-beamforming process of the SPARE-beamformer. Post-beamforming processes such as envelope detection and scan conversion are applied on the reconstructed data for the PA image display.

This theory applies to both fixed and dynamic focused beamformed ultrasound RF data with difference being that in dynamic focusing, the round-trip between the transmitter and the reflecting point in conventional ultrasound imaging must be considered along with the location of the virtual point source. Thus, in SPARE beamforming of dynamically focused data, the virtual point source depth, z_F, is considered to be dynamically varied by half of the photoacoustic beamforming focal point depth, z_R, as illustrated in Fig. 3d. Note that $z_R = 2z_F$ is always true in this special case.

2.3 Simulation Evaluation

The concept validation was performed through the ultrasound simulation tool, Field II [18]. A 128-element, 0.3 mm pitch, linear array transducer was assumed to be a receiver, which matches the setup of the experiment presented in Sect. 2.4. The standard delay-and-sum PA beamforming algorithm was applied to the simulated channel data in order to provide a ground-truth resolution value for this setup. Five-point targets were placed at depths of 10 mm to 50 mm with 10 mm intervals. To simulate defocused data, delay-and-sum with dynamic receive focusing and an aperture size of 4.8 mm was used to beamform the simulated channel data assuming ultrasound delays. The simulation results are shown in Fig. 4. The ultrasound beamformed RF data was defocused due to an incorrect delay function (Fig. 4b). The reconstructed PA images are shown in Figs. 4c–d. The measured full width at half maximum (FWHM) is shown in Table 1. The reconstructed point size was comparable to the point reconstructed using a 9.6 mm aperture on the conventional PA beamforming.

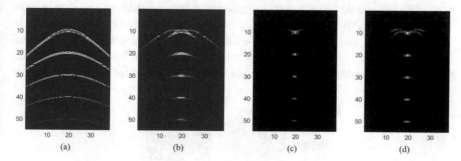

Fig. 4 Simulation results. **a** Channel data. **b** Ultrasound post-beamformed RF data. **c** Reconstructed PA image from channel data with an aperture size of 9.6 mm. **d** Reconstructed PA image through SPARE beamforming

Table 1 FWHM of the simulated point targets for corresponding beamforming methods

FWHM (mm)	Control using channel data	SPARE-beamforming
10 mm depth	0.60	0.63
10 mm depth	1.02	0.99
10 mm depth	1.53	1.43
10 mm depth	1.94	1.91
10 mm depth	2.45	2.42

2.4 Experimental Demonstration

The PA sensing system was employed for evaluating the LED-based PA imaging performance; a near-infrared pulsed LED illumination system (CYBERDYNE INC, Tsukuba, Japan) was used for PA signal generation. To collect the generated PA signals, a clinical ultrasound machine (Sonix Touch, Ultrasonix) with a 10 MHz linear ultrasound probe (L14-5/38, Ultrasonix) was used to display and save the received data. A line phantom made with fishing wire was imaged to evaluate the SNR and resolution performance. The ultrasound post-beamformed RF data with dynamic receive focusing was then saved. To validate the channel data recovery through inverse beamforming, the raw channel data was collected using a data acquisition device (DAQ). Figure 5 shows the experimental results imaging the cross section of a line phantom [19, 20]. The control data was reconstructed from channel data collected from the DAQ. The SPARE result used the ultrasound post-beamformed data collected from the ultrasound scanner as the input. The SPARE algorithm produced better imaging contrast and SNR when comparing the inherent resolution of the two methods. By quantifying the SNR change over the number of averaging, these two were correlated in a log-linear model for both with and without the use of channel data, depicted in Fig. 5b. In result, the gradient of the SPARE method was larger than conventional PA reconstruction from channel data, because the ultrasound beamformed data was summed already across the aperture once even with incorrect focus, and the random

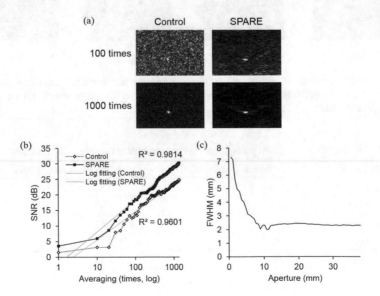

Fig. 5 Experiment results with LED light source imaging line phantom. **a** Comparison of control using channel data from DAQ and SPARE results using ultrasound post-beamformed data. **b** SNR analysis of both control and SPARE results. **c** Resolution analysis of SPARE results. Resolution improvement was hindered at FWHM of 2 mm due to the aperture size [19, 20]

Fig. 6 In vivo PA imaging of human fingers using LED light source. Experimental configuration of ultrasound and PA images of human fingers are shown. PA images were reconstructed using channel data from a DAQ device and beamformed RF data with the SPARE algorithm [19, 20]

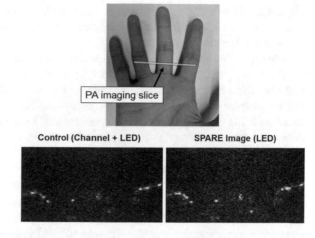

noise can be suppressed in this process. The control result showed better spatial resolution compared to the SPARE result because the ultrasound beamformed data was formed from a restricted aperture size (maximum 32 elements) due to restriction of the ultrasound scanner, while the channel data could utilize the complete aperture for reconstruction (Fig. 5c).

Human fingers were imaged using 850 nm LED bars for an in vivo experiment (Fig. 6). The channel data was collected first, then the ultrasound beamformed data was produced to compare standard and suggested solutions to beamforming. The raw channel data was averaged 3000 times to maximize the imaging contrast. It was verified that the SPARE approach could achieve comparable image quality to the channel data.

2.5 Discussion

The introduced SPARE method would work for any structures that have high optical absorption such as blood vessels that show strong contrast for near-infrared wavelength light excitation. Reconstruction artifacts such as side lobe and grating lobe could appear and influence non-point targets making the image quality of SPARE image was worse than standard PA image using channel data. The algorithm could also be incorporated with clinical ultrasound machines in real-time imaging schemes. Theoretically, the SNR of two beamformers should be similar, and this discrepancy could be attributed to the summation of axially distributed coherent information twice, once for each beamforming step. When the SNR of the channel signals is considerably low, the reconstructed image may contain a noise-related gradation artifact as the number of summations differs for each focal point. Hence, beamforming with the full aperture is more suitable in this high-noise situation. The image quality improvement strategies (apodization, transmit gain compensation, etc.) are expected to have a comparable impact on the SPARE image enhancement. Apodization improves the appearance of the reconstructed image, because it reduces the sidelobes in the ultrasound beam.

The suggested technique is superior than the speed of sound adjustment approach [11] and is applicable to steered beams (e.g. phased arrays) and to beam geometries that vary from linear arrays (e.g. curvilinear arrays). As formulated in Eqs. (3) and (4), the proposed beamformer applies a delay-and-sum assuming the PA signals are received at the virtual element. Therefore, even if the ultrasound beam is angled, the delay-and-sum algorithm is still applicable with the virtual element created by the angled beam.

Suppression of ultrasound transmission may be regarded as another system requirement. The ideal solution is to turn off the transmit events. However, if this function is not available, an option is to lower the transmission energy voltage. The use of an electrical circuit to regulate the timing of the laser transmission is another strategy. Subtracting the images with and without laser excitation would highlight the PA signals.

One system requirement for the SPARE beamformer is a high pulse repetition frequency (PRF) laser system. In order to maintain the frame rate, so that it is comparable to that of ultrasound B-mode imaging, the PRF of the laser transmission should be the same as the ultrasound transmission rate, in the range of at least several kHz. In fact, a high PRF laser system, such as a LED, is idealistic. Based on the assumption that the LED frame rate is 16,000 and the reception ultrasound has 128 lines of acquisition, Fig. 7 summarizes the estimated frame rate and laser energy by varying the number of averaging. Since SNR improvement under averaging is the square root of the number of averaging, outputting 1 mJ and 5 mJ light source energy requires 25 and 625 times averaging, respectively. The highest frame rate available when the DAQ unit is accessible is 625 and 25.6 frames per second, respectively. When a clinical ultrasound scanner was used for data acquisition, the frame rate becomes 5 and 0.2 frames per second, respectively. Using clinical ultrasound machine, the highest frame rate available is 125 without averaging.

The novelty of the SPARE algorithm suggested its potential for integration with clinical ultrasound scanners to become real-time imaging systems [21]. Most real-time photoacoustic imaging systems are currently based on open platform research systems [9]. However, the option of using a clinical ultrasound system already with FDA approval eases the transition of photoacoustic technology into the clinic. Potential applications include in vivo real-time photoacoustic visualization for brachytherapy monitoring [22–24], brain imaging [25–28], image-guided surgery [29, 30], interventional photoacoustic tracking [31], multispectral interventional imaging [32, 33], and cardiac radiofrequency ablation monitoring [34].

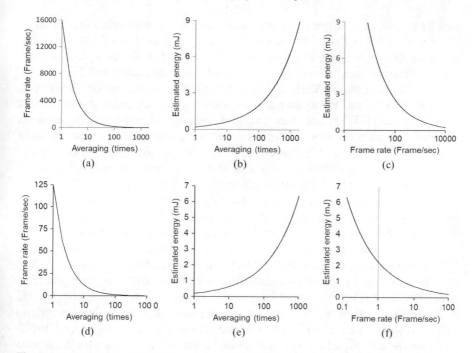

Fig. 7 Numerical estimation of frame rate using a LED system. Frame rate (**a**, **d**) and estimated energy (**b**, **e**) by varying the number of averaging, and the relationship between frame rate and estimated energy (**c**, **f**) are shown using a DAQ device (**a**–**c**), and using a clinical ultrasound system (**d**–**f**) for data collection

3 SNR Enhancement with Convolutional Neural Network

3.1 Problem Statement

The most classic and conventional strategy to improve the SNR with a low-power light source such as the LED-based scheme is averaging, obtaining multiple frames (ten, hundreds, or a few thousand) of the same sample, then averaging them over. When the noise has its distribution of ʊ, the noise distortion after the averaging of N times is expressed as

$$\sigma_{avg-N} = \frac{\sqrt{N}}{N}\sigma, \tag{5}$$

and the SNR improvement is proportional to the number of frames used for averaging. While using more frames to average earns an enhanced SNR, it decreases PA imaging's effective frame rate. Reduced frame rate makes it difficult to adapt this technology to moving objects, like the heart, and prone to motion artifacts. The signal processing approaches, such as adaptive denoising, empirical mode decomposition, wavelet transform or Wiener deconvolution could be used to tackle the limitation of averaging [7, 35]. Coded excitation is a strategy that increases the SNR without compromising the measurement time. In temporal encoding, the laser pulses are sent with a special encoded pattern without the need for waiting the acoustic TOF. The PA signals with an improved SNR are decoded from the received encoded RF signals. Golay codes [36] and m-sequence family (such as preferred pairs of m-sequences and Gold codes) [37, 38] have been proposed for temporal encoding. The limitation of coded excitation is that it presents its benefit only if the pulse interval is shorter than that of the acoustic TOF, thus ultra-high PRF lasers with hundreds kHz or several MHz pulsing capabilities are required. Therefore, a more generalized approach is needed to improve the SNR for the usage of LED light source.

A recently emerging approach based on deep convolutional neural networks is a powerful alternative. Deep neural networks have been introduced to image classification [39, 40], image segmentation [41], image denoising [42] and image super-resolution [43–46] and outperforms state-of-the-art signal processing approaches. The published image enhancement techniques are based on stacked denoising auto-encoder [42], densely connected convolutional net [46] or including perceptual loss to enhance the spatial structure of images [44]. Neural networks have been applied on PA imaging for image reconstruction [47–49] and removal of reflection artifacts [50]. This section focuses on the usage of deep convolutional neural network to differentiate the main signal from the background noise and to denoise a PA image with a reduced number of averaging.

The introduced architecture consists of two key components; one is convolutional neural networks (CNN) that extracts the spatial features, and the other one is recurrent neural networks (RNN) that leverages the temporal information in PA images. The CNN is built upon a state-of-the-art dense net-based architecture [46] that uses series of skip-connections to enhance the image content. Convolutional variant of short-long-term-memory [51, 52] is used for the RNN to exploit the temporal dependencies in a given PA image sequence. Skip-connections are integrated throughout the networks, including both CNN and RNN components, to effectively propagate features and eliminate vanishing gradients. While the full description of approaches can be found in Refs. [53, 54], this section provides digest of them.

3.2   Deep Convolutional Neural Network

A dense net-based CNN architecture to denoise PA images is introduced by Anas et al. [46, 53, 54]. The PA image with a limited number of averaging is used as the input, and the objective is to produce a high-quality PA image that provides an equivalent SNR compared to a PA image with a considerably high number of averaging. Figure 8 shows the deep neural network architecture [46]. The network focusing on improving the image quality of a single PA image is illustrated in Fig. 8a. The number of feature maps in each convolutional layer is defined as 'xx' in 'Conv xx'. The architecture consists of three dense blocks, and each dense block is composed of two densely

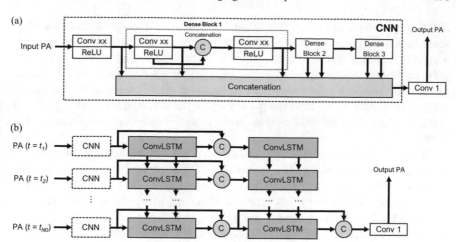

Fig. 8 A schematic of the introduced deep neural network-based approach (Reproduced from [53]). **a** The dense net-based CNN architecture to improve the quality of PA image. The architecture consists of three dense blocks, each dense block includes two 3 by 3 dense convolutional layers followed by rectified linear units. **b** The architecture that integrates CNN and ConvLSTM together to extract the spatial features and the temporal dependencies, respectively

connected convolutional layers and rectified linear units (ReLU). The benefit of using the dense convolutional layer is elimination of the vanishing gradient problem of deep networks [55] because all the features initially produced are inherited and succeeded in the following layers. The output image is produced by convoluting the feature map with all features from the concatenated dense blocks.

In addition to CNN, a recurrent neural network (RNN) [56, 57] is implemented to mitigate the temporal dependencies in a specific sequence. While several variants of RNN have been reported, and long-short-term-memory (LSTM) [51] showed the most successful performance in different applications. ConvLSTM [52] is an extension of LSTM that uses the convolution operation to extract temporal features from a series of 2D maps. The introduced architecture combining CNN and ConvLSTM to improve the denoising performance is shown in Fig. 8b. The architecture takes as inputs a series of PA images in different time points. It initially uses CNN to obtain the spatial features and then subsequently utilizes ConvLSTM to exploit the temporal dependencies. Two layers of ConvLSTM including skip connections are used for the recurrent connection. At the end, all the features generated in the previous layers are concatenated to compute the SNR improved PA image as the final output.

3.3 Experimental Demonstration

The concept was validated by training the network and assessing the SNR enhancement with a point target and proved further with human fingers in vivo. Two sets of

LED bar-type illuminators were placed on both sides of a linear ultrasound transducer array for the image setup. The LED's pulse repetition frequency was set at 1 kHz and PA data acquisition was synchronized with the LED excitation. PA images of the point target from the wire phantom were used to train the neural networks, assuming that those PA images with multiple point targets at different depths enable our network to learn how to improve the quality of the point spread function. The trained network with a point spread function can be applied to any arbitrary function of PA target.

The number of averaging was used to control the reconstructed image quality to produce input data consisting of low and high SNR target PA images for the training. For low SNR inputs, lower values of N in the range of 200–11,000 was chosen, with a step of 200. The averaging frame numbers in the sequence can be represented as $\{N_s; 2N_s; 3N_s; \ldots; N_0\}$ corresponding to time index $\{t_1; t_2; t_3; \ldots; t_{N0}\}$, where N_s was set to 200, and N_0 was 11,000. For each chosen value of N, the large set of 11,000 frames was split into several subsets, where each subset consists of N frames of PA signals. For each subset of N frames data, the PA signals are averaged first, followed by reconstruction to obtain one post-processed PA image. With the collection of 11,000 frames for one phantom sample, the greatest possible quality PA image can be achieved by reconstructing it from the averaged signal over all frames, which is regarded as the ground truth target image. Note that for each experiment, there is only one gold-standard target image that corresponds to more than one input sequences. Mean square losses are used as a loss function between the predicted and gold-standard target PA images. To minimize the loss function, TensorFlow library (Google, Mountain View, CA) with Adam [58] optimization technique is used. The quantitative assessment was performed with the independent test dataset. The peak-signal-to-noise ratio (PSNR) and structural similarity index (SSIM) were used as evaluation indices that compare the output of our networks with the highest quality target image [59].

The comparison of PSNR and SSIM of two techniques using deep neural networks (CNN-only and RNN + CNN) for different averaging frame numbers is shown in Fig. 9a, b. The solid line in the figure shows the mean value for each computing method calculated from 30 test samples. The shaded region reflects the corresponding standard deviation of each evaluation index. While both deep neural network approaches outperform the SNR enhancement over averaging, the approach of RNN + CNN presented the highest performance among them. The improvements of RNN + CNN in PSNRs of 5.9 dB and 2.9 dB was accomplished on average with respect to averaging and CNN-only techniques, respectively. Figure 9c presents the amount of frame rate enhancement two deep neural network approaches relative to averaging to attain certain PSNA. The gain is calculated with respect to the frame number of the averaging approach. For example, at a mean PSNR of 35.4 dB, the RNN + CNN, CNN-only and averaging techniques need 1360, 3680 and 11,000 averaging frames, respectively. When the averaging approach was treated as reference, the RNN + CNN and CNN-only achieved gains in the frame rate of 8.1 and 3.0 times, respectively. With the deep neural network approaches, the improved frame rate can be achieved without compromising the SNR.

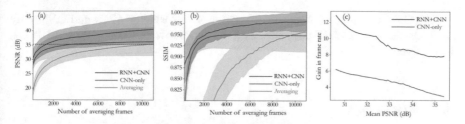

Fig. 9 A comparison of PSNR and SSIM of our (RNN + CNN) method with those from the simple averaging and CNN-only methods [53]. **a** PSNR versus averaging frame numbers. An improvement at all the averaging frame numbers is seen for our method compared to the two other methods. A higher improvement rate of the method is observed compared to the CNN-only method. **b** SSIM versus averaging frame numbers. Unlike CNN-only method, the trend of improvement is observed with the averaging frame numbers for our method. **c** Gain in frame rate versus mean PSNR

Figure 10 shows a qualitative comparison among all three comparative methods for proper digital arteries of three fingers of a volunteer (anatomy is shown at bottom in the figure). Three blood veins were noticeable for each finger, where enhanced blood vessel detections were observed for the RNN + CNN approach (highlighted by arrows). Note that the PA image averaged from the 5000 frames (high quality in the figure) includes some remaining noises and artifacts due to the movement during the scanning period.

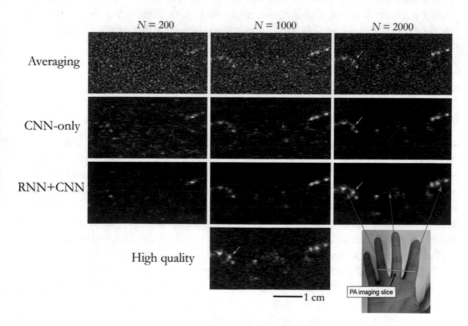

Fig. 10 A comparison of our method with the averaging and CNN-only techniques for an in vivo example [53]. Improvements are noticeable compared to those of other two methods in recovering the blood vessels (marked by arrows)

The GPU computation times are 15 and 4 ms for RNN + CNN and CNN-only methods, respectively. The corresponding run-times in the CPU are 190 and 105 ms, respectively.

3.4 Discussion

This section presented a deep neural networks approach to improve the quality of PA images in real-time while simultaneously reducing the scanning period. Besides using CNN to obtain the spatial features, RNN is used in the architecture to exploit the temporal information in PA images. The network was trained using a sequence of PA images from 32 phantom experiments. On the test from 30 samples, a gain in the frame rate of 8.1 times is achieved with a mean PSNR of 35.4 dB compared to the conventional averaging approach. A temporal PA sequence allows the neural networks to learn the image and noise contents more effectively than a single image-based CNN-only network does. In addition, for the CNN-only method, saturation in both image quality indices is observed for higher averaging frame numbers (Fig. 9a, b) indicating a decrease in the rate of improvement with a rise in the averaging frame number, as opposed to the higher improvement rate for the CNN + RNN method. Furthermore, the improved performance of the deep neural network approach was demonstrated through an in vivo example (Fig. 10). The key benefit of the technique is that it could improve the image quality from a reconstructed image with low averaging frame number, thus eliminating the potential effect of the artifacts.

4 Conclusions and Future Directions

In this chapter, we reviewed a paradigm on PA imaging using LED light source and image reconstruction with ultrasound post-beamformed RF data from a clinical ultrasound system. SPARE-beamforming takes the post-beamformed data and compensate the delay error by producing a PA image. Simulation and experimental studies presented that this approach can achieve an equivalent resolution compared to PA image generated from channel data. In addition, it was demonstrated that deep neural networks have a potential to exploit the temporal information in PA images for an improvement in image quality as well as a gain in the imaging frame rate.

Future directions along the line of this research include the exploration of developing beamforming algorithms utilizing more accessible data from clinical ultrasound machine such as post-beamformed B-mode images. It was reported that if the PA target is a point-like target, the post-beamformed B-mode image can be used as the source for PA image recovery [60]. For the image quality enhancement based on deep neural networks, more extensive in vivo evaluations are required to validate the clinical translatability. Other training architectures that do not take the final B-mode image but pre-beamformed or post-beamformed RF data may enhance both SNR and resolution of a PA image [61].

Acknowledgments The author would like to acknowledge the Medical UltraSound Imaging and Intervention Collaboration (MUSiiC) laboratory led by Dr. Emad Boctor for providing the opportunity to develop the technologies introduced in this book chapter. The author also thanks Mr. Doua Vang for proof-reading the manuscript and providing helpful comments. Financial support was provided by the Worcester Polytechnic Institute internal fund and the National Institute of Health grant DP5 OD0Z8162 for preparing the manuscript.

Compliance with Ethical Standards

Conflict of Interest All authors declare that they have no conflict of interest.

Ethical Approval All procedures performed in studies involving human participants were in accordance with the ethical standards of the institutional and/or national research committee and with the 1964 Helsinki declaration and its later amendments or comparable ethical standards. This article does not contain any studies with animals performed by any of the authors.

Informed Consent Additional informed consent to publish was obtained from all individual participants included in the study.

References

1. M. Xu, L.V. Wang, Photoacoustic imaging in biomedicine. Rev. Sci. Instrum. **77**, 041101 (2006)
2. S. Park, S.R. Aglyamov, S. Emelianov, Beamforming for photoacoustic imaging using linear array transducer, in *Proceedings of IEEE International Ultrasonics Symposium* (2007), pp. 856–859
3. B. Yin, D. Xing, Y. Wang, Y. Zeng, Y. Tan, Q. Chen, Fast photoacoustic imaging system based on 320-element linear transducer array. Phys. Med. Biol. **49**(7), 1339–1346 (2004)
4. C.K. Liao, M.L. Li, P.C. Li, Optoacoustic imaging with synthetic aperture focusing and cohehrence weighting. Opt. Lett. **29**, 2506–2508 (2004)
5. R.G.M. Kolkman, P.J. Brands, W. Steenbergen, T.G.V. Leeuwen, Real-time in vivo photoacoustic and ultrasound imaging. J. Biomed. Opt. **13**(5), 050510 (2008)
6. A. Hariri, J. Lemaster, J. Wang, A.S. Jeevarathinam, D.L. Chao, J.V. Jokerst, The characterization of an economic and portable LED-based photoacoustic imaging system to facilitate molecular imaging. Photoacoustics **9**, 10–20 (2018)
7. X. Dai, H. Yang, H. Jiang, In vivo photoacoustic imaging of vasculature with a low-cost miniature light emitting diode excitation. Opt. Lett. **42**(7), 1456–1459 (2017)
8. J.J. Niederhauser, M. Jaeger, M. Frenz, Comparision of laser-induced and classical ultrasound. Proc. SPIE **4960**, 118–123 (2003)
9. N. Kuo, H.J. Kang, D.Y. Song, J.U. Kang, E.M. Boctor, Real-time photoacoustic imaging of prostate brachytherapy seeds using a clinical ultrasound system. J. Biomed. Opt. **17**(6), 066005 (2012)
10. H.J. Kang, N. Kuo, X. Guo, D. Song, J.U. Kang, E.M. Boctor, Software framework of a real-time pre-beamformed RF data acquisition of an ultrasound research scanner. Proc. SPIE **8320**, 83201F (2012)
11. T. Harrison, R.J. Zemp, The applicability of ultrasound dynamic receive beamformers to photoacoustic imaging. IEEE Trans. Ultrason. Ferroelectr. Freq. Control **58**(10), 2259–2263 (2011)

12. H.K. Zhang, M. Bell, X. Guo, H.J. Kang, E.M. Boctor, Synthetic-aperture based photoacoustic re-beamforming (SPARE) approach using beamformed ultrasound data. Biomed Opt Express **7**(8), 3056–3068 (2016)
13. H.K. Zhang, X. Guo, B. Tavakoli, E.M. Boctor, Photoacoustic imaging paradigm shift: towards using vendor-independent ultrasound scanners. Med. Image Comput. Comput.-Assist. Interv. (MICCAI) **9900**, 585–592 (2016)
14. C.H. Frazier, W. D. O'Brien, Synthetic aperture techniques with a virtual source element. IEEE Trans. Ultrason. Ferroelec. Freq. Contr. **45**, 196–207 (1998)
15. S.I. Nikolov, J.A. Jensen, Virtual ultrasound sources in high resolution ultrasound imaging. Proc. SPIE, Progr. Biomed. Opt. Imaging **3**, 395–405 (2002)
16. J. Kortbek, J.A. Jensen, and K. L. Gammelmark, Synthetic aperture sequential beamforming, in *Proceedings of IEEE International Ultrasonics Symposium* (2008), pp. 966–969
17. K.E. Thomenius, Evolution of ultrasound beamformers, in *Proceedings of IEEE Ultrasonics Symposium*, vol. 2 (1996), pp. 1615–1622
18. J. A. Jensen and N. B. Svendsen, Calculation of pressure fields from arbitrarily shaped, apodized, and excited ultrasound transducers, in *IEEE Transactions on Ultrasonics, Ferroelectrics, and Frequency Control*, vol. 39 (1992), pp. 262–267
19. H. Zhang (2017) Enabling technologies for co-robotic translational ultrasound and photoacoustic imaging. Doctoral dissertation, Johns Hopkins University
20. H.K. Zhang, J. Kang, E.M. Boctor, In vivo low-cost photoacoustic imaging using LED light source and clinical ultrasound scanner. Proc. SPIE (presentation), 10494-177 (2018)
21. H.K. Zhang, H. Huang, C. Lei, Y. Kim, E.M. Boctor, Software-based approach toward vendor independent real-time photoacoustic imaging using ultrasound beamformed data, in *Photons Plus Ultrasound: Imaging and Sensing 2017*, vol. 10064 (International Society for Optics and Photonics, 2017), p. 100643O
22. M.A.L. Bell, N.P. Kuo, D.Y. Song, J. Kang, E.M. Boctor, In vivo visualization of prostate brachytherapy seeds with photoacoustic imaging. J. Biomed. Opt. **19**(12), 126011 (2014)
23. M.A.L. Bell, X. Guo, D.Y. Song, E.M. Boctor, Transurethral light delivery for prostate photoacoustic imaging. J. Biomed. Opt. **20**(3), 036002 (2015)
24. H.K. Zhang, A. Cheng, E.M. Boctor, Three-dimensional photoacoustic imaging using robotically tracked transrectal ultrasound probe with elevational synthetic aperture focusing, in *Proceedings of CLEO* (2016)
25. X. Wang, Y. Pang, G. Ku, X. Xie, G. Stoica, L.V. Wang, Noninvasive laser-induced photoacoustic tomography for structural and functional in vivo imaging of the brain. Nat. Biotechnol. **21**, 803–806 (2003)
26. W. Li, R. Chen, J. Lv, H. Wang, Y. Liu, Y. Peng et al., In vivo photoacoustic imaging of brain injury and rehabilitation by high-efficient near-infrared dye labeled mesenchymal stem cells with enhanced brain barrier permeability. Adv. Sci. (Weinh) **5**, 1700277 (2018)
27. J. Kang, H.K. Zhang, S.D. Kadam, J. Fedorko, H. Valentine, A.P. Malla, P. Yan, M. Harraz, J.U. Kang, A. Rahmim, A. Gjedde, L.M. Loew, D.F. Wong, E.M. Boctor, Transcranial recording of electrophysiological neural activity in the rodent brain in vivo using functional photoacoustic imaging of near-infrared voltage-sensitive dye. Front. Neurosci. **13**, 579 (2019)
28. J. Kang, E.M. Boctor, S. Adams, E. Kulikowicz, H.K. Zhang, K. Raymond, E. Graham, Validation of noninvasive photoacoustic measurements of sagittal sinus oxyhemoglobin saturation in hypoxic neonatal piglets. J. Appl. Physiol. **125**(4), 983–989 (2018)
29. L. Xi, G. Zhou, N. Gao, L. Yang, D.A. Gonzalo, S.J. Hughes, H. Jiang, Photoacoustic and fluorescence image-guided surgery using a multifunctional targeted nanoprobe. Ann. Surg. Oncol. **21**(5), 1602–1609 (2014)
30. M.A.L. Bell, A.K. Ostrowski, K. Li, P. Kazanides, E.M. Boctor, Localization of transcranial targets for photoacoustic-guided endonasal surgeries. Photoacoustics **3**(2), 78–87 (2015)
31. A. Cheng, H.J. Kang, H.K. Zhang, R.H. Taylor, E.M. Boctor, Ultrasound to video registration using a bi-plane transrectal probe with photoacoustic markers. Proc. SPIE **9786**, 97860J (2016)

32. W. Xia, D.I. Nikitichev, J. Mari, S.J. West, R. Pratt, A.L. David, S. Ourselin, P.C. Beard, A.E. Desjardins, Performance characteristics of an interventional multispectral photoacoustic imaging system for guiding minimally invasive procedures. J. Biomed. Opt. **20**(8), 086005 (2015)

33. H.K. Zhang, Y. Chen, J. Kang, A. Lisok, I. Minn, M.G. Pomper, E.M. Boctor, Prostate specific membrane antigen (PSMA)-targeted photoacoustic imaging of prostate cancer in vivo. J. Biophotonics, e201800021 (2018)

34. N. Dana, L.D. Biase, A. Natale, S. Emelianov, R. Bouchard, In vitro photoacoustic visualization of myocardial ablation lesions. Heart Rhythm **11**(1), 150–157 (2014)

35. S.A. Ermilov, T. Khamapirad, A. Conjusteau, M.H. Leonard, R. Lacewell, K. Mehta, T. Miller, A.A. Oraevsky, Laser optoacoustic imaging system for detection of breast cancer. J. Biomed. Opt. **14**(2), 024007 (2009)

36. M.P. Mienkina, C.-S. Friedrich, N.C. Gerhardt, M.F. Beckmann, M.F. Schiffner, M.R. Hofmann, G. Schmitz, Multispectral photoacoustic coded excitation imaging using unipolar orthogonal golay codes. Opt. Express **18**, 9076–9087 (2010)

37. H. Zhang, K. Kondo, M. Yamakawa, T. Shiina, Simultaneous multispectral coded excitation using gold codes for photoacoustic imaging. Jpn. J. Appl. Phys. **51**, 07GF03 (2012)

38. H.K. Zhang, K. Kondo, M. Yamakawa, T. Shiina, Coded excitation using periodic and unipolar M-sequences for photoacoustic imaging and flow measurement. Opt. Express **24**, 17–29 (2016)

39. A. Krizhevsky, I. Sutskever, G.E. Hinton, Imagenet classification with deep convolutional neural networks, in *Advances in Neural Information Processing Systems* (2012), pp. 1097–1105

40. K. He, X. Zhang, S. Ren, J. Sun, Deep residual learning for image recognition, in *IEEE CVPR* (2016), pp. 770–778

41. J. Long, E. Shelhamer, T. Darrell, Fully convolutional networks for semantic segmentation, in *Proceedings of the IEEE Conference on Computer Vision and Pattern Recognition* (2015), pp. 3431–3440

42. J. Xie, L. Xu, E. Chen, Image denoising and inpainting with deep neural networks, in *Advances in Neural Information Processing Systems* (2012), pp. 341–349

43. C. Dong, C.C. Loy, K. He, X. Tang, Image super-resolution using deep convolutional networks. IEEE Trans. Pattern Anal. Mach. Intell. **38**, 295–307 (2016)

44. J. Johnson, A. Alahi, L. Fei-Fei, Perceptual losses for real-time style transfer and super-resolution, in *European Conference on Computer Vision* (Springer, 2016), pp. 694–711

45. C. Ledig, L. Theis, F. Huszár, J. Caballero, A. Cunningham, A. Acosta, A. Aitken, A. Tejani, J. Totz, Z. Wang et al., Photo-realistic single image super-resolution using a generative adversarial network, arXiv preprint (2016)

46. T. Tong, G. Li, X. Liu, Q. Gao, Image super-resolution using dense skip connections, in *2017 IEEE International Conference on Computer Vision (ICCV)* (IEEE, 2017), pp. 4809–4817

47. S. Antholzer, M. Haltmeier, J. Schwab, Deep learning for photoacoustic tomography from sparse data. arXiv preprint arXiv:1704.04587 (2017)

48. J. Schwab, S. Antholzer, R. Nuster, M. Haltmeier, DALnet: high-resolution photoacoustic projection imaging using deep learning. arXiv preprint arXiv:1801.06693 (2018)

49. S. Antholzer, M. Haltmeier, R. Nuster, J. Schwab, Photoacoustic image reconstruction via deep learning, in *Photons Plus Ultrasound: Imaging and Sensing 2018*, vol. 10494 (International Society for Optics and Photonics, 2018), p. 104944U

50. D. Allman, A. Reiter, M.A.L. Bell, Photoacoustic source detection and reflection artifact removal enabled by deep learning. IEEE Trans. Med. Imaging **37**(6), 1464–1477 (2018)

51. S. Hochreiter, J. Schmidhuber, Long short-term memory. Neural Comput. **9**, 1735–1780 (1997)

52. S. Xingjian, Z. Chen, H. Wang, D.-Y. Yeung, W.-K. Wong, and W.-c. Woo, Convolutional LSTM network: a machine learning approach for precipitation nowcasting, in *Advances in Neural Information Processing Systems* (2015), pp. 802–810.

53. E.M.A. Anas, H.K. Zhang, J. Kang, E.M. Boctor, Enabling fast and high quality LED photoacoustic imaging: a recurrent neural networks based approach. Biomed. Opt. Express **9**(8), 3852 (2018)

54. E.M.A. Anas, H.K. Zhang, J. Kang, E.M. Boctor, Towards a fast and safe LED-based photoacoustic imaging using a deep convolutional neural networks. MICCAI (2018)
55. G. Huang, Z. Liu, K.Q. Weinberger, L. van der Maaten, Densely connected convolutional networks, in *Proceedings of the IEEE Conference on Computer Vision and Pattern Recognition*, vol. 1 (2017), p. 3
56. J.J. Hopfield, Neural networks and physical systems with emergent collective computational abilities, in *Spin Glass Theory and Beyond: An Introduction to the Replica Method and Its Applications* (World Scientific, 1987), pp. 411–415
57. D.E. Rumelhart, G.E. Hinton, R.J. Williams et al., Learning representations by back-propagating errors. Cogn. Model. **5**, 1 (1988)
58. D. Kingma, J. Ba, Adam: a method for stochastic optimization. arXiv preprint arXiv:1412.6980 (2014)
59. Z. Wang, A.C. Bovik, H.R. Sheikh, E.P. Simoncelli, Image quality assessment: from error visibility to structural similarity. IEEE Trans. Image Process. **13**, 600–612 (2004)
60. H.K. Zhang, X. Guo, H.J. Kang, E.M. Boctor, Photoacoustic image reconstruction from ultra-sound post-beamformed B-mode image. Proc. SPIE Photonics West BiOS 2016 **9708**(970837) (2016)
61. E.M.A. Anas, H.K. Zhang, C. Audigier, E.M. Boctor, Robust photoacoustic beamforming using dense convolutional neural networks. In *Simulation, Image Processing, and Ultrasound Systems for Assisted Diagnosis and Navigation* (Springer, Cham 2018), pp. 3–11

Deep Learning for Image Processing and Reconstruction to Enhance LED-Based Photoacoustic Imaging

Kathyayini Sivasubramanian and Lei Xing

Abstract Photoacoustic imaging is a rapidly growing imaging technique which combines the best of optical and ultrasound imaging. For the clinical translation of photoacoustic imaging, a lot of steps are being taken and different parameters are being continuously improved. Improvement in image reconstruction, denoising and improvement of resolution are important especially for photoacoustic images obtained from low energy lasers like pulsed laser diodes and light emitting diodes. Machine learning and artificial intelligence can help in the process significantly. Particularly deep learning based models using convolutional neural networks can aid in the image improvement in a very short duration. In this chapter we will be discussing the basics of neural networks and how they can be used for improving photoacoustic imaging. We will also discuss few examples of deep learning networks put to use for image reconstruction, image denoising, and improving image resolution in photoacoustic imaging. We will also discuss further the possibilities with deep learning in the photoacoustic imaging arena.

1 Introduction

1.1 Photoacoustic Imaging

Photoacoustic (PA) imaging is a hybrid imaging technique which has gained importance in the last several years [1–4]. It is based on the photoacoustic effect discovered by Alexander Graham Bell in 1881 [5]. He observed that light energy absorbed by a material results in an acoustic signal. He demonstrated this with an apparatus called photophone which he designed. Almost after a century, it started gaining importance

K. Sivasubramanian (✉) · L. Xing
Department of Radiation Oncology, School of Medicine, Stanford University, Stanford, CA, USA
e-mail: kathya19@stanford.edu

L. Xing
e-mail: lei@stanford.edu

© Springer Nature Singapore Pte Ltd. 2020
M. Kuniyil Ajith Singh (ed.), *LED-Based Photoacoustic Imaging*,
Progress in Optical Science and Photonics 7,
https://doi.org/10.1007/978-981-15-3984-8_9

as we discovered that it can be used for the purpose of imaging. The major advantage of photoacoustic imaging is that it combines the best features of two different imaging modalities: the contrast of optical imaging and the resolution of ultrasound imaging [1, 6]. Photoacoustic imaging is based on the principle that when a pulsed laser light (pulse width in the range of nanoseconds) falls on a sample, if the sample absorbs the light at the particular wavelength then it undergoes a small increase in temperature in the order of milli kelvin (mK). Following the heating, there occurs thermoelastic expansion of the sample, which leads to the generation of the pressure waves. The pressure waves can be detected as photoacoustic waves by ultrasound transducers. The sound waves captured by the transducers are then reconstructed to form images known as photoacoustic images [7, 8]. Contrast agents are very important for photoacoustic imaging because when a sample is irradiated with a particular wavelength of light, only if the contrast agent absorbs light at that wavelength, detectable photoacoustic waves will be emitted from the sample [9–11]. The imaging wavelength is usually in the visible and the near-infra red (NIR) region of the optical spectrum, very recently the second NIR window is being explored for PA imaging [12]. Therefore, availability of contrast agents in these wavelength regions is very critical for PA imaging. Fortunately, there are some intrinsic contrast agents like blood (hemoglobin), melanin, lipids, etc. present in the human body which provides great contrast in the visible and near infrared spectrum [13–16]. However, the contrast from these are only sufficient and suitable for imaging certain body parts and for certain applications. Therefore, for imaging other organs and for different applications, the use of extrinsic contrast agents becomes inevitable. Some of the most commonly used extrinsic contrast agents are organic dyes, inorganic dyes, nanoparticles, nanomaterials, etc. Constant research is being done to develop highly efficient photoacoustic contrast agents [17–25]. Different forms of photoacoustic imaging are available like photoacoustic microscopy, tomography, endoscopy etc. [16, 26–32]. The applications of photoacoustic imaging ranges from cellular level imaging to systems imaging. Photoacoustic imaging can be used for obtaining both structural and functional data from the sample. Some of the most explored applications of photoacoustic imaging includes sentinel lymph node imaging, brain imaging, blood vasculature imaging, tumor imaging and monitoring, oxygen saturation monitoring etc. [33–44].

The PA wave equation is given by

$$(\nabla^2 - v_s^{-2}\partial^2/\partial t^2)p(\vec{r}, t) = -(\beta/C_P)\partial H(\vec{r}, t)/\partial t$$

Here v_s refers to acoustic speed, $p(\vec{r}, t)$ refers to the acoustic pressure at location r and time t, β refers to the thermal expansion coefficient, C_P refers to the specific heat constant at constant pressure, and H denotes the heating function which can be described as the thermal energy converted per unit volume and per unit time. The left-hand side of this equation describes the wave propagation, whereas the right-hand side represents the source term.

Traditionally, large and bulky Nd:YAG or dye based lasers are used as illumination source for photoacoustic imaging. They often need an optical table for housing them and are non-portable. Even the smallest misalignment will alter the results greatly

[45]. Very recently portable, mobile Nd:YAG laser with optical parametric oscillator for tuning different wavelengths has been commercially available from opotek [37]. The biggest advantage of using these lasers is the high laser energy, which in turn translates to higher penetration and high-resolution images. The catch with this laser is that it is very difficult to combine the light to the ultrasound transducer, which makes the clinical use of these lasers very limited. However, in recent times compact, lightweight lasers are starting to be used for imaging like the pulsed laser diode (PLD) and the light emitting diodes (LED) [46, 47]. The pulsed laser diode is very small often palm size, and very light weight which makes it easily portable to use. It can also be integrated with the ultrasound transducer much easily than the OPO laser. The frequency of these lasers is very high therefore, they can provide large number of frames in a short period of time. The problem with this laser is that the pulse energy is very low and will often need averaging over multiple frames to obtain a high-resolution image. LEDs are similar to the PLDs but with lesser energy. The frequency of the LEDs is also very high. Multiple LEDs are placed in an array to generate light for imaging. But, even with an array of LEDs the pulse energy of the system is very low [48, 49]. The system requires a lot of averaging to obtain an acceptable photoacoustic image. One major disadvantage with these systems is that they are usually single wavelength and cannot be tuned. Therefore, cannot be used for spectroscopic studies like the Nd:YAG laser. However, in the last few years, multiple commercial systems have been developed for real-time photoacoustic imaging using different types of lasers like the Nd:YAG, PLD, LEDs etc. Ongoing research is being done on how to improve the resolution of images from the low energy laser sources like PLD, LEDs etc.

1.2 Photoacoustic Image Acquisition and Reconstruction

As much as the light plays a crucial role in photoacoustic imaging, equally important are the ultrasound transducers. The signal form the sample can be acquired using ultrasound transducer [50, 51]. There are many ways in which an ultrasound transducer is used for photoacoustic imaging, a single element transducer can be used for signal acquisition or a raster scan be performed to obtain a 2D or 3-D image or it can be rotated around the sample to obtain a cross-sectional image. However, it can be very time consuming to scan a big area. In order to complete scanning in a very short time multiple transducers can be combined to make an array of transducers to obtain images [33, 52, 53]. When using commercial systems, linear array, concave array and convex array-based transducers are also available for data acquisition. These transducers are supported by the data acquisition cards (DAQ) for image acquisition. Once the data is acquired using the ultrasound transducers, it goes through the reconstruction process to form the final image. Different types of reconstruction methods, such as filtered back-projection, Fourier transform, alternative algorithm, time reversal, inversion of the linear Radon transform, and delay and sum beamforming, have been developed under different assumptions and approximation for ultrasound and photoacoustic imaging [54–59]. The issue with these image reconstruction methods is that these methods assumes that the wave propagation is through a homogenous media, but in reality, that is often not the case. Another issue with using these image reconstruction methodologies is that these methods often generates artifacts like reflection artifact, etc. and cannot remove these artifacts. Post-processing of images is often used to remove some of the artifacts from the reconstructed images. However, the existing reconstruction and post-processing techniques are not sufficient to improve the quality of the images. There is a great need to improve the imaging resolution, reduce noise and remove artifacts of the photoacoustic images for clinical translation. Continuous research is required and being done on how to improve the image resolution from the perspective of reconstruction and post-processing.

Among the various image reconstruction methods, the delay-and-sum beamforming method is the most widely used algorithm for the reconstruction of both PA and US images. This algorithm works by summing the corresponding US signals while adjusting their time delays in accordance to the distance between the detectors and the sample. However, it has few drawbacks like low resolution, low contrast, and strong side lobes which results in artifact generation. Matrone et al. proposed a modification to the DAS algorithm leading to a novel beamforming algorithm, called the delay-multiply-and-sum (DMAS) beamformer, in order to help in overcoming the limitations of DAS in ultrasound imaging. The DMAS provides the high contrast and enhanced image quality, it also helps in obtaining narrow main lobes, and weaker side lobes in comparison to DAS. Owing to these advantages, several researchers extended the ultrasound DMAS algorithm to PA imaging also. Park et al. introduced a DMAS-based synthetic aperture focusing technique to PA microscopy. Alshaya et al. demonstrated the DMAS based PA imaging can be useful when using a linear array transducer also and additionally they introduced a subgroup of DMAS method to improve the signal to noise ratio (SNR) and the speed of image processing. To improve the quality of the image obtained from DMAS algorithm even more, Mozaffarzadeh et al. proposed using a double-stage DMAS operation, a minimum variance beamforming algorithm, or modified coherence factor [60–62]. In spite of all these advances, it has been difficult to use DMAS for image reconstruction clinically because of the heavy computation complexity involved in the incorporation of this algorithm to a clinical PA imaging system.

Another commonly used reconstruction technique for photoacoustic imaging is the back-projection (BP) method. This reconstruction technique and its derivatives like the filtered back projection (FBP) are one of the major reconstruction algorithms used for the photoacoustic computed tomography (PACT) specifically. This algorithm makes use of fact that the pressure propagating from an acoustic source reach the detectors at different time delays, which depends on a myriad of factors like the speed of sound, the distance between the source and the detectors, etc. The BP algorithm requires large number of signals collected from various view angles as its input. These signals can be collected by a single transducer or use an array of transducers rotating around the sample. Both the methods have their own pros and cons. This is a faster reconstruction technique, back-projection (BP) algorithms are capable of producing good images for common geometries (planar, spherical, cylindrical) in simulations and is also applied widely for volumetric image reconstruction in PA imaging. Constant development of BP algorithms leads to improved image quality, which has improved the possibilities with PA imaging and the capabilities of PA imaging in the various biomedical applications. The formulas of back projection techniques are implemented either in the spatio-temporal domain or in the Fourier domain. The BP algorithms are constantly modified to improve the applications and the image quality, one of the modifications is based on a closed-form inversion formula. This modified algorithm was very successful in detection of the position and shape of absorbing objects in turbid media.

Although filtered back-projection (FBP) reconstruction techniques has proven its use in solving for time-dependent partial differential equations through Fourier spectral methods, there are still many critical problems that needs addressing to further improve the quality of FBP-reconstructed images [63, 64]. One of the shortcomings of the conventional back-projection algorithm is that they are not exact in experimental setting and may lead to the generation of substantial artifacts in the reconstructed image, such as the accentuation of fast variations in the image, which is accompanied by negative optical-absorption values that otherwise have no physical interpretation [59, 65]. The presence of the artifacts has not restricted the use of BP algorithms for structural PA imaging, they do affect the quantification capacity, the image fidelity, and the accurate use of the method for functional and molecular imaging applications.

Time reversal is another reconstruction method used in photoacoustic imaging. In the typical time reversal imaging reconstruction method, the recorded pressure time series are enforced in time reversed order as a Dirichlet boundary condition as the position of detectors on the measurement surface [66–69]. If the array of detectors is placed sparsely to collect the measurement rather instead of a continuous surface, the time reversed boundary condition will be discontinuous. This can cause severe blurring in the reconstructed images. To solve the problem, Treeby et al., improved time reversal image reconstruction technique with the usage of interpolated sensor data. In the course, the interaction can be avoided by interpolating the recorded data onto a continuous rather than discrete measurement surface within the space grid used for the reconstruction. The edges of the reconstructed image are considerably sharper, and the magnitude has also been improved. After that, they used the enforced time reversal boundary condition to trap artifacts in the final image, and by truncating the data, or introducing an adaptive threshold boundary condition, this artifact trapping can be mitigated to some extent.

1.3 Types of Artifact

Artifacts are one of the major problems in photoacoustic imaging. The presence of artifacts limits the application of photoacoustic imaging from a clinical perspective and hampers the clinical translation of the imaging modality greatly. Reflection artifact is one of the most commonly observed artifacts in photoacoustic imaging [67,

70, 71]. These reflections are not considered by traditional beamformers which use a time-of-flight measurement to create images. Therefore, reflections appear as signals that are mapped to incorrect locations in the beamformed image. The acoustic environment can also additionally introduce inconsistencies, like the speed of sound, density, or attenuation variations, which makes the propagation of acoustic wave very difficult to model. The reflection artifacts can become very confusing for clinicians during diagnosis and treatment monitoring using PA imaging. Until these are corrected the possibility of clinical translation is very slim.

In order to minimize the effect of artifacts in photoacoustic imaging different signal processing approaches have been implemented to enhance signal and image quality. These signal processing techniques use singular value decomposition and short-lag spatial coherence. But these techniques are not so efficient in the removal of intense acoustic reflection artifacts. A technique called photoacoustic-guided focused ultrasound (PAFUSion) was developed which differs from other traditional photoacoustic artifact reduction methodologies as it uses ultrasound to mimic wavefields produced by photoacoustic sources in order to identify reflection artifacts for removal [72, 73]. A slight modification of this approach was developed which uses plane waves instead of focused waves, but the implementation was very similar. Both of these methods make the assumption that acoustic reception pathways are identical, which may not always be true. When performing simultaneous ultrasound and photoacoustic imaging in real-time it is not always possible to have an exact overlay of the image because of the motion induced artifact caused by the moving organs inside the body especially organs like heart, abdominal cavity, blood vessels etc. Certain reconstruction methods have been proposed to overcome these types of artifacts, but the problem is they don't account for the inter patient variability and sometimes variability in the same patient when imaging different body parts (Fig. 1).

1.4 LED Based Photoacoustic Imaging

LED based photoacoustic system can play a very important role in the clinical translation of photoacoustic imaging. LEDs are less expensive compared to the traditional lasers for photoacoustic imaging, they are very compact, and capable of imaging in multiwavelength (e.g., 750, 810, 930, and 980 nm) [47–49]. The energy output from the LED arrays is much lesser than the energy from powerful class-IV lasers, therefore these can be used for clinical applications easily. But they have very low energy and usually produce low resolution images and are noisier. They also have higher laser pulse width, which limits the spatial resolution of the images. In order to obtain better images, signal averaging in the order of 1000s is required to obtain one image, which increases the image acquisition time. But, in spite of all the shortcomings LED based photoacoustic imaging has gained a lot of momentum with different types of applications that are possible with the system [47–49, 75–77]. The LED based photoacoustic imaging system from Cyberdyne INC (Tsukuba, Japan) can be operated at multiple wavelengths in the visible and the near infrared region. It has a

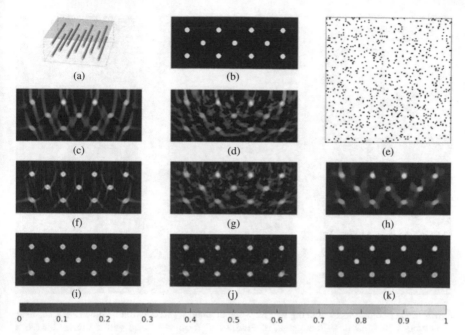

Fig. 1 Images of errors of different image reconstruction methods using a simple numerical phantom consisting of tubes. **a, b** Images of the numerical phantoms. **e** Illustration of the sub-sampling pattern. **c, d** Slice view of full and sub-sampled data respectively. **f–k** Slice views through the reconstructions of the tube phantom by different methods and for full or sub-sampled data. **f, i** non negative least squares (NNLS) of full data at different iterations. **g, j** NNLS of sub-sampled data at different iterations. **h, k** total variation (TV) of sub-sampled data at different iterations. Reprinted with permission from Ref. [74]

linear array transducer for image acquisition and a 128-channel data acquisition card. The system comes with inbuilt image reconstruction algorithms based on delay and sum model, the image further undergoes post-processing through various filters [48, 77–79]. Figure 2 shows the schematic and the photograph of the LED-photoacoustic imaging system (PLED-PA). It has been demonstrated using this system that it can be used for applications like blood vessel imaging, diagnosis of inflammatory arthritis, detection of head and neck cancer, etc. It has also been used for functional imaging of blood oxygen saturation.

With the current reconstruction techniques and post-processing methodologies in photoacoustic imaging it is really difficult to generate artifact and noise free images in shorter time, with lesser averaging and minimal post-processing. This is especially more relevant to the low energy laser sources like the PLD, LED etc. [80, 81]. In order to improve the image reconstruction and reduce noise in the images in a shorter duration, artificial intelligence can be made use of, specifically deep learning using convolution neural networks could be very useful for this purpose. In the rest of the chapter we will focus on how to make use of deep learning for photoacoustic imaging.

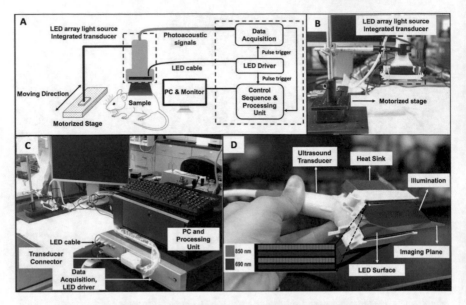

Fig. 2 LED photoacoustic imaging system. **A** Schematic representation of the PA system using LED array light source. **B** Photograph of PLED-PA probe associated with motorized stage. **C** Whole imaging setup. **D** PLED-PA probe with imaging plane and illumination source are shown schematically. LED array design is also shown in the inset—there were alternating rows of LEDs with different wavelengths. Reprinted with permission from Ref. [47]

2 Machine Learning and Artificial Intelligence

In the year 1950, Alan Turing proposed a 'Learning Machine' that could learn and become artificially intelligent. Research in neurology had shown that synapses worked like a network firing electric impulses, based on this idea, the construction of an electronic brain was suggested. Marvin Minsky and Dean Edmonds build the first neural network machine in the year 1951, it was called the Stochastic Neural Analog Reinforcement Calculator (SNARC) [82]. Starting from the 80s the golden age of machine learning began, in that period many ground-breaking discoveries were made but due to non-availability of the infrastructure for higher computing power and speed, further developments were hindered. In the year of 1981, the government of Japan funded a project with the goal to develop machines which could carry on conversations, translate languages, interpret pictures and reason like human beings, some of which are not realized even today. In 1997 IBMs computer 'Deep Blue' bet the world chess champion, Garry Kasparov and in 2005 a robot from Stanford was able to drive autonomously for 131 miles. There are countless other examples of the success of deep learning approaches in numerous fields [83–88].

Artificial intelligence (AI) is a technology that aims to make machines which tries to mimic human brain and the field has grown exponentially in the last few years and continues to impact the world significantly. The applications of artificial intelligence and machine learning are plenty, almost across all fields with a profound impact on improvement of human lives. Machine learning (ML) and deep learning (DL) has played a very significant role in the improvement of healthcare industry at different levels, right from diagnosis to patient monitoring. It is widely used in the areas of image processing, image analysis, diagnostics, treatment planning and follow-up, thus benefitting a large number of patients. It also helps the clinicians by reducing their workload and helps in making quicker decisions in many cases. Its impact on image processing and analysis especially are noteworthy.

Machine learning is a subset of AI which relies on pattern recognition and data analytics. In ML it is tested to identify if a computer can learn on its own from data without being programmed to perform different tasks. The iterative aspect of machine learning is critical because as models are exposed to new data, they are able to independently adapt. The system learns effectively from previous computations to make predictions and decisions.

2.1 Neural Networks

Artificial intelligence and machine learning are not complete without mentioning neural networks (NN). Neural networks are a set of algorithms, modeled inspired and based on human neural system, which are designed to recognize patterns [84, 89–91]. They interpret sensory data through a kind of machine perception, labeling or clustering raw input. Neural networks are capable of recognizing patterns from various input formats such as images, sound, text, etc. The information from the different types of inputs are translated into numerical values that can be understood by the machine. Some of the key areas in which neural networks help are in clustering and classification [88, 92–94]. When group of unlabeled data is presented to the neural networks, they are capable of grouping them according to the similarities between them. When the neural network is presented with a group of labelled data for training, they can classify data effectively. Neural networks are capable of extracting features which are then provided to the clustering and classification algorithms.

Deep learning is a subset of machine learning and is a rapidly growing field of research that targets to significantly enhance the performance of many pattern recognition and machine learning applications. Deep learning makes use of neural network designs for representing the nonlinear input to output map together with optimization procedures for adjusting the weights of the network during the training phase. In the last few years, deep learning-based algorithms were developed for achieving highly accurate reconstruction of tomography images.

2.2 Convolution Neural Network

The convolutional neural network (CNN), is a special neural network model that is designed to predominantly work with two-dimensional image data. Among neural networks, CNNs are primarily used for image recognition, images classification, Objects detections, recognition faces etc. [88, 93, 95, 96]. As the name suggests CNN derives its name from the convolutional layer and as it suggests this layer performs the "convolution" operation. In CNN, convolution is classified as a linear operation which involves the moving one or more convolution filters (with a set of assigned weights) across the input image [89, 97, 98]. Each of the weight from the convolution layer gets multiplied with the input data from the image on which it is scanned to yield a matrix that is smaller than the input image. The convolution filter is always chosen to be smaller than the input image data and element-wise multiplication followed by summation (dot product) is carried on between the convolution filter and the filter-sized patch of the input. The CNN always uses a filter smaller than the input image because, this filter can be used multiple times on the input data, and it can be moved across the entire input image at different times also. This can happen with data overlap or without overlap (top to bottom and left to right). Each of the convolution filter is designed in such a manner that they can detect a specific type of feature from the input image. As the filter is moved across the image it starts detecting the specific feature it is supposed to [89, 99, 100]. For more efficient and high-quality feature extraction, the filter can be passed through the input image multiple times. The result that is obtained after the filter performs the function of feature extraction is called the feature map, which is a two-dimensional array of the filtered input. Once the feature map is generated, it is passed through a nonlinearity like ReLU. Many numbers of convolution filters can be used on the same input image to extract and identify different types of feature maps. The more the feature maps for a given image, more accurate is the performance of the neural networks.

All CNN consists of at least three different types of layers, an input layer, an output layer and several hidden layers. Initial versions of CNNs were shallow (one input and one output layer with a hidden layer). Deep learning networks is classified as anything that has a minimum of three layers. In deep learning, each passing layer trains on features that are generated as the output from the previous layers and as the number of layers progresses, they are able to recognize more complex features which is known as feature hierarchy. This feature of deep-learning enables the networks to handle very large, high-dimensional data sets with a multitude of parameters. Some of the most commonly used layers in a CNN are discussed below. After the input layer, the very first layer of a CNN is the convolutional layer. The convolution layer performs the convolution operation on the input data. Once the convolution layer extracts the features to generate the feature map, it is passed through a Rectified linear units (ReLu), which are activation functions. Leaky ReLus allow a small, non-zero gradient when the unit is not active. Once the data passes through the ReLu, it goes through the pooling layers. The pooling layers combines the outputs from the neuron clusters at each layer into a single neuron in the next layer. Max Pooling is one of

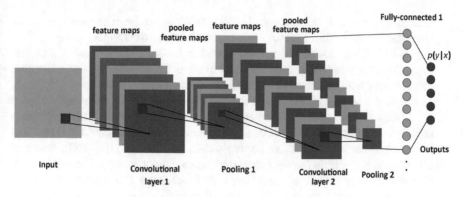

Fig. 3 Representation of a typical CNN, consisting of convolutional, pooling, and fully-connected layers. Reprinted with permission from Ref. [89]

the examples of pooling layers, this layer chooses and uses the maximum value from each cluster of neurons in the previous layer and sends only the maximum values of the cluster to the next layer. Next, upsampling layers performs an upsampling using nearest neighbor, linear, bi-linear and tri-linear interpolation. Finally, fully connected layers connect every neuron of one layer to every other neuron in another layer. Model of a traditional CNN is shown in Fig. 3.

Feature extraction is a highly time-consuming task and it can be very tedious to perform especially by humans. Deep-learning networks can perform feature extraction with very minimal or without human intervention. This comes in very handy in the medical community, especially in the field like radiology, where there are always limited personnel to scan all the diagnostic images of a patient. This is specifically important because diagnosis from the images is very crucial for the treatment planning and monitoring of a patient.

2.3 Learning by Neural Networks

There are two different ways in which neural networks learns, which are supervised learning and unsupervised learning.

1. Supervised learning: Supervised learning is a machine learning method and is widely used, in this method a large dataset is required with corresponding labels. A supervised learning algorithm is trained using ground truth images which is a set of labelled data. Therefore, the algorithm attempts to reproduce the label and calculates a loss function that measures the error between the output from the machine and the label. The algorithm then considers the error value, which is then further factored to modify the internal adjustable parameters (weights), to further minimize this error and improve the efficiency of the model. The performance rate of any machine learning algorithm is based on its how accurately

it handles previously unseen data [101–105]. This can be evaluated with some data that the algorithm has not been exposed during the training process. This data is called test set. The algorithm is said to be more generalized if it is able to predict closer to the ground truth on unseen data. In contrast, if an algorithm can perform with accuracy on previously exposed data but perform very badly on the new test data, it shows that the algorithm only tries to memorize known solutions without any abstraction and does not generalize well. This problem is called overfitting. Overfitting is one of the most frequently encountered problem in machine learning and it can be avoided in many ways [106–108]. Some of them are to either train the algorithm with more data or data augmentation or by completely using a different neural network. Choosing a different neural network works best when the current network cannot handle the complexity of the data.

2. Unsupervised learning: Unsupervised learning is a method in which the algorithms train themselves automatically as they are trained on unlabeled data. In this method each node in every layer of the network tries to learn the features automatically by repeatedly trying to reconstruct the data from the input set, it tries to minimize the variation between the guesses of the network and the probability distribution of the input data itself [109–111]. Also, in this process, the neural networks learn to identify similarities and relationships between certain relevant features and optimal results. The networks try to find connections between feature signals and what it represents, whether it be a full reconstruction, or with labeled data [112, 113]. A deep-learning network can first be trained on labeled data can then be applied to unlabeled data as well. This way, it gives the network access to much more input than just the machine-learning nets. The key to the performance of any deep learning model is data, the greater the amount of data a network trains on, the network's probability of accuracy improves likewise. The output layer of any deep-learning network is either a softmax or logistic layer, the classifier assigns a probability to a specific outcome or label, this type of network is predictive in nature. Neural network follows a corrective feedback loop, that rewards the weights which support the correct guesses, and punishing weights that leads to error. The network tests extensively which combination of input is significant as it tries to reduce error.

Gradient descent is a very commonly used function for optimization. It further adjusts the weights according to the error values obtained. The slope of a neural networks depicts the relationship between the allotted weights and the error function. As a neural network continues to learn, it gradually starts adjusting many weights so that they can map signal to meaning more accurately [114–117]. The relationship between each weight of the network and the error is a derivative, every weight of a network is just one factor in whole deep network which involves multitude of transforms; the signal of each weight passes through activations and gets summed over several layers. The basic crux of a deep learning network is to constantly adjust and modify its weights in response to the error calculated in each iteration. This continues to happen until the error can't be reduced any more. The activation function layer of a network determines the possible output from a given node, based on the input data [100, 118]. The activation function is set in the layer level and gets applied to all neurons present in that layer. Every output node produces binary output (0 or 1) as the two possible outcomes, as it determines whether an input variable either deserves a label or it does not. Neural networks working on labeled data only produces a binary output, as the input they receive is often continuous. That is, the signals that the network receives as input will be over a range of values and include any number of metrics, depending on the problem it is attempting to solve. The mechanism that is used for the conversion of continuous signals into binary output is known as logistic regression. It calculates the probability that a set of inputs match the label. For continuous inputs to be expressed as probabilities, they must output positive results, since there is no such thing as a negative probability.

(a) Training the network: The neural network starts by randomly initializing weights to the model and calculates the output from the first image. The obtained output image is compared with the ground truth with the help of a loss function. The loss is then back propagated to update and modify the weights of the network. This process is performed multiple times to optimize the performance of the networks.

(b) Testing of the network: After training, a network, testing will be done to evaluate the networks performance. In the testing data no labels are used. The network with the previously trained weights is evaluated on new data that was not encountered by the network previously. These weights determine the prediction of the network.

The Cost functions minimum is searched and an easy way to find a minimum is using gradient descent. Hence, the cost function needs to be differentiable. To perform the adjustment of the weights that are calculated by the gradient descent, the machine learning algorithm computes a gradient vector that, for each weight, which gives an indication on the error amount would increase or decrease if the weight were modified (increased or decreased) by a small amount. By updating the weights step-by-step the cost functions minimum is approached.

The learning rate parameter is introduced to improve the working efficiency of the algorithm. The learning rate is multiplied to the cost function, which thereby decides the step size for each iteration. If for a given algorithm the learning rate is chosen too low, then the algorithm takes a long time to converge to the minima, in contrary if the learning rate is chosen too large, then there is a possibility for the algorithm to overshoot the minima. In the state-of-the-art deep learning algorithms, the learning rate is made flexible which adapts continuously [119]. A lot of work has been done to optimize gradient descent algorithms in recent years [118]. One state-of-the-art algorithm called Adam [120] that is based on adaptive estimates of lower order moments, it is made with high computational efficiency and can deal with large datasets with ease [120].

2.4 Backpropagation

Backpropagation is a methodology in which the contribution of each neuron towards error is calculated after the completion of processing of a batch of data. Using backpropagation, after the calculation of loss function and propagation of the error backwards, the weight of the neurons can be modified. The recently developed networks using back propagation are faster than earlier approaches, thus enabling the neural networks to be used for solving problems which were previously unsolvable. Backpropagation based algorithm is the most commonly used optimization approach in neural networks [121]. Using backpropagation, the networks weights are continuously adapted and thereby facilitating the network to learn the best parameters [121]. Back propagation-based algorithms are being used extensively in medical image processing.

2.5 Improving the Networks Performance

As the applications of the neural networks keeps growing, it becomes very important to constantly improve the performance of the network to optimize their functions better and to improve their efficiency. Some of the ways in which the neural networks performance be improved are as follows.

1. Batch computing: Batch computing is used to improve the computational performance of a neural network. A group of data is consolidated and grouped to form a batch which helps in improving the computational performance as most of the libraries are optimized better for array computing [122].
2. Data Augmentation: Data augmentation is commonly used to increase the amount of data on which the algorithms arc bcing trained on. When there is an increase in the amount of data on which the algorithms learn it leads to an increase in prediction accuracy of the algorithm. Therefore, data augmentation can improve an algorithms performance [123].
3. GPU Computing: GPU computing is a technique used to increase computational speed of processing by using a graphics processing unit (GPU), this unit traditionally handles only computations for computer graphics but it can also be used to compute tasks that are normally carried out using the central processing unit (CPU). The GPU is usually designed with more cores than a CPU and are capable of processing far more graphical data per second than the handling capacity of a CPU. Thus, if the data is transferred to the GPU instead of the CPU and processed there, it can lead to a significant speedup of the computing time.

2.6 *Evaluation Indices*

For the quantitative evaluation of a neural networks based on their performance on the test set, some of the most commonly used evaluation parameters includes signal-to-noise-ratio (SNR), peak-signal-to-noise-ratio (PSNR) and structural similarity index (SSIM). These are calculated for each of the test set data for comparison.

SNR

Signal to noise ratio (SNR) can be defined as the ratio of peak signal intensity from the sample to standard deviation of the background intensities represented in decibels. It is based on absolute signal strength and noise statistics of a given image. SNR can be mathematically represented as follows:

$$SNR = 20 \log_{10}(\mu I / \sigma b)$$

where, μI and σb represent the peak signal amplitude of the target area and the standard deviation of the background, respectively.

PSNR

The term peak signal-to-noise ratio (PSNR) can be defined as the ratio between the maximum possible value of a signal in a given image and the power of distorting noise which affects the image quality. Because a variety of signals have a very wide dynamic range, (ratio between the largest and smallest possible values of a changeable quantity) the PSNR is usually expressed in terms of the logarithmic decibel scale.

The mathematical representation of the PSNR is as follows:

$$PSNR = 20 \log_{10} \left(\frac{MAX_f}{\sqrt{MSE}} \right)$$

The PSNR is a conventional measurement of the image quality in decibels (dB) based on the mean square differences between the estimated and reference images as:

where the MSE (Mean Squared Error) is given by:

$$MSE = \frac{1}{mn} \sum_{0}^{m-1} \sum_{0}^{n-1} \| f(i,j) - g(i,j) \|^2$$

Here, f represents the matrix data from the original image, g represents the matrix data from the degraded image, m represents the numbers of rows of pixels of the images and i represents the index of that row n represents the number of columns of pixels of the image and j represents the index of that column and MAXf is the maximum signal value that exists in the ground truth image.

SSIM

The Structural Similarity Index (SSIM) metric that is used to quantify the image quality degradation which can be caused due to the image processing tools like as data compression or by loss due to data transmission. It is a reference metric which requires two images from the same image capture namely the reference image and the processed image. The processed image is usually the compressed version.

SSIM measures the perceived quality of a digital image; a higher SSIM (in a scale of 1.0) indicates a better representation of an estimated image in terms of perception.

2.7 Training Data

Training data is very important for deep learning for photoacoustic imaging. Generating training data and ground truth images for training algorithms is very crucial as this data determines the efficiency of the model. Also, the number of training data available, the quality of the images and the variety of images in the training data pool helps the neural network model to learn more effectively and be able to handle any kind of images that it might come across in real-time scenario. For different imaging modalities the training data can be acquired in different ways. For the most commonly used clinical imaging techniques like MRI, CT etc. there are multiple open source libraries with thousands of data. We can choose the dataset which is most appropriate for our application and train the neural network with the dataset. However, in case of certain applications where relevant data set might not be available online or for imaging modalities that are not so commonly used for clinical imaging, training data needs to be custom generated. This can be done in two ways, the first is to use different imaging systems to acquire high quality images for the specific application, these types of images are more realistic, and it is easy to get a good ground truth image. But it can be very expensive to acquire enough number of training data to train model and it can also be very time consuming. The other method to generate training data is through simulation models. Simulation is a cost-effective way to generate images and the ideal case scenario can be obtained through simulation images. We can also add any type of artifact on the image to help the model to perform better for a specific application. Simulation images make good training data, but the shortcoming with this method is that it can sometimes be very far from reality that when the model comes across a real image, it may not be trained to work on the image.

2.8 Neural Networks for Medical Imaging

Now that what neural networks are and what they do is clear let's explore its applications especially in the field of medical imaging [93, 124–128]. Neural networks are starting to have a huge impact on different aspects of medical imaging like segmentation, detection, classification etc. especially in the field of radiology. Classification is one of the most important tasks in radiology, it typically consists of predicting some target class like a lesion category or condition in the patient from an image or region of interest in a dataset [99, 104, 129, 130]. This task is used for a wide range of applications, right from determining the presence or absence of a disease to identifying the type of malignancy. Deep learning is very frequently used for the segmentation task which can be defined as the identification of pixels or voxels composing an organ or structure of interest [88, 131, 132]. For a machine learning algorithm, it can be considered as a pixel-level classification task, where the end goal is to determine whether a given pixel belongs to the background or to a target class (e.g., prostate, liver, lesions). For this, from image classification tasks, image masks can be used to perform various quantitative analyses such as virtual surgery planning, radiation therapy planning, or quantitative lesion follow-up. Detection is another common task for the deep learning, it can be used to identify focal lesions such as lung nodules, hepatic lesions, or colon polyps. This can be used as a screening technique before a radiologist can take a look at it [105, 133]. Detection is a subset of the classification task however, classification only aims to predict labels, detection tasks aim to predict the location of potential lesions, often in the form of points, regions, or bounding boxes of interest. All of the three tasks are extremely useful for diagnosis, treatment planning of a disease condition. The labeling of the images varies based on the task it performs. Classification of images requires image labeling. Detection of images requires marking the region of interest, such as a boxplot. Segmentation of images requires pixel-wise delineation of the desired object.

2.8.1 Deep Learning for Radiology

Among the various clinical imaging techniques, radiology is one place where deep learning is being explore more extensively. Radiology is one of the most important and widely used clinical imaging tool for diagnosis of many diseases and clinicians depend on it every day. Therefore, using deep learning in radiology can have more impact in the clinics than any other imaging technique [134–140]. In this section we will see how deep learning has come to play in the hospitals.

CheXNeXt is a convolutional neural network that was developed by a team of researchers at Stanford, it has the potential to concurrently detect up to 14 different pathologies, including pneumonia, pleural effusion, pulmonary masses, and nodules in frontal-view chest radiographs. The CheXNeXt CNN was trained and validated internally on dataset of ChestX-ray8 images [141]. A set of 420 images were used for training and kept for validation purpose including images of all the original pathology labels. 3 board-certified cardiothoracic specialist radiologists voted on the images which served as reference standard. The performance of the CheXNeXt's was compared with the performance of 9 radiologists using the area under the receiver operating characteristic curve (AUC) on the validation dataset. It was observed that the performance of CheXNeXt was similar to the level of radiologists on 11 different pathologies but was not able to the achieve performance level of radiologist on 3 pathologies. The radiologists significantly higher performance on three different pathologies (cardiomegaly, emphysema, and hiatal hernia). CheXNeXt has performed significantly better than radiologists in detecting atelectasis. For the other 10 pathologies there was no statistical significance in differences between radiologists and the CheXNeXt. For the radiologists, the average time to interpret the validation set (420 images) was significantly longer than CheXNeXt. Radiologists took about 240 min but the CheXNeXt took only 1.5 min. One of the drawbacks in this study was that both the CNN and the radiologists were not given any patient history. Another limitation is that all the data acquired for this study was from a single institution only. So, the performance of the algorithm may be biased and limited by it. Figure 4 shows the performance of the algorithm for various disease models in comparison to a doctor. Figure 5 shows the predictions of the algorithm for disease conditions.

This is one example of how a CNN can aid the physicians in the field of radiology. Similarly, different types of algorithms are attempting to solve different types of problems in radiology.

2.8.2 Deep Learning for Ultrasound Imaging

Ultrasound imaging is a commonly used imaging technique in the clinics for patient diagnosis. There are many different types of artifacts present in ultrasound imaging, which needs efficient methods for artifact reduction or elimination. Deep learning is being explored for image classification, segmentation and artifact removal problems in ultrasound [142–145]. One such example of a classification problem of thyroid nodules is discussed below.

In the ultrasound images, thyroid nodules appear very heterogeneous in nature with unclear boundaries with various internal components, this makes it very difficult for physicians to discriminate between the benign thyroid nodules and malignant ones. A study was proposed for the diagnosis of thyroid nodules using a hybrid method. The model was developed using a combination of two different pre-trained convolutional neural networks. The two CNNs have different convolutional layers and fully connected layers. Initially, the two which are pretrained with the ImageNet database are trained individually. After individual training the two neural networks, the feature maps are learned by the trained convolutional filters, pooling and normalization operations of the two CNNs. After this the two obtained feature maps are fused and a softmax classifier is used to diagnose (classify) the thyroid nodules. This method was validated on 15,000 ultrasound images obtained from two different hospitals.

For CNN1 and CNN2, a single testing was performed on the training step. A multi-view was adapted to improve the performance of the network. For the input of the trained CNNs 256 views of the thyroid nodule images were cropped and was sampled randomly and used. The output was the average of the result of 256 views.

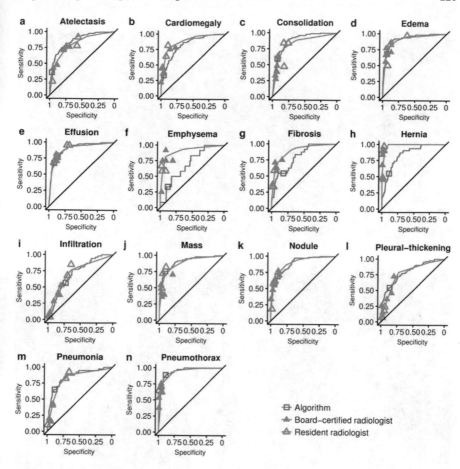

Fig. 4 ROC curves of radiologists and algorithm for each pathology on the validation set. Each plot illustrates the ROC curve of the deep learning algorithm (purple) and practicing radiologists (green) on the validation set, Individual radiologist (specificity, sensitivity) points are also plotted. The ROC curve of the algorithm is generated by varying the discrimination threshold. Reprinted with permission from Ref. [141]

The two fused pretrained CNN used the fused feature maps that was generated by the two CNNs in multi-view testing as shown in Fig. 6. The softmax layer was trained for thyroid nodule classification. To compare the performance of the CNNs a well-established classification method called SVM was also implemented. The SVM with radial basis function (RBF) kernel was used for experiments [146].

The accuracy of the classification algorithm was tested and represented graphically in Fig. 7; this graph compares the classification accuracy of different methods used in this study. It can be noted from the results that CNN based methods outperform the various other methods significantly in the classification of thyroid nodules. Especially, the combination of CNN1 and CNN2 achieved a classification accuracy

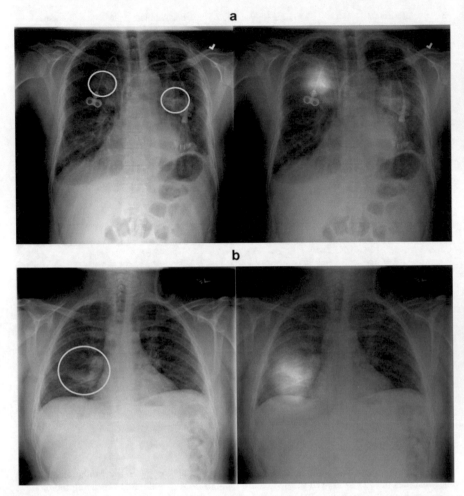

Fig. 5 In the normal chest radiograph images (left), the pink arrows and circles highlight the locations of the abnormalities. **a** Frontal chest radiograph (left) demonstrates 2 upper-lobe pulmonary masses in a patient with both right- and left-sided central venous catheter. The algorithm correctly classified and localized both masses as indicated by the heat maps. **b** Frontal chest radiograph demonstrates airspace opacity in the right lower lobe consistent with pneumonia. The algorithm correctly classified and localized the abnormality. Reprinted with permission from Ref. [141]

as 83.02% ± 0.72%, sensitivity as 82.41% ± 1.35%, and specificity as 84.96% ± 1.85%. These demonstrate the potential clinical applications of this method.

Photoacoustic imaging is very similar to ultrasound imaging and the techniques from ultrasound can be easily adapted for photoacoustic with minimal modifications.

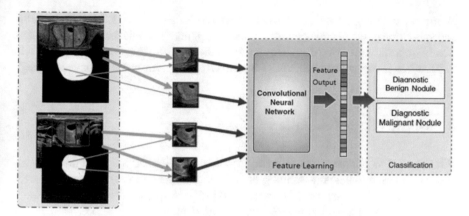

Fig. 6 An overview testing of CNNs. This CNN based approach first extract multiple nodule patches to capture the wide range of nodule variability from 2D ultrasound images. The obtained patches are then fed into the networks simultaneously to compute discriminative features. Finally, a softmax is applied to label the input nodule. Reprinted with permission from Ref. [146]

Fig. 7 Box plots of performance measures for classifying between benign and malignant thyroid nodules. In each box plot, the center red line is the median and the edges of the box are the 25th and 75th percentiles, the whiskers extend to the most extreme data points not considered outliers, and outliers are plotted individually. Reprinted with permission from Ref. [146]

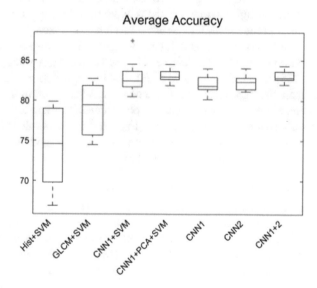

2.8.3 Deep Learning for Photoacoustic Imaging

Photoacoustic imaging is not being used in clinics yet. It aspires to become a clinical tool for diagnosis. Image quality and easy interpretability is very crucial for that to happen. Improvement in reconstruction and post processing of images is just one part of it. As evident from the other types of clinical imaging modalities, deep learning can be used for the improvement of photoacoustic imaging for artifact removal and reduction [74, 147–154]. One of the major limitations for using deep learning for photoacoustic imaging is that there are not much clinically recorded data for training and validation of the neural networks. Therefore generating data through other means is very important for using deep learning for photoacoustic imaging. Using data from simulations is one possible solution for generating data for photoacoustic imaging.

3 Monte Carlo Simulation

With respect to photoacoustic imaging monte carlo simulations for light propagation can be used to generate training data. Using monte carlo simulations, the light absorbance by the sample can be calculated [155–158]. The absorbance is usually directly proportional to the photoacoustic signal intensity. Therefore, we can get an idea of how the photoacoustic image will look like. In monte carlo simulation a sample object (of desired shape and size) is simulated in medium like tissue or water, with the properties of the tissue specified. The absorption coefficient and the transmission coefficient of the sample and tissue are predetermined from literature and the number of layers are also mentioned [157–161]. Photon packet is launched from the light source and the movement of the photon is tracked as it propagates through the tissue. It loses weight as it passes through each layer where it either gets absorbed or transmitted. It loses weight as it moves across the tissue and some of the photons might hit the sample of interest and can get reflected, transmitted or absorbed [155, 162, 163]. To obtain a high-resolution image millions of photons are launched from the light source simultaneously and at the end of it, the light absorbance by the sample is calculated. This is reconstructed to form the absorbance map. These are equivalent to the photoacoustic images that are obtained from the imaging systems.

While building a MC simulation model, a large number of photons are modelled to propagate through the simulation medium (tissue). While passing through any medium photons undergo either reflection or refraction or absorption or scattering or a combination of these. The path that the photon takes is determined by the optical properties of the medium such as refractive index (n), absorption coefficient (μ_a), scattering coefficient (μ_s), and scattering anisotropy (g). Absorption coefficient (μ_a) of a sample can be defined as the probability absorption by the photon in the medium per unit (infinitesimal) path length. This physical quantity is measured by Beer's law. Similarly, scattering coefficient (μ_s) can be defined as the probability scattering of light in a medium per unit (infinitesimal) path length. Scattering anisotropy (g) is defined as the mean of the cosine of the scattering angle. In biological tissues, the typical values for the various optical parameters are as follows, $\mu_a = \sim 0.1$ cm^{-1}, $\mu_s = \sim 100$ cm^{-1}, g = 0.9, and n = 1.4. The flow chart for MC for an embedded sphere as object is shown in Fig. 8.

The images generated from the monte carlo simulation of light propagation through tissues can be used to train the neural networks. An example of absorbance maps generated from monte carlo simulation for a spherical object is shown in Fig. 9. For training networks on artifact detection and correction, monte carlo simulations can generate artifacts on the images as well. Many different types of artifacts can

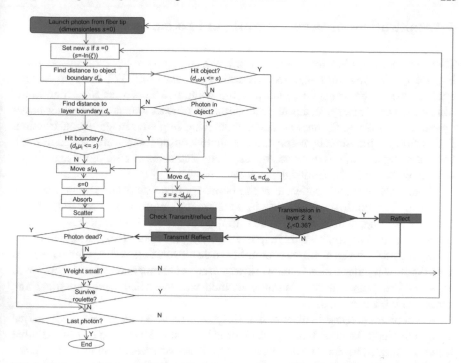

Fig. 8 Flow chart of Monte Carlo with embedded sphere (MCES). Reprinted with permission from Ref. [164]

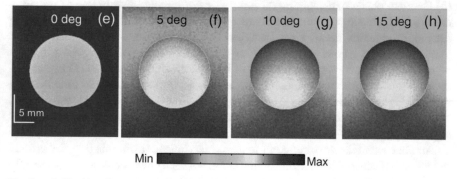

Fig. 9 e–h The absorbance maps of sphere at depth 0.5 cm for illumination angles 0°, 5°, 10°, and 15°. Reprinted with permission from Ref. [164]

be generated in the images to train the neural networks model appropriately. Images from the simulation can also be used to test a neural network. The major advantage of using monte carlo simulation for photoacoustic imaging is that a large amount of training data can be obtained very easily, and the data can be customized based on the problem.

4 Applications of Deep Learning in Photoacoustic Imaging

There are two ways in which deep learning can be applied for photoacoustic imaging. First, is during the image reconstruction process itself. When the traditional photoacoustic reconstruction techniques are being used, deep learning algorithms can be used on the raw data during reconstruction process to make the images better by reducing artifacts. Second, as a post-processing step after image reconstruction. In this case the traditionally reconstructed photoacoustic images are passed through deep learning algorithms to reduce the artifacts. Examples of both of these methods will be discussed in the following sections.

Photoacoustic signals collected at the boundary of a tissue surface and are most often band limited. In a recent work, in an attempt to improve the bandwidth of the photoacoustic signal detected from the sample, a deep neural network was proposed. Using the neural network would help in improving the quantitative accuracy of the reconstructed PA images. A least square-based deconvolution method which involves the Tikhonov regularization framework was used for comparison with the proposed network. The proposed deep learning method was evaluated with numerical and experimental data as well.

The network proposed contains five fully connected layers, out of the five, one layer is the input layer and one other layer is the output layer. The rest of the three layers are hidden layers. The architecture of this network very similar to that of the decoder network. Three different numerical phantoms (different from the training data) were used to evaluate the performance of the network: (a) a blood vessel network is frequently used as PA numerical phantom for imaging blood vasculature, (b) Derenzo phantom containing different sizes of circular distribution of pressure, and (c) PAT phantom to simulate sharp edges. The bandwidth enhancement using the proposed neural network can be evidently observed from the images as shown in Fig. 10. Here, frequency response of the signal calculated using the proposed neural networks was very similar to full bandwidth signal response. These results indicate that the proposed method using neural networks are capable of enhancing detected PA signal's bandwidth [165]. This further improves the contrast recovery and quality of reconstructed PA images without increasing any computational complexity significantly.

Another example of using deep learning is photoacoustic imaging for artifact reduction is discussed here. In one of the recent works, a novel technique with the help of a deep learning neural network which are trained layer-by-layer to reconstruct 3D photoacoustic images with high resolution was proposed. This network incorporates the physical model into the reconstruction procedure to iteratively reduce artefacts [74]. In this method a U-Net was used to post process data from direct reconstruction, the limitation of using neural networks for post processing is that the result from the neural networks are highly dependent on the quality of the initially reconstructed photoacoustic image. The U-Net is one of the commonly used deep neural network for image denoising, it is a state-of-the-art deep learning technique. The U-Net consists of equal number of contracting and expansive layers. In this network, the number of

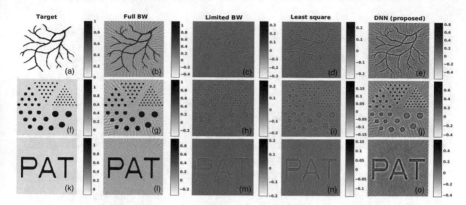

Fig. 10 Numerical phantoms used for evaluation: **a** blood vessel network, **f** Derenzo phantom, and **k** PAT phantom. Reconstructed (backprojected) initial pressure images with 100 detectors using **b, g, l** full BW signal, **c, h, m** limited BW signal, **d, i, n** predicted signal from least square deconvolution method, and **e, j, o** predicted signal from the proposed DNN. The SNR of the data is at 40 dB. Reprinted with permission from Ref. [165]

feature channels is the same in the first and last layer, similarly, the number of feature channels in the second layer is two times the first layer and the same is true for the second-last layer. The resolution is being halved in of each contraction step and gets doubled in each of the expansion step. Every single layer in the neural network has a large number of feature channels, this aspect of it allows the propagation of context information to higher resolution layers in the network. Because of this the network assumes a symmetry and providing a U-shape architecture. For the down sampling, the contracting layers of the network consists of unpadded convolutions which are followed by rectified linear units and a pooling operation. Number of feature channels gets doubled in each of the down sampling step [166]. The expansive layers consist of up sampling feature map followed by an up-convolution where the number of feature channels gets halved, each of this is followed by a rectified linear unit. Owing to the high complexity of the photoacoustic forward operator, the training and computation of the gradient information was separated. This network used data from a set of segmented vessels from lung computed tomography scans for training and testing. The network was then applied to in-vivo photoacoustic data measurement.

Use of directly reconstructed images on the neural networks to remove artifacts is a valid approach in many applications, specifically if the goal is to achieve fast and real-time reconstructions. This approach only needs an initial direct reconstruction and one application of the trained network. In the case of a full-view data, this is a promising approach, but it has been demonstrated that even with limited-view images this technique performs very well. A comparison of DGD and U-Net for simulated data is shown in Fig. 11 (top row). The final image is cleaned up and many vessels are properly reconstructed although, some of the minor details are missing in the image and could not be recovered from the initially reconstructed data. The difference to the true target is also shown in Fig. 11 (bottom row). The

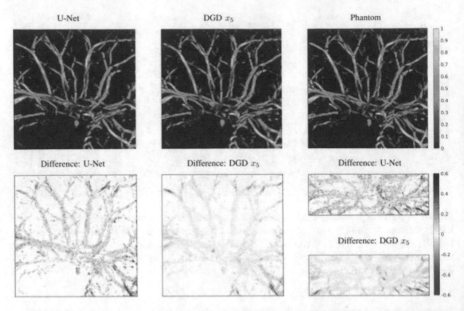

Fig. 11 Comparison of reconstructions for a test image from the segmented CT data. Left: top and bottom shows the result by applying U-Net to the initialization x0 and the difference to the phantom, maximal value of difference is 0.6012. Middle: shows the result of the DGD after 5 iterations and the difference to the phantom, maximal value of difference is 0.4081. Right bottom: difference images as side projections for the results of DGD and U-Net. Reprinted with permission from Ref. [74]

differences are most pronounced in the outer parts of the domain as a consequence of the limited view geometry. In comparison the reconstruction by DGD has a much smaller overall error, but this is especially true in the center of the domain. The maximal error of the U-net reconstruction is 0.6012 (on the scale of [0, 1]) and of the DGD reconstruction 0.4081 as can be observed form Fig. 12. In conclusion we can say that the U-net architecture performs very well and is even capable of removing

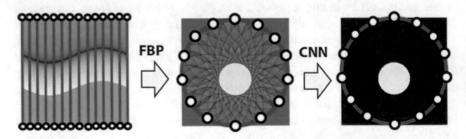

Fig. 12 Illustration of the proposed network for PAT image reconstruction. In the first step, the FBP algorithm is applied to the sparse data. In a second step, a deep CNN is applied to the intermediate reconstruction which outputs an almost artefact-free image. Reprinted with permission from Ref. [150]

some limited-view artifacts but is ultimately limited by the information contained in the initial reconstruction.

In another work, deep learning approach was used for photoacoustic imaging from sparse data. In this approach, linear reconstruction algorithm was first applied to the sparsely sampled data and the results were further applied to a CNN with weights adjusted based on the training data set. Evaluation of the neural networks is non-iterative process and it takes similar numerical effort as a traditional FBP algorithm for photoacoustic imaging. This approach consists of two steps: In the first step, a linear image reconstruction algorithm was applied to the photoacoustic images, this method provides an approximate result of the original sample including under-sampling artifacts. In the next step, a deep CNN is applied for mapping the intermediate reconstruction to form an artifact-free end image.

The neural network is first trained using simulated ellipse shaped phantoms samples. 1000 pairs of images were generated and used for training. One part of the training data includes pressure data without any noise and the second part of the data random noise was introduced to the simulated pressure data. The neural network was evaluated on similar simulated images of ellipse samples which was not introduced to the network during training. The network performed well by eliminating all the artifacts from the test images. The network was further tested on Shepp-Logan type phantoms and as expected the network was not able to remove all the artifacts from the image as it was not trained on this data [167]. Hence, additional CNNs were trained on 1000 randomly generated ellipse phantoms and 1000 randomly generated Shepp–Logan type phantoms. The newly retrained network was once again tested on the Shepp-Logan type phantoms. It is evident from the images in Fig. 13 that when the neural network is trained with appropriate and correct training data, the performance of the neural networks improves significantly.

4.1   Deep Learning for LED Based Photoacoustic Imaging

As discussed earlier, the image quality of the LED based photoacoustic imaging system is not great. To improve the image resolution, improve artifact removal and reduce the averaging for these images would greatly help in the clinical use of this system. Deep learning can be applied to the photoacoustic images from LED systems to improve the overall system efficiency. One of the recent works uses deep neural networks-based performance improvement of the system for improving the quality of the images and also to reduce the average scanning time (averaging) of LED-based PA images. The proposed architecture of the neural networks consists of two important components; the first is a CNN which is used for the spatial feature extraction, and the second one is the recurrent neural networks (RNN) to leverage the temporal information from the PA images. RNN is a form of neural networks in which the output of each step is fed as input to the next step. It varies from the traditional neural networks in the sense that, in CNNs the input and output through different steps are independent of each other. The most unique and important feature of the

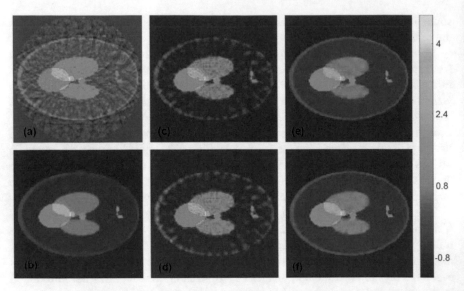

Fig. 13 Reconstruction results for a Shepp–Logan type phantom from data with 2% Gaussian noise added. **a** FBP reconstruction; **b** reconstruction using TV minimization; **c** proposed CNN using wrong training data without noise added; **d** proposed CNN using wrong training data with noise added; **e** proposed CNN using appropriate training data without noise added; **f** proposed CNN using appropriate training data with noise added. Reprinted with permission from Ref. [150]

RNN is the hidden state, which helps the network to remember the information about a sequence. The neural networks are built based on the state-of-the-art algorithm of densenet-based architecture which uses a series of skip-connections to enhance the image content. For the RNN component, convolutional variant of short-long-term-memory was used to make use of the temporal dependencies in a given PA image sequence. Skip connections was introduced in the both the networks, CNN and RNN for effective feature propagation and elimination of vanishing gradient.

Figure 14a shows the densenet-based CNN architecture. The neural network accepts a low-quality PA image as input and as output generates high quality PA image. The number of feature maps are shown in Fig. 14. The architecture of the network consists of three dense blocks, where each dense block consists of two convolutional layers followed by a ReLU. One of the major advantages of using the dense convolutional layer is that it utilizes all the generated features from previous layers as inputs through skip connections. This enables the propagation of features more effectively through the network which leads to the elimination of the vanishing gradient problem. Finally, to obtain the output image, all the features from the dense blocks are concatenated, a single convolution with one feature map is performed at the end.

In order to train the network experimental study was done using the LED based photoacoustic system and evaluate the performance. The experiment including

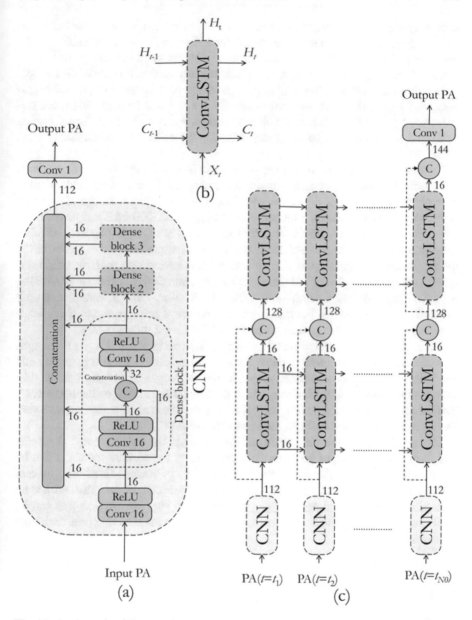

Fig. 14 A schematic of the neural network. **a** The densenet-based CNN architecture to improve the quality of a single PA image. **b** A schematic of ConvLSTM cell. In addition to current input X_t, it exploits previous hidden and cell states to generate current states. **c** The architecture that integrates CNN and ConvLSTM together to extract the spatial features and the temporal dependencies, respectively. Reprinted with permission from Ref. [168]

acquiring images from phantoms and also in vivo human fingers. For the phantom experiments, PA signal was acquired for a time period of 11 s leading to the generation of 11,000 frames of pre-beamformed signals. To obtain a noise free image through averaging having a steady set up without any motion is critical. This is possible with phantoms whereas, maintaining a steady position for in-vivo imaging is very challenging therefore was only done for 5 s. After data acquisition, PA signals were averaged over certain number of frames, followed by beamforming using delay-and-sum technique, subsequently detecting the envelope to reconstruct the PA image.

Two different types of phantoms were used in this study, wire and magnetic nanoparticle phantoms because of their high optical absorption coefficients. For the wire phantom, a total of 62 sets of PA data from 62 different image planes was acquired, and each of the data set consists of a total of 11,000 frames. The phantom was built with fives cylindrical tubes that are placed at multiple depths. The tubes were varied in concentration and depth to perform a comprehensive evaluation of the performance of the neural networks. This helps in evaluating the sensitivity of the system at various depths. Tubes 1–3 were of same concentration but placed in decreasing depths. At the maximum depth along with tube 3, tubes 4 and 5 were placed with decreasing concentrations. For the phantom experiment suing nanoparticle tubes, a total of 10 sets of PA data was acquired from 10 different image planes.

For effective training of the neural networks, different qualities of input PA images was used. As stated previously, greater the averaging, better is the image quality and resolution. This aspect was made use of to obtain images of varying quality. The averaging works well with phantom data as the three is not motion artifacts involved. The number of frames to be averaged (N) was chosen from a range starting with very low value and increased to the highest possible value (11,000).

Figure 15 depicts the photoacoustic images from the two different phantoms and in-vivo human finger here. The performance of the various networks can be clearly observed from the difference in the quality of images from the different neural networks for all the different samples and at various depths. This work is an example of how a neural network can be trained on very simple data that can be easily acquired to improve the image quality and reduce the scanning time for image acquisition.

5   Limitations of Deep Learning

Deep learning has been very successful in the recent times for a variety of applications. In spite of its success, there are many limitations associated with the application of the technique. Firstly, deep learning is not the best machine learning technique for all the different types of data analysis problems. For various issues in which the data is already well structured or if optimal features are well-defined, instead of deep learning a lot of other simple machine learning methods like logistic regression, support vector machines, and random forests can be applied to solve it. It will be much

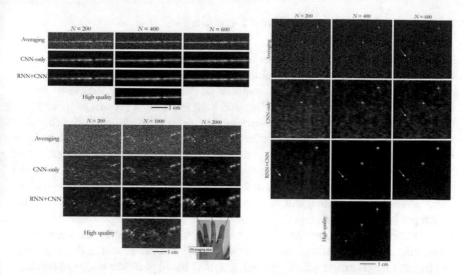

Fig. 15 Qualitative comparison of our method with the simple averaging and CNN-only techniques for a wire phantom, in vivo example. The in vivo data consists of proper digital arteries of three fingers of a volunteer. Example effect of depth on the PA image quality on nanoparticles. Reprinted with permission from Ref. [168]

easier to apply and are also usually more effective with such datasets. CNNs have become very dominant in the field of computer vision, there are some limitations there as well. One of the most significant limitation is that deep learning is a technology that requires a large amount of data; for the network to learn the weights from scratch for a large network requires a huge number of labeled examples to achieve accurate classification. Deep learning scales very well with large datasets. Therefore, computing resources, time needed for training a deep learning model is very high. Also, obtaining so much of labelled training data is very difficult.

Transfer learning is receiving more research for moving to an effective way of reducing the data requirements. In recent transfer learning approaches, it reuses weights from networks trained on ImageNet (a labeled collection of low-resolution 2D color images). For most applications in radiology, higher-resolution volumetric images are required, for which pretrained networks are not yet available. As a result, creating a large labelled medical image library is really important step for further progress in applying deep learning, which is not easy due to cost, privacy etc. Also, with more future breakthroughs in deep learning, data requirements can be significantly reduced for training of deep learning systems.

6 Future Directions for Deep Learning

Deep learning models has shown expert-level or better performance at few tasks. Deep learning algorithms are capable of extracting or identifying more features than humans. Data availability and curation of data into repositories is becoming more organized now for better handling and usage of data. This will further help in developing better models for deep learning as there will be more availability of a variety of training data including different scenarios. In the recent past there have been approaches where they use data from one imaging modality to train a network for better performance on another imaging modality. This will help in boosting the performance of the neural networks as they train on better ground truth images. The importance of deep learning will keep increasing in the days to come in the hospitals.

For photoacoustic imaging, deep learning will have a more important role to play. Deep learning for photoacoustic is not much explored till now, so the potential of it has fully not been understood. Some of the major areas in which deep learning can be used for photoacoustic in general and LED based photoacoustic systems also includes better and faster reconstruction algorithms, reduction of artifacts in images, reduction in averaging to produce a high resolution image, decreasing the data acquisition time, possibility of reducing the laser power used for image acquisition and lesser exposure time. Further research in all the above-mentioned area will greatly improve the performance of photoacoustic imaging system and may make the clinical translation and utilization of photoacoustic for diagnosis and real-time monitoring more feasible in the near future.

7 Conclusion

In this chapter we discussed the limitations of the current image reconstruction and denoising techniques in photoacoustic imaging. The basic concepts of machine learning and artificial intelligence was established with a focus on deep learning. The applications of deep learning in various medical imaging techniques was discussed. Based on this, the use of deep learning in photoacoustic imaging was analysed especially for improvement in areas of image reconstruction, image denoising and image resolution. Although, deep learning has a lot of potential applications for improving photoacoustic imaging, it comes with certain limitations, especially in terms of training data. Upon overcoming the limitations, deep learning will definitely help in clinical translation and utilization for various clinical applications in the near future.

Next section of this book will focus on preclinical imaging applications and early clinical pilot studies using LED-based photoacoustics.

References

1. P. Beard, Biomedical photoacoustic imaging. Interface Focus **1**, 602–631 (2011)
2. M. Erfanzadeh, Q. Zhu, Photoacoustic imaging with low-cost sources: a review. Photoacoustics **14**, 1–11 (2019)
3. P.K. Upputuri, M. Pramanik, Recent advances toward preclinical and clinical translation of photoacoustic tomography: a review. J. Biomed. Opt. **22**(4), 1–19 (2016)
4. V. Ntziachristos et al., Looking and listening to light: the evolution of whole-body photonic imaging. Nat. Biotechnol. **23**(3), 313–320 (2005)
5. S. Manohar, D. Razansky, Photoacoustics: a historical review. Adv. Opt. Photonics **8**(4), 586–617 (2016)
6. M. Xu, L.V. Wang, Photoacoustic imaging in biomedicine. Rev. Sci. Instrum. **77**(4), 041101 (2006)
7. J.L. Su et al., Advances in clinical and biomedical applications of photoacoustic imaging. Expert Opin. Med. Diagn. **4**(6), 497–510 (2010)
8. I. Steinberg et al., Photoacoustic clinical imaging. Photoacoustics **14**, 77–98 (2019)
9. J. Weber, P.C. Beard, S.E. Bohndiek, Contrast agents for molecular photoacoustic imaging. Nat. Methods **13**, 639 (2016)
10. Q. Fu et al., Photoacoustic imaging: contrast agents and their biomedical applications. Adv. Mater. **31**(6), 1805875 (2019)
11. S.W. Yoo et al., Biodegradable contrast agents for photoacoustic imaging. Appl. Sci. **8**(9), 1567 (2018)
12. P.K. Upputuri, M. Pramanik, Photoacoustic imaging in the second near-infrared window: a review. J. Biomed. Opt. **24**(4), 1–20 (2019)
13. G. Xu et al., Photoacoustic spectrum analysis for microstructure characterization in biological tissue: analytical model. Ultrasound Med. Biol. **41**(5), 1473–1480 (2015)
14. Y. Wang et al., Toward in vivo biopsy of melanoma based on photoacoustic and ultrasound dual imaging with an integrated detector. Biomed. Opt. Express **7**(2), 279–286 (2016)
15. P. Hai et al., High-throughput, label-free, single-cell photoacoustic microscopy of intratumoral metabolic heterogeneity. Nat. Biomed. Eng. **3**(5), 381–391 (2019)
16. H.D. Lee et al., Label-free photoacoustic microscopy for in-vivo tendon imaging using a fiber-based pulse laser. Sci. Rep. **8**(1), 4805 (2018)
17. M. Mathiyazhakan et al., In situ synthesis of gold nanostars within liposomes for controlled drug release and photoacoustic imaging. Sci. China Mater. **59**(11), 892–900 (2016)
18. W. Li, X. Chen, Gold nanoparticles for photoacoustic imaging. Nanomedicine **10**(2), 299–320 (2015)
19. D. Wu et al., Contrast agents for photoacoustic and thermoacoustic imaging: a review. Int. J. Mol. Sci. **15**(12), 23616–23639 (2014)
20. M. Pramanik et al., In vivo photoacoustic (PA) mapping of sentinel lymph nodes (SLNs) using carbon nanotubes (CNTs) as a contrast agent, in *Proc SPIE*, San Francisco (2009)
21. C.-W. Wei et al., In vivo photoacoustic imaging with multiple selective targeting using bioconjugated gold nanorods, in *Proc SPIE* (SPIE, Bellingham, 2008)
22. D. Das et al., On-chip generation of microbubbles in photoacoustic contrast agents for dual modal ultrasound/photoacoustic in vivo animal imaging. Sci. Rep. **8**(1), 6401 (2018)
23. K. Sivasubramanian et al., Near-infrared light-responsive liposomal contrast agent for photoacoustic imaging and drug release applications. J. Biomed. Opt. **22**(4), 041007 (2016)
24. G.P. Luke et al., Silica-coated gold nanoplates as stable photoacoustic contrast agents for sentinel lymph node imaging. Nanotechnology **24**(45), 455101 (2013)
25. W. Lu et al., Photoacoustic imaging of living mouse brain vasculature using hollow gold nanospheres. Biomaterials **31**(9), 2617–2626 (2010)
26. R. Ansari et al., All-optical forward-viewing photoacoustic probe for high-resolution 3D endoscopy. Light Sci. Appl. **7**(1), 75 (2018)
27. J.-M. Yang et al., Photoacoustic endoscopy. Opt. Lett. **34**(10), 1591–1593 (2009)

28. L.V. Wang, Multiscale photoacoustic microscopy and computed tomography. Nat. Photonics **3**(9), 503 (2009)
29. D. Cai et al., Dual-view photoacoustic microscopy for quantitative cell nuclear imaging. Opt. Lett. **43**(20), 4875–4878 (2018)
30. H. Zhang et al., Functional photoacoustic microscopy for high-resolution and noninvasive in vivo imaging. Nat. Biotechnol. **24**, 848 (2006)
31. S. Jeon et al., Review on practical photoacoustic microscopy. Photoacoustics **15**, 100141 (2019)
32. W. Liu, J. Yao, Photoacoustic microscopy: principles and biomedical applications. Biomed. Eng. Lett. **8**(2), 203–213 (2018)
33. L. Li et al., Single-impulse panoramic photoacoustic computed tomography of small-animal whole-body dynamics at high spatiotemporal resolution. Nat. Biomed. Eng. **1**, 0071 (2017)
34. A. Horiguchi et al., Pilot study of prostate cancer angiogenesis imaging using a photoacoustic imaging system. Urology **108**, 212–219 (2017)
35. P.K. Upputuri et al., A high-performance compact photoacoustic tomography system for in vivo small-animal brain imaging. J. Visualized Exp. **124**, e55811 (2017)
36. S. Wang et al., Recent advances in photoacoustic imaging for deep-tissue biomedical applications. Theranostics **6**(13), 2395 (2016)
37. J. Kim et al., Programmable real-time clinical photoacoustic and ultrasound imaging system. Sci. Rep. **6**, 35137 (2016)
38. R. Li et al., Assessing breast tumor margin by multispectral photoacoustic tomography. Biomed. Opt. Express **6**(4), 1273–1281 (2015)
39. A. Garcia-Uribe et al., Dual-modality photoacoustic and ultrasound imaging system for non-invasive sentinel lymph node detection in patients with breast cancer. Sci. Rep. **5**, 15748 (2015)
40. E.I. Galanzha, V.P. Zharov, Circulating tumor cell detection and capture by photoacoustic flow cytometry in vivo and ex vivo. Cancers **5**(4), 1691–1738 (2013)
41. M. Jeon, J. Kim, C. Kim, Photoacoustic cystography. J. Visualized Exp. **76**, e50340 (2013)
42. M. Pramanik, L.V. Wang, Thermoacoustic and photoacoustic sensing of temperature. J. Biomed. Opt. **14**(5), 054024 (2009)
43. P.K. Upputuri et al., Recent developments in vascular imaging techniques in tissue engineering and regenerative medicine. BioMed Res. Int. **2015** (2015)
44. K. Sivasubramanian et al., Hand-held, clinical dual mode ultrasound—photoacoustic imaging of rat urinary bladder and its applications. J. Biophotonics **11**, e201700317 (2018)
45. X. Wang et al., Noninvasive laser-induced photoacoustic tomography for structural and functional in vivo imaging of the brain. Nat. Biotechnol. **21**(7), 803–806 (2003)
46. K. Sivasubramanian, M. Pramanik, High frame rate photoacoustic imaging at 7000 frames per second using clinical ultrasound system. Biomed. Opt. Express **7**(2), 312–323 (2016)
47. A. Hariri et al., The characterization of an economic and portable LED-based photoacoustic imaging system to facilitate molecular imaging. Photoacoustics **9**, 10–20 (2018)
48. Y. Zhu et al., Light emitting diodes based photoacoustic imaging and potential clinical applications. Sci. Rep. **8** (2018)
49. T.J. Allen, P.C. Beard, High power visible light emitting diodes as pulsed excitation sources for biomedical photoacoustics. Biomed. Opt. Express **7**(4), 1260–1270 (2016)
50. S.K. Kalva, M. Pramanik, Experimental validation of tangential resolution improvement in photoacoustic tomography using a modified delay-and-sum reconstruction algorithm. J. Biomed. Opt. **21**(8), 086011 (2016)
51. R.A. Kruger et al., Photoacoustic ultrasound (PAUS)—reconstruction tomography. Med. Phys. **22**(10), 1605–1609 (1995)
52. Z. Deng, W. Li, C. Li, Slip-ring-based multi-transducer photoacoustic tomography system. Opt. Lett. **41**(12), 2859–2862 (2016)
53. S.K. Kalva, Z.Z. Hui, M. Pramanik, Calibrating reconstruction radius in a multi single-element ultrasound-transducer-based photoacoustic computed tomography system. J. Opt. Soc. Am. A **35**(5), 764–771 (2018)

54. R.A. Kruger et al., Photoacoustic ultrasound (PAUS)—reconstruction tomography. Med. Phys. **22**, 1605 (1995)
55. C. Huang et al., Full-wave iterative image reconstruction in photoacoustic tomography with acoustically inhomogeneous media. IEEE Trans. Med. Imaging **32**, 1097 (2013)
56. H. Jiang, Z. Yuan, X. Gu, Spatially varying optical and acoustic property reconstruction using finite-element-based photoacoustic tomography. J. Opt. Soc. Am. A **23**, 878 (2006)
57. M. Xu, Y. Xu, L.V. Wang, Time-domain reconstruction-algorithms and numerical simulations for thermoacoustic tomography in various geometries. IEEE Trans. Biomed. Eng. **50**, 1086 (2003)
58. P. Omidi et al., A novel dictionary-based image reconstruction for photoacoustic computed tomography. Appl. Sci. **8**(9), 1570 (2018)
59. J. Wang, Y. Wang, Photoacoustic imaging reconstruction using combined nonlocal patch and total-variation regularization for straight-line scanning. BioMed. Eng. OnLine **17**(1), 105 (2018)
60. M. Mozaffarzadeh et al., The double-stage delay-multiply-and-sum image reconstruction method improves imaging quality in a LED-based photoacoustic array scanner. Photoacoustics **12**, 22–29 (2018)
61. M. Mozaffarzadeh, A. Mahloojifar, M. Orooji, Medical photoacoustic beamforming using minimum variance-based delay multiply and sum, in *SPIE Digital Optical Technologies*, vol. 10335 (SPIE, Bellingham, 2017)
62. M. Mozaffarzadeh et al., Linear-array photoacoustic imaging using minimum variance-based delay multiply and sum adaptive beamforming algorithm. J. Biomed. Opt. **23**(2), 1–15 (2018)
63. H. Huang et al., An adaptive filtered back-projection for photoacoustic image reconstruction. Med. Phys. **42**(5), 2169–2178 (2015)
64. Z. Ren, G. Liu, Z. Huang, Filtered back-projection reconstruction of photo-acoustic imaging based on an modified wavelet threshold function, in *International Symposium on Optoelectronic Technology and Application 2016*, vol. 10155 (SPIE, Bellingham, 2016)
65. C. Zhang, Y. Zhang, Y. Wang, A photoacoustic image reconstruction method using total variation and nonconvex optimization. BioMed. Eng. OnLine **13**(1), 117 (2014)
66. E. Bossy et al., Time reversal of photoacoustic waves. Appl. Phys. Lett. **89**(18), 184108 (2006)
67. B.T. Cox, B.E. Treeby, Artifact trapping during time reversal photoacoustic imaging for acoustically heterogeneous media. IEEE Trans. Med. Imaging **29**(2), 387–396 (2010)
68. X. Minghua, X. Yuan, L.V. Wang, Time-domain reconstruction algorithms and numerical simulations for thermoacoustic tomography in various geometries. IEEE Trans. Biomed. Eng. **50**(9), 1086–1099 (2003)
69. Y. Xu, L.V. Wang, Time reversal and its application to tomography with diffracting sources. Phys. Rev. Lett. **92**, 033902 (2004)
70. M.A. Lediju Bell, J. Shubert, Photoacoustic-based visual servoing of a needle tip. Sci. Rep. **8**(1), 15519 (2018)
71. H.N.Y. Nguyen, A. Hussain, W. Steenbergen, Reflection artifact identification in photoacoustic imaging using multi-wavelength excitation. Biomed. Opt. Express **9**(10), 4613–4630 (2018)
72. M.K.A. Singh, W. Steenbergen, Photoacoustic-guided focused ultrasound (PAFUSion) for identifying reflection artifacts in photoacoustic imaging. Photoacoustics **3**(4), 123–131 (2015)
73. M.K.A. Singh et al., Reflection-artifact-free photoacoustic imaging using PAFUSion (photoacoustic-guided focused ultrasound), in *SPIE BiOS*, vol. 9708 (SPIE, Bellingham, 2016)
74. A. Hauptmann et al., Model-based learning for accelerated, limited-view 3-D photoacoustic tomography. IEEE Trans. Med. Imaging **37**(6), 1382–1393 (2018)
75. M. Tabei, T.D. Mast, R.C. Waag, A k-space method for coupled first-order acoustic propagation equations. J. Acoust. Soc. Am. **111**(1), 53 (2002)
76. Y. Adachi, T. Hoshimiya, Photoacoustic imaging with multiple-wavelength light-emitting diodes. Jpn. J. Appl. Phys. **52**(7S), 07HB06 (2013)

77. Y. Zhu et al., LED-based photoacoustic imaging for monitoring angiogenesis in fibrin scaffolds. Tissue Eng. Part C Methods **25**(9), 523–531 (2019)
78. J. Jo et al., Detecting joint inflammation by an LED-based photoacoustic imaging system: a feasibility study. J. Biomed. Opt. **23**(11), 1–4 (2018)
79. J. Leskinen et al., Photoacoustic tomography setup using LED illumination, in *European Conferences on Biomedical Optics*, vol. 11077 (SPIE, Bellingham, 2019)
80. P.K. Upputuri et al., Pulsed laser diode photoacoustic tomography (PLD-PAT) system for fast in vivo imaging of small animal brain, in *SPIE BiOS*, vol. 10064 (SPIE, Bellingham, 2017)
81. P.K. Upputuri, M. Pramanik, Performance characterization of low-cost, high-speed, portable pulsed laser diode photoacoustic tomography (PLD-PAT) system. Biomed. Opt. Express **6**(10), 4118–4129 (2015)
82. D. Crevier, *AI: The Tumultuous History of the Search for Artificial Intelligence* (Basic Books, Inc., New York, 1993), p. 386
83. O. Vinyals, M. Fortunato, N. Jaitly, *Pointer Networks*. arXiv e-prints (2015)
84. W. Samek, T. Wiegand, K.-R. Müller, *Explainable Artificial Intelligence: Understanding, Visualizing and Interpreting Deep Learning Models*. arXiv e-prints (2017)
85. H. Chen et al., The rise of deep learning in drug discovery. Drug Discovery Today **23**(6), 1241–1250 (2018)
86. H. Chen et al., Low-dose CT via convolutional neural network. Biomed. Opt. Express **8**(2), 679–694 (2017)
87. J.X. Wang et al., *Learning to Reinforcement Learn*. arXiv e-prints (2016)
88. F. Isensee et al., *Brain Tumor Segmentation Using Large Receptive Field Deep Convolutional Neural Networks* (Springer, Berlin, 2017)
89. S. Albelwi, A. Mahmood, A framework for designing the architectures of deep convolutional neural networks. Entropy **19**(6), 242 (2017)
90. G.I. Parisi et al., Continual lifelong learning with neural networks: a review. Neural Networks **113**, 54–71 (2019)
91. J. Schmidhuber, Deep learning in neural networks: an overview. Neural Networks **61**, 85–117 (2015)
92. O.I. Abiodun et al., State-of-the-art in artificial neural network applications: a survey. Heliyon **4**(11), e00938 (2018)
93. G. Litjens et al., A survey on deep learning in medical image analysis. Med. Image Anal. **42**, 60–88 (2017)
94. J. Zhou et al., *Graph Neural Networks: A Review of Methods and Applications*. arXiv e-prints (2018)
95. X. Hu, W. Yi, L. Jiang, S. Wu, Y. Zhang, J. Du, T. Ma, T. Wang, X. Wu, Classification of metaphase chromosomes using deep convolutional neural network. J. Comput. Biol. **26**(5), 473–484 (2019)
96. A. Kensert, P.J. Harrison, O. Spjuth, Transfer learning with deep convolutional neural networks for classifying cellular morphological changes. SLAS DISCOVERY Advancing Life Sci R&D **24**(4), 466–475 (2019)
97. K.H. Jin et al., Deep convolutional neural network for inverse problems in imaging. IEEE Trans. Image Process. **26**(9) (2017)
98. Y. Ren, X. Cheng, Review of convolutional neural network optimization and training in image processing, in *10th International Symposium on Precision Engineering Measurements and Instrumentation (ISPEMI 2018)*, vol. 11053 (SPIE, Bellingham, 2019)
99. W. Wang et al., Development of convolutional neural network and its application in image classification: a survey. Opt. Eng. **58**(4), 1–19 (2019)
100. X. Zhou, Understanding the convolutional neural networks with gradient descent and backpropagation. J. Phys. Conf. Ser. **1004**, 012028 (2018)
101. M.C. Belavagi, B. Muniyal, Performance evaluation of supervised machine learning algorithms for intrusion detection. Procedia Comput. Sci. **89**, 117–123 (2016)
102. A.N. Dalrymple et al., A supervised machine learning approach to characterize spinal network function. J. Neurophysiol. **121**(6), 2001–2012 (2019)

103. S. Klassen, J. Weed, D. Evans, Semi-supervised machine learning approaches for predicting the chronology of archaeological sites: a case study of temples from medieval Angkor, Cambodia. PLoS ONE **13**(11), e0205649 (2018)
104. M. Rucco et al., A methodology for part classification with supervised machine learning. Artif. Intell. Eng. Des. Anal. Manuf. **33**(1), 100–113 (2018)
105. H. Yao et al., MSML: a novel multilevel semi-supervised machine learning framework for intrusion detection system. IEEE Internet Things J. **6**(2), 1949–1959 (2019)
106. D.M. Hawkins, The problem of overfitting. J. Chem. Inf. Comput. Sci. **44**(1), 1–12 (2004)
107. S. Salman, X. Liu, *Overfitting Mechanism and Avoidance in Deep Neural Networks*. arXiv e-prints (2019)
108. R. Hingorani, C.L. Hansen, Can machine learning spin straw into gold? J. Nucl. Cardiol. **25**(5), 1610–1612 (2018)
109. M. Ceriotti, Unsupervised machine learning in atomistic simulations, between predictions and understanding. J. Chem. Phys. **150**(15), 150901 (2019)
110. A. Dik, K. Jebari, A. Ettouhami, An improved robust fuzzy algorithm for unsupervised learning. J. Intell. Syst. **29**(1) (2018)
111. M. Usama et al., Unsupervised machine learning for networking: techniques applications and research challenges. IEEE Access **7**, 65579–65615 (2019)
112. S. Becker, Unsupervised learning procedures for neural networks. Int. J. Neural Syst. **02**(01n02), 17–33 (1991)
113. S. Guan et al., Application of unsupervised learning to hyperspectral imaging of cardiac ablation lesions. J. Med. Imaging **5**(4), 1–12 (2018)
114. P. Baldi, Gradient descent learning algorithm overview: a general dynamical systems perspective. IEEE Trans. Neural Networks **6**(1), 182–195 (1995)
115. N. Cui, Applying gradient descent in convolutional neural networks. J. Phys. Conf. Ser. **1004**, 012027 (2018)
116. Q. Mercier, F. Poirion, J.-A. Désidéri, A stochastic multiple gradient descent algorithm. Eur. J. Oper. Res. **271**(3), 808–817 (2018)
117. D. Newton, F. Yousefian, R. Pasupathy, Stochastic gradient descent: recent trends, in *Recent Advances in Optimization and Modeling of Contemporary Problems* (INFORMS, Aliso Viejo, 2018), pp. 193–220
118. S. Ruder, *An Overview of Gradient Descent Optimization Algorithms*. arXiv e-prints (2016)
119. R.A. Jacobs, Increased rates of convergence through learning rate adaptation. Neural Networks **1**(4), 295–307 (1988)
120. D.P. Kingma, J. Ba, *Adam: A Method for Stochastic Optimization*. arXiv e-prints (2014)
121. Y. LeCun, Y. Bengio, G. Hinton, Deep learning. Nature **521**, 436 (2015)
122. D. Masters, C. Luschi, *Revisiting Small Batch Training for Deep Neural Networks*. arXiv e-prints (2018)
123. L. Perez, J. Wang, *The Effectiveness of Data Augmentation in Image Classification using Deep Learning*. arXiv e-prints (2017)
124. S.M. Anwar et al., Medical image analysis using convolutional neural networks: a review. J. Med. Syst. **42**(11), 226 (2018)
125. J. Jiang, P. Trundle, J. Ren, Medical image analysis with artificial neural networks. Comput. Med. Imaging Graph. **34**(8), 617–631 (2010)
126. D. Shen, G. Wu, H.-I. Suk, Deep learning in medical image analysis. Annu. Rev. Biomed. Eng. **19**(1), 221–248 (2017)
127. Z. Shi et al., Survey on neural networks used for medical image processing. Int. J. Comput. Sci. **3**(1), 86–100 (2009)
128. I. Wolf et al., The medical imaging interaction toolkit. Med. Image Anal. **9**(6), 594–604 (2005)
129. T.J. Brinker et al., Skin cancer classification using convolutional neural networks: systematic review. J. Med. Internet Res. **20**(10), e11936 (2018)
130. P. Chang et al., Deep-learning convolutional neural networks accurately classify genetic mutations in gliomas. Am. J. Neuroradiol. **39**(7), 1201 (2018)

131. Y. Guo et al., A review of semantic segmentation using deep neural networks. Int. J. Multimedia Inf. Retrieval **7**(2), 87–93 (2018)
132. G. Wang et al., Interactive medical image segmentation using deep learning with image-specific fine tuning. IEEE Trans. Med. Imaging **37**(7), 1562–1573 (2018)
133. D.A. Ragab et al., Breast cancer detection using deep convolutional neural networks and support vector machines. PeerJ **7**, e6201 (2019)
134. F. Pesapane, M. Codari, F. Sardanelli, Artificial intelligence in medical imaging: threat or opportunity? Radiologists again at the forefront of innovation in medicine. Eur. Radiol. Exp. **2**(1), 35 (2018)
135. L. Saba et al., The present and future of deep learning in radiology. Eur. J. Radiol. **114**, 14–24 (2019)
136. G. Chartrand et al., Deep learning: a primer for radiologists. RadioGraphics **37**(7), 2113–2131 (2017)
137. A. Fourcade, R.H. Khonsari, Deep learning in medical image analysis: a third eye for doctors. J. Stomatology Oral Maxillofac. Surg. **120**(4), 279–288 (2019)
138. S. Soffer et al., Convolutional neural networks for radiologic images: a radiologist's guide. Radiology **290**(3), 590–606 (2019)
139. D. Ueda, A. Shimazaki, Y. Miki, Technical and clinical overview of deep learning in radiology. Jpn. J. Radiol. **37**(1), 15–33 (2019)
140. G. Zaharchuk et al., Deep learning in neuroradiology. Am. J. Neuroradiol. **39**(10), 1776 (2018)
141. P. Rajpurkar et al., Deep learning for chest radiograph diagnosis: a retrospective comparison of the CheXNeXt algorithm to practicing radiologists. PLoS Med. **15**(11), e1002686 (2018)
142. L.J. Brattain et al., Machine learning for medical ultrasound: status, methods, and future opportunities. Abdom. Radiol. (New York) **43**(4), 786–799 (2018)
143. S. Liu et al., Deep learning in medical ultrasound analysis: a review. Engineering **5**(2), 261–275 (2019)
144. A.S. Becker et al., Classification of breast cancer in ultrasound imaging using a generic deep learning analysis software: a pilot study. Br. J. Radiol. **91**(1083), 20170576 (2018)
145. Q. Huang, F. Zhang, X. Li, Machine learning in ultrasound computer-aided diagnostic systems: a survey. Biomed. Res. Int. **2018**, 5137904 (2018)
146. J. Ma et al., A pre-trained convolutional neural network based method for thyroid nodule diagnosis. Ultrasonics **73**, 221–230 (2017)
147. D. Allman, A. Reiter, M.A.L. Bell, Photoacoustic source detection and reflection artifact removal enabled by deep learning. IEEE Trans. Med. Imaging **37**(6), 1464–1477 (2019)
148. D. Allman, A. Reiter, M.A.L. Bell, A machine learning method to identify and remove reflection artifacts in photoacoustic channel data, in *2017 IEEE International Ultrasonics Symposium (IUS)* (2017)
149. S. Antholzer et al., Photoacoustic image reconstruction via deep learning, in *Photonics West*, USA (2018)
150. S. Antholzer, M. Haltmeier, J. Schwab, Deep learning for photoacoustic tomography from sparse data. Inverse Prob. Sci. Eng. **27**(7), 987–1005 (2019)
151. C. Cai et al., End-to-end deep neural network for optical inversion in quantitative photoacoustic imaging. Opt. Lett. **43**(12), 2752–2755 (2018)
152. N. Davoudi, X.L. Deán-Ben, D. Razansky, Deep learning optoacoustic tomography with sparse data. Nat. Mach. Intell. **1**(10), 453–460 (2019)
153. J. Schwab et al., *Real-Time Photoacoustic Projection Imaging Using Deep Learning* (2018)
154. D. Waibel et al., Reconstruction of initial pressure from limited view photoacoustic images using deep learning, in *SPIE BiOS*, vol. 10494 (SPIE, Bellingham, 2018)
155. S.L. Jacques, Coupling 3D Monte Carlo light transport in optically heterogeneous tissues to photoacoustic signal generation. Photoacoustics **2**(4), 137–142 (2014)
156. V. Periyasamy, M. Pramanik, Monte Carlo simulation of light transport in tissue for optimizing light delivery in photoacoustic imaging of the sentinel lymph node. J. Biomed. Opt. **18**(10), 1–8 (2013)

157. V. Periyasamy, M. Pramanik, Monte Carlo simulation of light transport in turbid medium with embedded object—spherical, cylindrical, ellipsoidal, or cuboidal objects embedded within multilayered tissues. J. Biomed. Opt. **19**(4), 1–10 (2014)
158. V. Periyasamy, M. Pramanik, Advances in Monte Carlo simulation for light propagation in tissue. IEEE Rev. Biomed. Eng. **10**, 122–135 (2017)
159. Y. Liu, Z. Yuan, Monte Carlo simulation predicts deep-seated photoacoustic effect in heterogeneous tissues, in *Biomedical Optics 2016* (Optical Society of America, Fort Lauderdale, Florida, 2016)
160. G.S. Sangha, N.J. Hale, C.J. Goergen, Adjustable photoacoustic tomography probe improves light delivery and image quality. Photoacoustics **12**, 6–13 (2018)
161. A. Sharma et al., Photoacoustic imaging depth comparison at 532-, 800-, and 1064-nm wavelengths: Monte Carlo simulation and experimental validation. J. Biomed. Opt. **24**(12), 1–10 (2019)
162. G. Paltauf, P.R. Torke, R. Nuster, Modeling photoacoustic imaging with a scanning focused detector using Monte Carlo simulation of energy deposition. J. Biomed. Opt. **23**(12), 1–11 (2018)
163. L. Wang, S.L. Jacques, L. Zheng, MCML—Monte Carlo modeling of light transport in multilayered tissues. Comput. Methods Programs Biomed. **47**(2), 131–146 (1995)
164. K. Sivasubramanian et al., Optimizing light delivery through fiber bundle in photoacoustic imaging with clinical ultrasound system: Monte Carlo simulation and experimental validation. J. Biomed. Opt. **22**(4), 041008 (2016)
165. S. Gutta et al., Deep neural network-based bandwidth enhancement of photoacoustic data. J. Biomed. Opt. **22**(11), 1–7 (2017)
166. O. Ronneberger, P. Fischer, T. Brox, *U-Net: Convolutional Networks for Biomedical Image Segmentation*. arXiv e-prints (2015)
167. L.A. Shepp, B.F. Logan, Reconstructing interior head tissue from X-ray transmissions. IEEE Trans. Nucl. Sci. **21**(1), 228–236 (1974)
168. E.M.A. Anas et al., Enabling fast and high quality LED photoacoustic imaging: a recurrent neural networks based approach. Biomed. Opt. Express **9**(8), 3852–3866 (2018)

Preclinical Applications, Clinical Translation, Trends and Challenges

Light Emitting Diodes Based Photoacoustic and Ultrasound Tomography: Imaging Aspects and Applications

Kalloor Joseph Francis, Yoeri E. Boink, Maura Dantuma, Mithun Kuniyil Ajith Singh, Srirang Manohar, and Wiendelt Steenbergen

Abstract Tomographic photoacoustic and ultrasound imaging is essential for isotropic spatial resolution and to obtain a full view of the target tissue. However, tomographic systems with pulsed laser sources and custom made transducer arrays are expensive. Additionally, there are other factors that limit the wide use of photoacoustic and ultrasound tomographic systems which include the size of the tomographic systems that use pulsed laser and the laser safety issues. A cost-effective, compact and safe photoacoustic and ultrasound tomographic system can find several imaging applications both in clinics and small animal labs. LED-based photoacoustic imaging has shown the potential to bring down the cost, enable faster imaging with high pulse repetition rate and is safer when compared to pulsed lasers. The conventional US system can be adopted for photoacoustic imaging by adding a light source to it. Hence, linear transducer arrays are preferred as they are cheaper and allow faster

K. J. Francis (✉) · Y. E. Boink · M. Dantuma · W. Steenbergen
Biomedical Photonic Imaging Group, Technical Medical Center,
University of Twente, Enschede, The Netherlands
e-mail: f.kalloorjoseph@utwente.nl

Y. E. Boink
e-mail: y.e.boink@utwente.nl

M. Dantuma
e-mail: m.dantuma@utwente.nl

W. Steenbergen
e-mail: w.steenbergen@utwente.nl

K. J. Francis · M. Dantuma · S. Manohar
Multi-Modality Medical Imaging Group, Technical Medical Center,
University of Twente, Enschede, The Netherlands
e-mail: s.manohar@utwente.nl

Y. E. Boink
Department of Applied Mathematics, University of Twente, Enschede, The Netherlands

M. Kuniyil Ajith Singh
Research & Business Development Division, Cybedyne Inc.,
Cambridge Innovation Center, Rotterdam, The Netherlands
e-mail: mithun_ajith@cyberdyne.jp

© Springer Nature Singapore Pte Ltd. 2020
M. Kuniyil Ajith Singh (ed.), *LED-Based Photoacoustic Imaging*,
Progress in Optical Science and Photonics 7,
https://doi.org/10.1007/978-981-15-3984-8_10

imaging. The combination of LED-based illumination and linear transducer array-based tomographic imaging can be a cost-effective alternative to current tomographic imaging, especially in point-of-care applications.

1 Introduction

In photoacoustic (PA) imaging, pulsed (nanosecond) light induces thermoelastic expantion in optical absorbers, resulting in acoustic signal generation. Detection of these optically excited acoustic signals for imaging enables optical absorption contrast at ultrasound (US) resolution [1, 2]. This imaging modality can surpass the high optical scattering with the detection of less scattered acoustic signals. Hence, it is a preferred modality where optical contrast, larger imaging depth, and high resolution are required [3]. PA tomography is one of the widely used configurations for deep tissue imaging where the target biological tissue allows a larger view angle [4, 5]. PA tomography has been demonstrated for applications such as breast cancer imaging [6, 7], thyroid cancer imaging [8], finger joint imaging [9], brain functional imaging [10], and small animal whole-body imaging [11]. An additional advantage of PA imaging is that it can be combined with US imaging as the same transducer can be used for both modalities [3, 12–14]. The visualization of PA images in the context of well known US images, with their complementary nature combining structural and functional information is of added value for clinical applications [3, 15]. Major limitations of tomographic systems are its cost, large size and safety issues with the use of pulsed laser sources [16]. Additionally, the large number of transducers and acquisition channels in tomographic systems also contribute to its size and cost. Hence, there is an urgency for cost-effective, compact and safe to use systems for an increasing number of clinical applications [16].

System cost and size of PA tomography systems can be considerably reduced by replacing pulsed Q-switched lasers with alternative sources such as laser diodes or LEDs [17]. Recent developments in LED-based PA imaging has shown promise in developing systems which are portable, safe to eye and having high frame rate [18]. A commercial system with LED-based illumination and linear US transducer is available for research applications [18]. This system is an example which shows that conventional US systems can be combined with PA imaging by simply plugging in a light source [18, 19]. An advantage of such a configuration is that with a software update, widely used US systems can incorporate PA imaging and can find faster clinical acceptance [19]. Another aspect is the use of linear transducer array instead of custom-developed transducers for tomographic imaging. The wide use of linear array and its production in large numbers resulted in high yield and low cost. Additionally, it enables faster US imaging with switching on and off subsets in an array for transmission [20]. While for PA imaging the whole array can simply be in the receive mode [20]. The usage of linear array also allows the use of less computational complex reconstruction methods such as the Fourier domain algorithm

[21]. Using the conventional US systems with linear array and low-cost illumination sources such as LEDs for tomographic imaging can enable cost-effective, compact and faster imaging systems.

2 Tomographic Imaging Using Linear Array

Linear transducer arrays are one of the first choices for tomographic imaging from its very beginning, starting with Oraevsky et al. in [22]. The factors which enabled the widespread clinical use were its availability in various frequency ranges, low cost, and ease of integration [23]. However, due to limited aperture and directional sensitivity of the linear transducer array, imaging suffers from loss of structural information, anisotropic resolution and artifacts [23]. In order to overcome the limited view problem, several methods were suggested. The use of a reverberant cavity by Cox et al. [24] and two planar acoustic reflectors at an angle by Li et al. [25] are some of the examples. However, the detected field of view in these proposed methods is limited and multiple artifacts degrade the image quality making it challenging in a practical setting. Scanning the transducer still remains a viable option for high-quality tomographic imaging at the expense of scanning time.

A combined PA and US imaging by circular scanning of the linear transducer was first studied by Kruger et al. in [26]. Two transducer scanning modes were used for tomographic PA and US imaging, as shown in Fig. 1 [27]. Rotating the long axis of the transducer about a center for imaging results in 2D imaging and rotating the shorter axis provides 3D imaging [27]. In the 2D imaging case, multiple views of the same imaging plane in a target tissue can be obtained by circular scanning of the transducer. In this case, the tomographic image can be formed either by a combined acoustic reconstruction from all the elements or by compounding B-scan images from individual views [23, 28]. In case of 3D imaging, the cylindrical focusing of the linear transducer array must be taken into account. This focus along the elevation direction

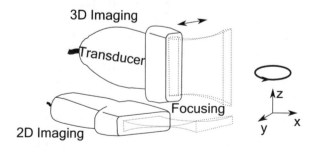

Fig. 1 Transducer configuration for 2D and 3D tomographic imaging. In the 2D imaging the long axis of the transducer is rotated around the sample. The imaging plane of the transducer with the cylindrical focusing is shown with doted lines. In the 3D imaging consists of linear scan at every angular location

results in imaging a different plane in each scan location. Combining these non-overlapping images from subsequent scan locations can result in discontinuities and smearing of structures [29]. To overcome these discontinuities in the 3D imaging, a translate-rotate scanning was proposed, where at each angular location a linear scan is performed [30, 31]. In this chapter, we are focusing on the 2D imaging configuration.

3   Imaging Aspects Using a Linear Array

The characteristics of the US transducer should be considered while designing the tomographic imaging configuration. Commercially available transducer arrays have acoustic focusing and directional sensitivity. The focus along the axial direction and the focus-zone where the transducer response is uniform should be considered to define the center of rotation and imaging area [26]. Further, the directivity of the transducer should be considered when defining the angular steps for the scanning. In this section, we present ultrasound transducer characterization results, followed by numerical simulations to determine an optimal number of angular views for tomographic imaging. We also present our image reconstruction approach and analysis of resolution improvement with tomographic imaging.

3.1   Transducer Characterization

AcousticX system (Cyberdyne Inc., Japan) was used in this work for both PA and US imaging. We have considered a linear transducer array with 128 elements having a center frequency of 7 MHz and a −6 dB bandwidth of 4–10 MHz, with a pitch of 0.315 mm. The transducer was first characterized to measure the directivity and focus. The characterization was performed in an acoustic receive mode. A 30 μm thick black suture wire (Vetsuture, France) was used as a PA target with LED-based illumination. We have used an LED array, illuminating at 850 nm wavelength with a pulse energy of 200 μJ and a pulse duration of 70 ns. The PA source was fixed and using motorized stages the transducer was moved to various axial and lateral locations. In the first experiment, the focus of the transducer was measured. The PA source was aligned to the short axis of one of the center elements of the transducer. The raw PA signal from the line target was acquired and the peak-to-peak value was plotted against the distance between the transducer and the PA source. The measurements were repeated for varying the axial distance between the transducer and the PA source. Figure 2a shows the peak-to-peak value of the PA signal at different axial positions. From the plot, the maximum PA signal value (3.5×10^4) was observed at 20 mm which is the focus of the transducer. On either side of the focus, a uniform drop of PA signal intensity to a value of 2.8×10^4 was observed. A similar pattern was observed by Kruger et al. [26]. They considered the focus of the transducer as the center of rotation for tomographic imaging and the uniform region for imaging. The same approach

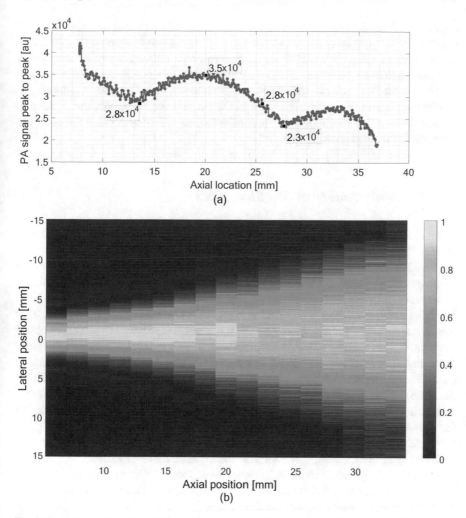

Fig. 2 Transducer characterization. **a** Axial response of the transducer. **b** Acoustic field of a single element

was adopted in our study. We have considered the focus distance (20 mm) to define the center of rotation with a ±5 mm window within the uniform region (15–25 mm) in the axial response of the transducer for larger and smaller samples.

In the second experiment, the directional sensitivity of a single element in a transducer array was measured. The transducer was scanned in both lateral and axial direction and peak PA signal values were recorded for one of the center elements. Figure 2b shows the PA intensity at different lateral and axial locations. At each depth with a Gaussian fit on the measured data, the Full Width Half Maximum (FWHM) was calculated. The angle made by the FWHM point with the center of the transducer element was calculated to be 26.8 ± 0.2°. To obtain the combined directivity of the transducer, the acoustic field of individual elements were combined together. The opening angle on either side of the transducer was calculated in a similar fashion using the FWHM points. The combined directivity of the transducer was estimated to be $16.6° \pm 1.5°$. The opening angle of a single element need to be considered for PA mode as well as conventional line-by-line B-mode US imaging as all elements are in the receiving mode. In the plane wave US mode, the directivity of the whole transducer needs to be considered, as US transmission from all the elements is involved in this mode.

3.2 Optimal Number of Angular Views

A target structure placed at an angle with the transducer is detectable if the normal from it falls within the opening angle of a transducer element [32]. Hence, with the opening angle of $26.8° \pm 0.25°$ to detect a structure with an arbitrary angle, a minimum of 14 views are required to cover the entire 360°. While in the case of plane wave US imaging, with the opening angle of the transducer being $16.6° \pm 1.57°$, 24 angular views are required to have full-view tomographic imaging. This theoretical estimation is a minimum as the transducer directivity is not a sharp cutoff function. Hence, we performed an acoustic simulation to study the number of angular views required to obtain an adequate image quality in the tomographic setting. A phantom was designed for the simulation study as shown in Fig. 3a. Considering the tomographic imaging we would like to test whether structures having different orientations can be reconstructed well. Angular dependency was incorporated in the phantom with 24 line targets placed at 15° angular steps as shown in Fig. 3a. Next, we considered targets with different sizes to test the resolution enhancement with tomographic imaging. Line targets and circular disks of different sizes were considered in the phantom to make it resolution dependent. Given a center-frequency (f_0) of the transducer and the corresponding wavelength (λ_0), a typical resolution of $\lambda_0/2$ is expected from the transducer. In our experiments with the center frequency of 7 MHz and a bandwidth of 4–10 MHz, the second set of structures were incorporated in the phantom with different thickness varying from $\lambda_0/4$, $\lambda_0/2$, λ_0, $2\lambda_0$ to $4\lambda_0$. The circular targets of diameter $\lambda_0/2$, λ_0, $3\lambda_0/2$ and $2\lambda_0$ were also placed in the phantom. Further, four different levels of initial pressures were also included making it ideal phantom for image quality based study.

Acoustic wave propagation was performed using k-Wave toolbox of MATLAB [34]. A homogeneous acoustic medium mimicking water with a speed of sound of 1502 m/s and a density of 1000 kg/m^3 was used in the forward model. The transducer elements were modeled to have finite size with spatial averaging of point detectors with a directivity mask [34]. Gaussian noise with a signal to noise ratio of 50 dB, with the signal level taken as the Root Mean Square (RMS) value of the raw acoustic signal was generated and added as the measurement noise. The simulation was repeated for different number of angular views with the linear array around the phantom. Individual B-scan images were formed from each angular view using the Fourier

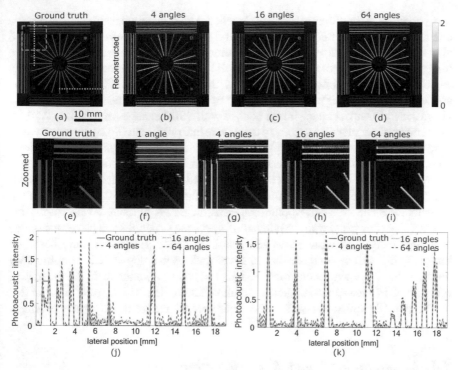

Fig. 3 Image quality versus angular views. **a** Ground truth photoacoustic phantom. **b, c** Reconstructed images from 4, 16 and 64 angular views. **e** Zoomed in region of the phantom, marked with the green box in (**f–i**) and the corresponding reconstructed images from 1, 4, 16 to 64 angles respectively. **j, k** Line profiles from a vertical (yellow) and a horizontal (blue) region, comparing ground truth and reconstructed photoacoustic pressure (Reproduced with permission [33])

domain reconstruction algorithm [35]. The tomographic image was formed by spatial compounding of B-scan images from all angles. We have chosen structural similarity (SSIM) index [36] and peak signal-to-noise ratio (PSNR) to measure the image quality of the reconstructed image with respect to the ground truth. We have selected SSIM to check how well the structures are reconstructed and PSNR to measure the signal to the noise level in the image. Figure 3b–d shows reconstructed images from 4, 16 and 64 angular views respectively. A zoomed-in region (marked in green) of the ground truth and reconstructed images from 1, 4, 16 to 64 angles are provided in Fig. 3e–i. In imaging using a single view only horizontal line targets are reconstructed. This limited view problem can also be observed in the circular targets, as only the top and bottom boundaries were reconstructed. Calculated SSIM was 0.09 indicating low structural similarity. Additionally, the reconstructed image contains artifacts and measurement noise providing a poor PSNR value of 7. An additional view from 18° can only provide a small improvement in the image quality as no additional angular information is available. With 4 angular views, SSIM improved to 0.3 and PSNR to 11, as all the vertical and horizontal structures are reconstructed. It can be observed that structures with different intensity levels are also distinguishable. However, the circular targets are not fully reconstructed and the noise level is still high. From 4 to 16 views there is a linear increase in reconstructed image quality. In case of tomographic image with 16 number of angular views all the structures at different angles, intensity levels and sizes are resolved well with providing an SSIM of 0.53 and the noise and artifacts are much lower with a PSNR value of 15.3. Further increasing the number of views has little impact on the image quality with SSIM change from 0.52 to 0.56 for the number of angular views from 16 to 64. The line profiles in Fig. 3j, k compares the ground truth to the reconstructed photoacoustic intensity. The profile in Fig. 3j shows line target with 5 different thickness, the smallest circular target and three line targets placed at different orientations. The reconstructed PA pressure levels are compared with that of ground truth. The profile in Fig. 3k shows targets with 5 different initial pressure levels and the corresponding reconstructed PA pressure levels. These line profiles also shows that from 16 number of views the structures can be reconstructed to a large extent and more views add less information. From this study based on image quality, we can conclude that 16 angular views in a step of 22.5° is optimal for tomographic PA image reconstruction. This is a good agreement with the theoretical estimate of 14 views made earlier. A small oversampling with angular steps of 20° resulting in 18 number of angular views were considered for the experimental study.

3.3 Tomographic Image Reconstruction

Tomographic image formation can be performed by combined acoustic reconstruction [23] or by spatial compounding of the B-scan images [28, 37]. We have considered multi-angle spatial compounding to form tomographic photoacoustic and ultrasound images [28]. The PA and US signals were acquired using the linear transducer array by scanning around the sample as shown in Fig. 5a, b. The PA and US data from each angle are first reconstructed to form B-scan images. The B-scan images formed directly from the system using an in-built real-time reconstruction algorithm. This can also be performed off-line using a similar Fourier domain algorithm [35, 38]. Additionally, the system can perform both plane wave and B-mode US imaging. The B-scan images were then rotated to the corresponding angle they were acquired. An image rotation was performed to correct for the angle it was acquired. The rotated images are then averaged to obtain the tomographic image.

3.4 Resolution Improvement

Imaging using a linear array with a single view of the target results in an asymmetric resolution along the axial and lateral direction. A higher resolution can be achieved along the axial direction compared to the lateral, as the resolution along axial direction depends only on the bandwidth of the transducer [39]. There are two factors that

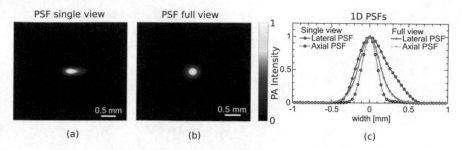

Fig. 4 Photoacoustic Point Spread Function (PSF) obtained from 30 μm target. **a** Normalized PSF from a single view with the linear array. **b** Normalized PSF from tomographic imaging obtained from 18 angular views. **c** Comparison of 1D PSF extracted from (**a**) and (**b**) along the lateral and axial direction

degrade the lateral resolution. First, the finite size of the element results in a spatial averaging of the received acoustic signals [39]. Second, the reconstruction from limited view measurements causing artifacts along the direction where detectors are absent [40]. Reconstruction from a full-view tomographic measurement can improve the resolution with projections from all the angles. Resolution improvement in tomographic imaging using a linear array was reported by Kruger et al. [41] with simple compounding of B-scan images from different angles. A small improvement in resolution was claimed with combined image reconstruction from all the measurements [42] and with the use of multi-view Hilbert transform [23].

We have performed Point Spread Function (PSF) measurements in PA mode with a single view and with tomographic imaging from 18 views. A suture wire (Vetsuture, France) with a diameter of 30 μm was used as PA target and an 850 nm LED array as illumination source. Figure 4a shows PSF from a single view. Measured axial Full-Width Half Max (FWHM) was 0.22 mm and lateral FWHM was 0.47 mm. PSF from tomographic reconstruction using spatial compounding of 18 angular views is shown in Fig. 4b. The PSF is symmetric and the axial and lateral resolution were measured to be 0.31 mm. A detailed analysis of 1D PSF along the lateral and axial direction for both single view and the tomographic case is shown in Fig. 4c. It can be observed that spatial compounding provides an improvement in lateral resolution at the expense of degrading the axial resolution. However, it is possible to obtain an isotropic resolution and limited view artifacts can be removed.

4 LED-Based Illumination for Tomography

In this section we present the illumination configurations using LED arrays. We also performed simulations to study the light propagation from the LEDs into soft tissue and present tomographic imaging results using the illumination configuration. The AcousticX system used in this study can drive four LED units simultaneously. While

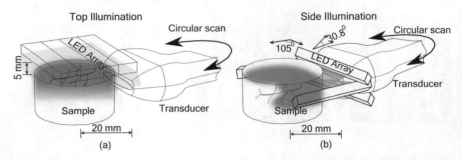

Fig. 5 Tomographic imaging configurations with **a** illumination from the top of the sample and **b** illumination from the sides of the sample (Reproduced with permission [33])

in the normal hand-held configuration only two LED units are used. Tomographic imaging demands more energy for better penetration depth. Hence a combination of all four LED units were considered for illumination. Each LED unit consists of 144 elements arranged in an array of 36×4 with an active area of 55×7 mm with an energy of $200 \, \mu J$ per pulse. Two LED configurations are presented in this work. The first one is for applications where the top of the sample can be accessed for illumination, such as brain functional imaging [43]. The second configuration is for samples that can only be accessed from its sides, such as finger joint imaging or small-animal whole-body imaging. A schematic of the illumination configurations is provided in Fig. 5.

To study the fluence distribution in a soft tissue using these two illumination configurations, combined optics and acoustic simulations were performed. The light propagation from LED array into the tissue was modeled using Monte Carlo simulation with Monte Carlo eXtreme (MCX) photon transport simulator [44]. A cylindrical phantom with a 25 mm diameter in water having average soft tissue optical tissue properties ($\mu_a = 0.56 \, \text{mm}^{-1}$, $\mu_s = 9.9 \, \text{mm}^{-1}$, g = 0.90, n = 1.4) [45] was considered in these simulations. The LED elements were modeled with a solid opening angle of $120°$. In the top illumination case, four adjacent LED units were positioned at a distance of 5 mm above the phantom as depicted in Fig. 5a. While for illumination from the side, two LED bars were placed above and below the transducer's active part, with an angle of $30.8°$ such that the illumination intersect the focus of the transducer (at 20 mm) in a non-scattering medium. Additionally, two more LED bars with an angle of $105°$ relative to the transducer array were placed in the imaging plane. To reflect the acoustic signals away from the transducer and to minimize artifact, the LED arrays were placed in an angle of $5°$ with the imaging plane.

The fluence maps from the Monte Carlo simulations were then coupled with the acoustic simulation. To obtain the initial pressure, the ground truth was multiplied with normalized fluence map. A plane 5 mm inside cylindrical phantom was considered for the fluence map. The Grüneisen parameter was not considered here, as we were interested only in the spatial variation of initial pressure and not in its absolute value. For the ground truth a vascular structure from a retinal image in the

DRIVE database was used [46]. To mimic soft tissue acoustic properties, a speed-of-sound of 1580 m/s and density of 1000 kg/m^3 were used for the phantom. For the coupling medium we assigned a speed-of-sound of 1502 m/s and a density of 1000 kg/m^3. The acoustic attenuation was modeled as power-law with pre-factor chosen to be 0.75 dB/(MHz$^{1.5}$ cm). The directivity and bandlimited nature of the transducer was also modeled as explained in the previous simulation (Sect. 3.2). The first-order k-space model was used for forward acoustic wave propagation. To mimic measurement noise, Gaussian noise with SNR of 30 dB (with respect to the RMS value of the PA signal) was added to the RF signals. For a realistic scenario the spatial variations in the acoustic properties are unknown. Hence, for the reconstruction, we assumed an uniform acoustic properties. The tomographic images were formed by spatial compounding of reconstructed B-scan images from all the angles. A normalization was performed on the reconstructed images such that the total intensity of the vascular structures are the same as that of the ground truth [47]. The normalization was performed by segmenting the pixels in the region of the vascular structures and normalizing such that the sum of pixel values in the segmented region is equal to that of the ground truth. Tomographic images obtained from top and side illuminations were compared validated against the ground truth.

A uniform illumination is desired to view the whole sample from all the angles and to form a perfect tomographic image. In this simulation study, we compare the difference in reconstructed image from a fixed top illumination and rotating side illumination. Figure 6a and b show optical fluence map from the top and side illumination. A line profile of the fluence map extracted from the center of the phantom is also shown below images. In the top illumination, the fluence is mostly uniform. An asymmetric drop of the fluence at the edges of the phantom in the vertical direction can be observed compared to the horizontal. This asymmetry is due to the rectangular (50 mm × 40 mm) illuminated region, as we stacked LED units. From the center of the phantom to the edge a maximum drop in fluence of 46% was observed. In the case of side illumination, the fluence dropped to 30% at the center compared to the one at the boundary.

The ground truth vascular phantom is shown in Fig. 6c. The initial pressure map from top and side illumination, obtained by multiplying the ground truth with the fluence map (Fig. 6a, b) is shown in Fig. 6d, e. Tomographic images obtained from 16 angular views are shown in Fig. 6f, g. In both the configurations, all the structures in the phantom were reconstructed. Figure 6f shows that with top illumination, the reconstructed pressure level is lower towards the boundary of the phantom compared to the center. While for side illumination in Fig. 6g, the reconstructed pressure is lower at the center of the phantom, compared to the rim. This is further evident in the line profiles in Fig. 6h, i, extracted from the reconstructed images along the vertical (white) and horizontal (green) lines through the center of the phantom. The speed-of-sound is considered uniform in the reconstruction. As a result of this assumption, a lateral shift in the peaks compared to the ground truth can be observed in the line profiles. In the tomographic imaging utilizing spatial compounding of B-scan images, this can result in smear artifacts from multiple angles and a change in the size of the structures. To concluded, both configurations can be used for tomographic

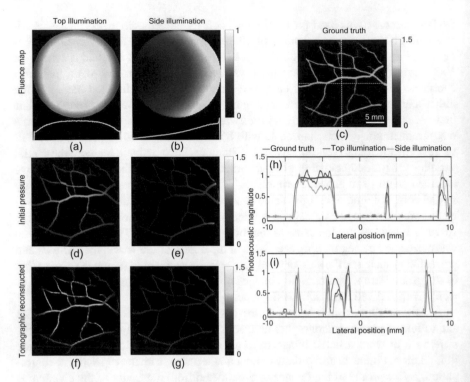

Fig. 6 A simulation study comparing top and side illumination configurations. **a, b** Normalized optical fluence maps in the two cases respectively. A line profile of the fluence map extracted from the center of the phantom is shown below the image. **c** Ground truth vascular phantom. **d, e** Initial pressure obtained from top and side illumination respectively. **f, g** Reconstructed and normalized tomographic images from 16 angular views. **h, i** Comparison of line profiles between the ground truth (**c**) and the reconstructed images (**f**) and (**g**), along horizontal (green) and vertical (white) lines passing through the center of the phantom respectively (Reproduced with permission [33])

imaging. However, with top illumination, a region of interest around the center can be reconstructed well. This aspect can be helpful in applications like small animal brain imaging. In the case of side illumination, there is a significant amount of overlap in the illuminated regions between angular views. This overlapping illumination region allows multiple view of the same structure enabling tomographic reconstruction. The lower fluence at the center of the sample can result in a lower reconstructed pressure level in this region. However, with more number of transducer elements observing the center higher SNR can be obtained with averaging. Using the side illumination the rim of the object is illuminated fairly uniform. Hence side illumination can be potentially used for finger joint tomographic imaging.

4.1 Imaging Experiments

The illumination configurations developed in Sect. 4, and scanning steps for tomographic imaging using a linear array determined in Sect. 2, were incorporated in the experimental setup. In this section, we present imaging using the two LED array configurations. We also discuss the imaging speed our tomographic system.

4.2 Tomographic Imaging Using Top Illumination

In the first configuration, all the four LED units are stacked together to form a large array of 576 elements, placed 5 mm above the sample for uniform illumination. In this case, the illumination is static and only the transducer is rotated around the sample for tomographic imaging. Uniform illumination of the entire sample for all angular views is ideal for tomographic imaging as this enables the transducer to view the same structure from different angles. To a large extent, this is possible with the illumination from the top of the sample.

Figure 7a shows a photograph of a leaf skeleton stained with India ink and included in a 3% agar phantom. The leaf phantom was selected as it has structures of different thickness and orientation for a resolution study in the tomographic setting. Additionally, the structures in the leaf mimic typical vascularization such as the one in small animal brain [43]. PA imaging of these structures indicates the applicability of this imaging system. Figure 7b–g shows B-scan PA and US images from three angular views. These images show the structure of the leaf with the limited view from a linear array. Structures which are having smaller angles with the transducer are reconstructed well. This shows the need for tomographic imaging to obtain a full view of the sample. Figure 7h, i show the PA and US tomographic images respectively. A zoomed-in image of the leaf in Fig. 7j shows four levels of structures based on the thickness. It can be observed in Fig. 7k–m, that with increasing number of angular views from 4, 12 to 18, finer structures are visible. Three levels of structures are reconstructed with 18 angular views leaving only the smallest structures in the leaf undetected. The smallest structures were not reconstructed as the bandwidth of the transducer is limited and the high frequency components in the PA signals were not detected. The tomographic US image formed with the plane-wave mode in Fig. 7c shows the boundary of the leaf and the larger veins. It should be noted that the number of angular views of 18 is insufficient in the plane wave mode for a complete tomographic imaging and the specular nature of the imaging is not appropriate to differentiate these structures.

It is also possible to image ex vivo tissue samples of small size using this configuration. A typical example is surgically excised tissue to look for malignancy before pathological inspection [39]. In another experiment, we have imaged an ex vivo mouse knee sample. In this case, we also compared the difference in tomographic US imaging using the plane wave and conventional line-by-line scanned B-mode.

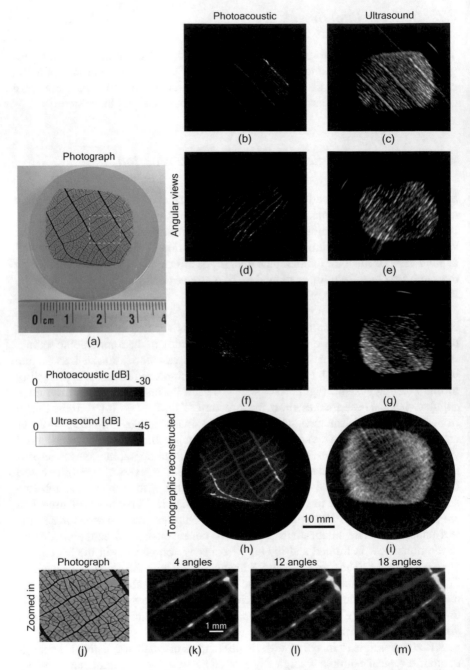

Fig. 7 Tomographic imaging of a leaf skeleton. **a** Photograph of the leaf. **b, d, f** Photoacoustic B-scan image from three angles. **c, e, g** Ultrasound B-scan image from three angles. **h** Photoacoustic tomographic image from 18 angular views. **i** Ultrasound tomographic image from 18 angular views. **j** Zoomed in region of the leaf skeleton indicated by green box and (**k**)–(**m**) are PA tomographic image of the zoomed in region using 4, 12 and 18 angular views (Reproduced with permission [33])

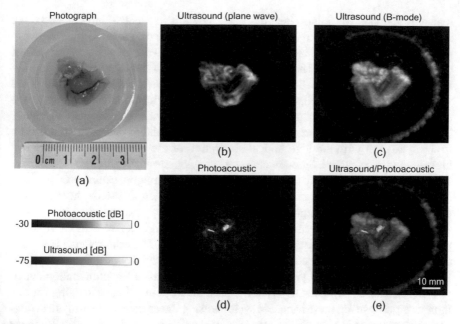

Fig. 8 Mouse knee imaging. **a** Photograph of the ex vivo mouse knee sample. US tomographic images obtained from **b** plane wave imaging and **c** B-mode imaging. **d** PA image of the sample. **e** Combined PA and US image (Reproduced with permission [33])

Figure 8a shows the mouse knee sample embedded in a 3% agar phantom. Figure 8b shows tomographic US imaging using the plane wave mode and Fig. 8c using B-mode. Tomographic imaging using B-mode US shows superior quality in its compared with the case of plane wave. Two bones forming the joint and the tissue around it are visible in the tomographic image. Adding more angular views can improve the plane wave ultrasound-based tomographic image. A major blood vessel running through the joint is visible in the photograph in Fig. 8a. The PA image in Fig. 8d shows the blood vessel and several branches from it. There are some discontinuities as the blood vessel was not completely in the imaging plane. Additionally, clotted blood near the point where the joint was dissected is also visible as a high absorbing spot. The coregistered PA and US tomographic image in Fig. 8e shows both joint and blood vessels. The capability to visualize both vascular development and joint damage can be useful for early detection of rheumatoid arthritis.

4.3 Finger Joint Imaging Using Side-Illumination

Rheumatoid Arthritis (RA) is an autoimmune disorder affecting joints, which leads to disabilities [48]. If not detected and treated at an early stage, the disease can permanently damage the joints. Imaging the synovium to identify inflammation and angiogenesis can indicate the disease activity. US and magnetic resonance imaging (MRI) are used for imaging joints. US imaging is less sensitive to early stage changes in the synovium. The vascularization of the synovium is imaged using Ultrasound power Doppler (US-PD) [49]. However, lack of sensitivity to small blood vessels limits the applicability of US-PD. MRI can provide good visuaization of the synovium. However, it is largely inaccessable, expensive and needs contrast agents [49]. In this scenario, PA imaging has shown the potential to detect vascularization in the synovium with the high optical absorption of the blood [49]. Physiological biomarkers in the synovial tissues such as neoangiogenesis, hypoxia and hyperemia can be detected using PA imaging [50]. Therefore, PA imaging can be used for joint imaging for RA and other rheumatologic conditions such as osteoarthritis, crystal deposition diseases, seronegative spondyloarthropathies, and systemic lupus [48]. There are many efforts to use low-cost and compact light sources in PA imaging systems, specifically for the application of RA [48, 49]. The primary goal is to have a compact PA system as a point-of-care device in the vicinity of a rheumatologist. Patient studies using LED [51] and laser diode [49] based PA illumination added to conventional US systems have shown promise in point-of-care imaging for RA monitoring. However, these handheld systems using linear transducer arrays can miss early signatures of RA due to the asymmetric resolution and limited view. Hence, a tomographic system is needed to obtain a full view of the target tissue [52, 53].

We have developed a finger joint imager using the side illumination configuration explained in Sec. 4. A holder for the transducer and LED units with the configuration design developed in Sect. 4 was 3D printed. A schematic of the finger joint imager is shown in Fig. 9a. A translational stage with an accuracy of 100 μm with a maximum range of 157.7 mm and a rotational stage with a 0.1° accuracy for 360° rotation was used for scanning the imaging probe. Additionally, hand rest and a fingertip positioner were used to minimize movement during the scan. A photograph of the finger joint imager is shown in Fig. 9b. In a proof-of-concept experiment, a finger joint of a healthy female volunteer was imaged. Figure 9c shows combined PA and US images of a finger showing the interphalangeal joint and the blood vessels around it. Tomographic PA and US images were acquired at locations $p1$ at the joint, and $p2$ which is 5 mm away from the finger joint. Figure 9d–f are tomographic PA, US and combined images at $p1$. Figure 9g–i are tomographic PA, US and combined images at $p2$. The ultrasound images show a hypoechogenic region for both bone and blood vessels. The skin and the blood vessels are visible in the PA images. The interphalangeal joint in the US image ($p1$) shows two distinct regions in the bone with a narrow separation possibly from the curved region of the joint. Figure 9j shows tomographic PA images with increasing number of angular views from a partial view of 90° to a full-view acquired from 360°. This images shows the need for a full-view tomographic imaging. In this proof-of-concept study to minimize the imaging time for in vivo measurement plane-wave, US image was used. The use of B-mode and more angular views can improve image quality. The initial result shows that the

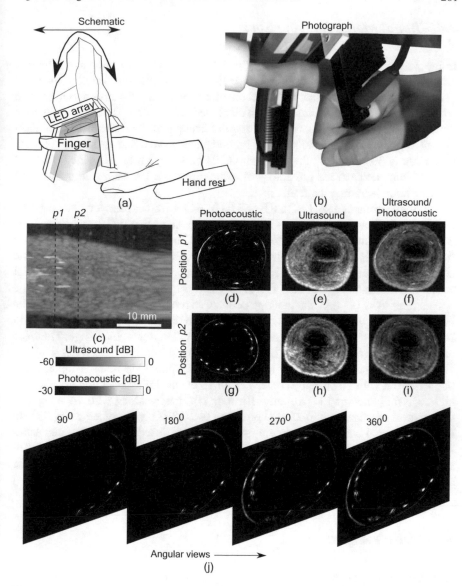

Fig. 9 Finger joint imaging. **a** Schematic and **b** photograph of the imager. **c** Combined PA and US image from a linear scan of the finger joint. **d–f** Photoacoustic, ultrasound and co-registered image respectively at position $p1$. **g–i** Photoacoustic, ultrasound and co-registered image respectively at position $p2$. **j** Tomographic photoacoustic images obtained from increasing number of angular views (Reproduced with permission [33])

tomographic US and LED-based PA system can be used to image the vascularisation around the finger joint and the bone structure, which shows its potential. To test the system for its applicability in RA imaging, an extensive patient study is required.

4.4   Tomographic Imaging Speed

In the AcousticX system, RF data from all US elements are acquired at a sampling rate of 40 MHz for US and at 20 MHz for PA imaging and transferred to the GPU board. Acquired data (US and PA) is reconstructed using an in-built Fourier-domain based reconstruction algorithm and then displayed in real-time. The system can drive the LED arrays as well as transmit and acquire data parallelly from all 128 elements of the US probe to generate interleaved PA and plane wave US images at a maximum frame rate of 30.3 Hz, with the maximum LED pulse repetition frequency (PRF) of 4 KHz which delivers the possibility of averaging more PA frames to attain good SNR without degrading the frame rate.

In our experiments at an LED PRF of 4 KHz, 64 PA frames are averaged on-board within the DAQ and then the data is transferred to AcousticX PC through the USB interface. This data is averaged 6 times in the PC and then reconstructed using a frequency-domain algorithm implemented in the GPU. One US (planewave) frame is acquired between every 64 PA frames to generate US and PA overlaid images at an interleaved frame rate of 10.3 Hz. After acquiring one PA and US image at the first angular view (97 ms), the imaging probe was rotated by 22.5° in 2.56 s and the next view was acquired. This was continued for 16 different angular views to attain 360° view, which is then used for generating a PA/US tomographic image. Thus, the total time required for generating a full view tomographic US/PA image is 42.5 s.

The system is also capable of performing conventional US line-by-line acquisition (transmit with 10 channels, receive with 16 channels) by scanning each line of an image for generating high quality B-mode US images along with PA images at the expense of frame rate (6.25 Hz compared to 10.3 Hz in the plane wave US). In the small animal cadaver experiment, this setting was used since high-speed scanning is not a prerequisite in this case. The total time required for generating a full view tomographic US/PA image is 43.5 s, which is almost a second slower than the acquisition involving plane-wave US imaging.

5   Future Perspectives

Our simulation, phantom, ex vivo and in vivo results give a direct confirmation that LED-based tomographic PA imaging using linear array US probes holds strong potential in multiple clinical and preclinical applications. However, several improvements are required for translating this technology to preclinical labs and clinics. The use of multispectral tissue illumination for obtaining blood oxygen saturation images is one of the key applications of PA imaging. By utilizing LED arrays of different wavelengths, we would like to explore this in our future studies in a tomographic setting. It is foreseen that 3D images of vasculature and oxygen saturation combined with anatomical information offered by US imaging would be a valuable tool for RA detection and staging in both preclinical and point-of-care clinical settings. Currently,

LED arrays are designed in such a way to fit well on both sides of a linear array US probe with an aperture close to 4 cm. In this work for tomographic imaging, we used a pseudo-arc-shaped illumination with commercially available LED matrices. However, a custom developed arc or ring-shaped LED array would be more appropriate for tomographic imaging applications. We would like to explore custom-developed LED arrays for specific applications such as small animal imaging in the future.

Another improvement we would like to incorporate into the system is on the US transducer. We have used a 7 MHz linear array probe in this work. For imaging deeper structures, we would like to use a transducer with low center frequency (1–3 MHz) with a wide bandwidth. From the PA images, we observed that blood vessels were visualized as double-layered features (only lumen) with no information inside the vessels. This is mainly because of the bandwidth limitation of the US probe. For accurate PA imaging, it is of paramount importance to design high bandwidth US probes in the future. This may help to visualize entire blood vessels, instead of walls as in conventional linear array US probe-based 2D PA imaging. Quantitative PA imaging will also be more accurate with an improved US transducer with high bandwidth.

From the application perspective, an update to the commercial system with circular scanning and tomographic image rendering would be useful for many labs to use the proposed tomographic imaging. This will enable the use of an LED-based system (AcousticX) in animal imaging facilities and pre-clinical labs for tomographic imaging. With all the above improvements in light delivery, acoustic detection, and circular scanning, we foresee to target small animal imaging (brain and abdomen) and several clinical applications (RA monitoring, breast imaging) in the future. An affordable, point-of-care full view tomographic imaging system with structural, functional, and molecular contrast is expected to have a significant impact in pre-clinical/clinical diagnostic imaging and treatment monitoring.

6 Conclusion

In this chapter, we have demonstrated that using LED-based illumination and a linear transducer array, tomographic photoacoustic and ultrasound imaging system can be performed. A method to determine the optimal number of angular views for a full view tomographic imaging is presented and it was validated with experimental results. In terms of penetration depth, the pulse energy from an LED is a limitation for tomographic imaging. Hence, we have demonstrated the use of 576 LED elements in two configurations in the tomographic setting. Both the configurations developed here have great potential biomedical imaging applications. We have successfully demonstrated the applicability of our method in joint imaging both in ex vivo samples and in vivo human finger. However, we restricted ourselves to imaging applications involving smaller samples ($\leq$30 mm), as the pulse energy is still a bottleneck for larger tissues. A tomographic imaging time using our system is mainly limited by the scanning time. Although the imaging takes less than a minute, this speed can still be a limitation in some applications. Imaging speed is a trade-off factor that we considered to reduce the system cost. Provided that the longer pulse duration (10s of nanoseconds) is acceptable for photoacoustic imaging, this inexpensive, compact and safe to use tomographic system demonstrated in this work can find applications especially in point-of-care imaging.

References

1. S. Manohar, D. Razansky, Photoacoustics: a historical review. Advances in optics and photonics **8**(4), 586–617 (2016)
2. M. Xu, L.V. Wang, Photoacoustic imaging in biomedicine. Rev. Sci. Instrum. **77**(4), 041101 (2006)
3. K.J. Francis, S. Manohar, Photoacoustic imaging in percutaneous radiofrequency ablation: device guidance and ablation visualization. Phys. Med. Biol. **64**(18), 184001 (2019)
4. P.K. Upputuri, M. Pramanik, Recent advances toward preclinical and clinical translation of photoacoustic tomography: a review. J. Biomed. Opt. **22**(4), 041006 (2016)
5. L.V. Wang, Prospects of photoacoustic tomography. Med. Phys. **35**(12), 5758–5767 (2008)
6. S. Mallidi, G.P. Luke, S. Emelianov, Photoacoustic imaging in cancer detection, diagnosis, and treatment guidance. Trends Biotechnol. **29**(5), 213–221 (2011)
7. S. Manohar, M. Dantuma, Current and future trends in photoacoustic breast imaging. Photoacoustics (2019)
8. A. Dima, V. Ntziachristos, In-vivo handheld optoacoustic tomography of the human thyroid. Photoacoustics **4**(2), 65–69 (2016)
9. J. Jo, G. Xu, M. Cao, A. Marquardt, S. Francis, G. Gandikota, X. Wang, A functional study of human inflammatory arthritis using photoacoustic imaging. Sci. Rep. **7**(1), 15026 (2017)
10. D. Wang, Y. Wu, J. Xia, Review on photoacoustic imaging of the brain using nanoprobes. Neurophotonics **3**(1), 010901 (2016)
11. J. Xia, L.V. Wang, Small-animal whole-body photoacoustic tomography: a review. IEEE Trans. Biomed. Eng. **61**(5), 1380–1389 (2013)
12. K.J. Francis, E. Rascevska, S. Manohar, Photoacoustic imaging assisted radiofrequency ablation: illumination strategies and prospects, in *TENCON 2019 IEEE Region 10 Conference* (IEEE, New York, 2019)
13. N. Rao, K.J. Francis, B. Chinni, Z. Han, V. Dogra, Innovative approach for including dual mode ultrasound and volumetric imaging capability within a medical photoacoustic imaging camera system, in *Optical Tomography and Spectroscopy* (Optical Society of America, 2018), pp. OW4D–2
14. E. Rascevska, K.J. Joseph Francis, S. Manohar, Annular illumination photoacoustic probe for needle guidance in medical interventions, in *Opto-Acoustic Methods and Applications in Biophotonics IV*, vol. 11077 (International Society for Optics and Photonics, 2019), p. 110770L
15. J.L. Su, B. Wang, K.E. Wilson, C.L. Bayer, Y.S. Chen, S. Kim, K.A. Homan, S.Y. Emelianov, Advances in clinical and biomedical applications of photoacoustic imaging. Expert Opin. Med. Diagn. **4**(6), 497–510 (2010)
16. A. Fatima, K. Kratkiewicz, R. Manwar, M. Zafar, R. Zhang, B. Huang, N. Dadashzadesh, J. Xia, M. Avanaki, Review of cost reduction methods in photoacoustic computed tomography. Photoacoustics p. 100137 (2019)
17. M. Erfanzadeh, Q. Zhu, Photoacoustic imaging with low-cost sources: a review. Photoacoustics (2019)
18. Y. Zhu, G. Xu, J. Yuan, J. Jo, G. Gandikota, H. Demirci, T. Agano, N. Sato, Y. Shigeta, X. Wang, Light emitting diodes based photoacoustic imaging and potential clinical applications. Sci. Rep. **8**(1), 9885 (2018)

19. X. Wang, J.B. Fowlkes, J.M. Cannata, C. Hu, P.L. Carson, Photoacoustic imaging with a commercial ultrasound system and a custom probe. Ultrasound Med. Biol. **37**(3), 484–492 (2011)

20. L.V. Wang, Multiscale photoacoustic microscopy and computed tomography. Nat. Photonics **3**(9), 503 (2009)

21. C. Lutzweiler, D. Razansky, Optoacoustic imaging and tomography: reconstruction approaches and outstanding challenges in image performance and quantification. Sensors **13**(6), 7345–7384 (2013)

22. A.A. Oraevsky, V.A. Andreev, A.A. Karabutov, R.O. Esenaliev, Two-dimensional optoacoustic tomography: transducer array and image reconstruction algorithm, in *Laser-Tissue Interaction X: Photochemical, Photothermal, and Photomechanical*, vol. 3601, pp. 256–267 (International Society for Optics and Photonics, 1999)

23. G. Li, L. Li, L. Zhu, J. Xia, L.V. Wang, Multiview hilbert transformation for full-view photoacoustic computed tomography using a linear array. J. Biomed. Opt. **20**(6), 066010 (2015)

24. B.T. Cox, S.R. Arridge, P.C. Beard, Photoacoustic tomography with a limited-aperture planar sensor and a reverberant cavity. Inverse Prob. **23**(6), S95 (2007)

25. G. Li, J. Xia, K. Wang, K. Maslov, M.A. Anastasio, L.V. Wang, Tripling the detection view of high-frequency linear-array-based photoacoustic computed tomography by using two planar acoustic reflectors. Quant. Imaging Med. Surgery **5**(1), 57 (2015)

26. R.A. Kruger, W.L. Kiser Jr., D.R. Reinecke, G.A. Kruger, Thermoacoustic computed tomography using a conventional linear transducer array. Med. Phys. **30**(5), 856–860 (2003)

27. M. Oeri, W. Bost, S. Tretbar, M. Fournelle, Calibrated linear array-driven photoacoustic/ultrasound tomography. Ultrasound Med. Biol. **42**(11), 2697–2707 (2016)

28. K.J. Francis, B. Chinni, S.S. Channappayya, R. Pachamuthu, V.S. Dogra, N. Rao, Multiview spatial compounding using lens-based photoacoustic imaging system. Photoacoustics **13**, 85–94 (2019)

29. S. Agrawal, C. Fadden, A. Dangi, S.R. Kothapalli, Light-emitting-diode-based multispectral photoacoustic computed tomography system. Sensors **19**(22), 48–61 (2019)

30. J. Gateau, M.Á.A. Caballero, A. Dima, V. Ntziachristos, Three-dimensional optoacoustic tomography using a conventional ultrasound linear detector array: whole-body tomographic system for small animals. Med. Phys. **40**(1), 013302 (2013)

31. M. Omar, J. Rebling, K. Wicker, T. Schmitt-Manderbach, M. Schwarz, J. Gateau, H. López-Schier, T. Mappes, V. Ntziachristos, Optical imaging of post-embryonic zebrafish using multi orientation raster scan optoacoustic mesoscopy. Light Sci. Appl. **6**(1), e16186 (2017)

32. Y. Xu, L.V. Wang, G. Ambartsoumian, P. Kuchment, Reconstructions in limited-view thermoacoustic tomography. Med. Phys. **31**(4), 724–733 (2004)

33. K.J. Francis, Y. Boink, M. Dantuma, M.K.A. Singh, S. Manohar, W. Steenbergen, Tomographic imaging with an led-based photoacoustic and ultrasound system. Biomed. Opt. Express **11**(4) (2020) https://doi.org/10.1364/BOE.384548

34. B.E. Treeby, B.T. Cox, k-wave: Matlab toolbox for the simulation and reconstruction of photoacoustic wave fields. J. Biomed. Opt. **15**(2), 021314 (2010)

35. M. Jaeger, S. Schüpbach, A. Gertsch, M. Kitz, M. Frenz, Fourier reconstruction in optoacoustic imaging using truncated regularized inverse k-space interpolation. Inverse Prob. **23**(6), S51 (2007)

36. Z. Wang, A.C. Bovik, H.R. Sheikh, E.P. Simoncelli et al., Image quality assessment: from error visibility to structural similarity. IEEE Trans. Image Process. **13**(4), 600–612 (2004)

37. H.J. Kang, M.A.L. Bell, X. Guo, E.M. Boctor, Spatial angular compounding of photoacoustic images. IEEE Trans. Med. Imaging **35**(8), 1845–1855 (2016)

38. K.J. Francis, B. Chinni, S.S. Channappayya, R. Pachamuthu, V.S. Dogra, N. Rao, Two-sided residual refocusing for an acoustic lens-based photoacoustic imaging system. Phys. Med. Biol. **63**(13), 13NT03 (2018)

39. K.J. Francis, B. Chinni, S.S. Channappayya, R. Pachamuthu, V.S. Dogra, N. Rao, Characterization of lens based photoacoustic imaging system. Photoacoustics **8**, 37–47 (2017)

40. S. Ma, S. Yang, H. Guo, Limited-view photoacoustic imaging based on linear-array detection and filtered mean-backprojection-iterative reconstruction. J. Appl. Phys. **106**(12), 123104 (2009)
41. R.A. Kruger, P. Liu, Y.R. Fang, C.R. Appledorn, Photoacoustic ultrasound (paus)–reconstruction tomography. Med. Phys. **22**(10), 1605–1609 (1995)
42. D. Yang, D. Xing, S. Yang, L. Xiang, Fast full-view photoacoustic imaging by combined scanning with a linear transducer array. Opt. Express **15**(23), 15566–15575 (2007)
43. J. Gamelin, A. Maurudis, A. Aguirre, F. Huang, P. Guo, L.V. Wang, Q. Zhu, A real-time photoacoustic tomography system for small animals. Opt. Express **17**(13), 10489–10498 (2009)
44. L. Yu, F. Nina-Paravecino, D.R. Kaeli, Q. Fang, Scalable and massively parallel monte carlo photon transport simulations for heterogeneous computing platforms. J. Biomed. Opt. **23**(1), 010504 (2018)
45. S.L. Jacques, Optical properties of biological tissues: a review. Phys. Med. Biol. **58**(11), R37 (2013)
46. J. Staal, M.D. Abràmoff, M. Niemeijer, M.A. Viergever, B. Van Ginneken, Ridge-based vessel segmentation in color images of the retina. IEEE Trans. Med. Imaging **23**(4), 501–509 (2004)
47. Y.E. Boink, M.J. Lagerwerf, W. Steenbergen, S.A. Van Gils, S. Manohar, C. Brune, A framework for directional and higher-order reconstruction in photoacoustic tomography. Phys. Med. Biol. **63**(4), 045018 (2018)
48. G. Xu, J.R. Rajian, G. Girish, M.J. Kaplan, J.B. Fowlkes, P.L. Carson, X. Wang, Photoacoustic and ultrasound dual-modality imaging of human peripheral joints. J. Biomed. Opt. **18**(1), 010502 (2012)
49. P.J. van den Berg, K. Daoudi, H.J.B. Moens, W. Steenbergen, Feasibility of photoacoustic/ultrasound imaging of synovitis in finger joints using a point-of-care system. Photoacoustics **8**, 8–14 (2017)
50. J.R. Rajian, X. Shao, D.L. Chamberland, X. Wang, Characterization and treatment monitoring of inflammatory arthritis by photoacoustic imaging: a study on adjuvant-induced arthritis rat model. Biomed. Opt. Express **4**(6), 900–908 (2013)
51. J. Jo, G. Xu, Y. Zhu, M. Burton, J. Sarazin, E. Schiopu, G. Gandikota, X. Wang, Detecting joint inflammation by an led-based photoacoustic imaging system: a feasibility study. J. Biomed. Opt. **23**(11), 110501 (2018)
52. P. van Es, S.K. Biswas, H.J.B. Moens, W. Steenbergen, S. Manohar, Initial results of finger imaging using photoacoustic computed tomography. J. Biomed. Opt. **19**(6), 060501 (2014)
53. M. Nishiyama, T. Namita, K. Kondo, M. Yamakawa, T. Shiina, Ring-array photoacoustic tomography for imaging human finger vasculature. J. Biomed. Opt. **24**(9), 096005 (2019)

Functional and Molecular Photoacoustic Computed Tomography Using Light Emitting Diodes

Sumit Agrawal and Sri Rajasekhar Kothapalli

Abstract Photoacoustic Computed Tomography (PACT) has been widely explored for inexpensive non-ionizing functional and molecular imaging of small animals and humans. In order for light to penetrate into deep tissue, a bulky and high-cost tunable laser is typically employed. Light Emitting Diodes (LEDs) have recently emerged as smaller and cost-effective alternative illumination sources for photoacoustic (PA) imaging. We recently developed a portable, low-cost multispectral three-dimensional PACT system using multi-wavelength LED arrays, referred to as LED-PACT, enabling similar functional and molecular imaging capabilities as standard tunable lasers. In this chapter, first the capabilities of commercial LED array-based B-mode PA and Ultrasound (US) imaging system, referred to as LED-PAUS, to perform functional and molecular imaging with both in vivo and phantoms studies are presented. We also present the details of the development of LED-PACT system with essential hardware components, acquisition and reconstruction software needed for the implementation. This chapter also covers simulations and experimental results comparing the capabilities of LED-PACT system with commercial LED-PAUS system. LED-PACT and LED-PAUS system together demonstrate the potential of LED based photoacoustic imaging for pre-clinical and clinical applications.

1 Introduction

Photoacoustic tomography (PAT) is an emerging biomedical imaging technology that combines rich optical contrast with the high spatial resolution of ultrasound [1, 2]. In PAT, a biological specimen is illuminated with a sufficiently short laser pulse, which

S. Agrawal (✉) · S. R. Kothapalli
Department of Biomedical Engineering, Pennsylvania State University, University Park, State College, PA 16802, USA
e-mail: sua347@psu.edu

S. R. Kothapalli
Penn State Cancer Institute, Hershey, PA 17033, USA
e-mail: srkothapalli@psu.edu

© Springer Nature Singapore Pte Ltd. 2020
M. Kuniyil Ajith Singh (ed.), *LED-Based Photoacoustic Imaging*,
Progress in Optical Science and Photonics 7,
https://doi.org/10.1007/978-981-15-3984-8_11

causes an increase in temperature due to energy absorption. This temperature increase leads to thermoelastic expansion of the specimen, which induces ultrasonic pressure generation. The generated pressure then propagates to transducers surrounding the specimen. The received ultrasound signal is converted into an optical-absorption map of the specimen with different image formation methods. PAT has shown great potential in biomedicine for several applications ranging from clinical breast angiography [3, 4] to pre-clinical whole body imaging of small animals [5–7].

Based on the image formation methods, PAT setups can be grouped into two major categories. In the first method, a single element transducer mechanically scans the imaging object in two dimensions. Photoacoustic received signal at each acquisition provides a one-dimensional image along the acoustic focus of the transducer element. Elementary stitching of these one-dimensional images along the two dimensions of the mechanically scanned region of interest gives a three-dimensional optical absorption map of the specimen. PAT setups using this direct method of image formation are called as photoacoustic microscopy (PAM) setups [8–10]. In the second method, different geometries of multi-element transducer arrays are used to electronically scan the imaging object. Each element of the transducer has a larger acceptance angle compared to the previous case. Photoacoustic image formation involves complex computational reconstruction algorithms that can effectively merge the data from all transducer elements. PAT setups using these complex reconstruction methods are called as photoacoustic computed tomography (PACT) setups [11–15].

Common PACT setups comprise of a cumbersome Q-switched Nd:YAG laser [11–15]. These lasers typically have a nanosecond pulse width with hundreds of mJ pulse energy. Such high pulse energies can easily provide sufficient signal to noise ratios for applications ranging from organelles to small animals to human imaging. However, these laser sources, widely explored in a typical research setup, are not suitable for clinical applications due to their massive cost and substantial footprints.

Recently, many researchers have employed both laser diodes [16–22] as well as LEDs [22, 23] in the development of PA imaging systems. However, the emission from laser diodes still remain coherent and are still under the category of class-IV lasers. LEDs, unlike the class-IV lasers, offer a unique opportunity to operate with more flexibility and ease of use. The challenging aspect of using LEDs compared to laser sources for PA imaging is their low power pulses (about two orders of magnitude smaller than the typical class-IV lasers) struggling to provide sufficient signal from deeper tissue regions. To overcome this, sizable signal averaging is performed utilizing the orders of magnitude higher repetition rates of the LEDs.

Several groups have explored the LED based PA imaging applications utilizing the higher repetition rates to achieve acceptable signal to noise ratios. Further, to improve the pulse energies researchers have also explored the arrayed arrangement of LEDs [24–29]. In the arrayed format, the typical pulse energies increases from few μJ to hundreds of μJ and thus improves the penetration depth of LED-based PA imaging.

The previous studies [24–29] utilized the arrayed format of LED arrays only in a typical B-mode fashion, where they place two LED arrays adjacent to a linear ultrasound probe and acquire B-mode photoacoustic images along the depth dimension

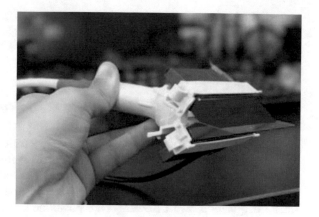

of the specimen. There are two disadvantages of using the LED arrays in the above format. First, in order for light and ultrasound to fall in the common imaging plane, the angled arrangement of LED arrays (as shown in Fig. 1) leads to a standoff region of about 8–10 mm.

Second, for deep tissue imaging applications, the signal to noise ratio reduces drastically with the lower light energy reaching to the deeper regions. We have addressed the above-mentioned problems with a novel approach of placing multiple LED arrays around the imaging specimen. Our setup not only provides a close-to-zero standoff region but also helps to increase the penetration depths in deep tissue imaging applications to about 35 mm, similar to the big lasers.

With this approach, ours is the first LED based three-dimensional PACT system that uses multiple LED arrays and a linear ultrasound probe to generate volumetric 3-D PACT images of the imaging object. Similar to the high-cost tunable lasers, our LED-PACT system allows for the multi-wavelength LED arrays housed in a cylindrical geometry to enable the functional and molecular imaging capabilities.

The rest of the chapter is organized as follows. Section 2 presents several phantom as well as in vivo studies demonstrating the capabilities of the commercial LED-PAUS system. Section 3 first describes our proposed LED-PACT system with the details of the development of the hardware as well as the reconstruction of PACT images. It also presents simulations and validation studies over different tissue mimicking phantoms comparing the capabilities of our proposed system with the LED-PAUS system and proposing solutions to overcome its limitations towards the goal of developing a clinically applicable, low-cost LED-PACT system.

2 LED-Based PAUS Imaging

The commercially available LED-based combined ultrasound (US)/photoacoustic (PA) B-mode imaging system, referred to as LED-PAUS system, explained below in Sect. 2.1 is capable of acquiring 2-D/3-D B-mode US/PA images in real time. In this section, we first introduce the commercially available LED based PAUS system and then present different studies involving LED based PAUS system.

2.1 Commercial LED-PAUS Imaging System

A commercially available LED-PAUS system (AcousticX, Cyberdyne Inc., Ibaraki, Japan) capable of acquiring interleaved PA and US B-mode images in real time was presented previously [26, 27]. This system uses LED arrays for PA excitation and a linear US transducer for ultrasonic excitation and detection. For B-mode 2-D/3-D US/PA acquisition, two LED arrays are positioned on either side of the US probe as shown in Fig. 1. Each of these LED arrays consists of four rows of 36 1 mm × 1 mm LEDs. These LED arrays are capable of delivering a maximum optical energy of 200 μJ per pulse and can be driven with a repetition rate of 1–4 kHz with pulse duration of 30–150 ns. The ultrasound probe is a lead zirconate titanate (PZT) 128-element linear array transducer having a pitch of 0.3 mm and total length of 38.4 mm. The central frequency of the transducer is 7 MHz and the measured −6 dB bandwidth is 75%. The ultrasound and photoacoustic modalities have sampling rates of 20 MHz and 40 MHz respectively. The US probe has an elevation focus of 15 mm, achieved with the help of an acoustic lens incorporated on top of the transducer array. In the following subsections, we have covered several phantom as well as in vivo studies published using this LED-PAUS system.

2.2 Capability of LED-PAUS System to Image Exogenous Contrast Agents

Exogenous contrast agents can be targeted for specific molecules or cells for preclinical and clinical applications. Photoacoustic contrast agents have significant feasibility to assist in monitoring and diagnosis of diseases [30–36]. This work from Hariri et al. [24] characterized the detection limit of some common small molecules used in photoacoustic imaging: ICG [31], MB [32, 33], and DiR [36]. These are NIR-sensitive, Food and Drug Administration (FDA)-approved contrast agents for both Fluorescent and photoacoustic imaging. Various concentrations were scanned, and the detection limits were calculated at three standard deviations above baseline. Figure 2a, e, i show MIP images of high concentrations of ICG (640, 320, and 160 μM), MB (6, 3, and 1.5 mM), and DiR (592, 296, and 148 μM). Figure 2b, f, j shows the average photoacoustic intensity along all ten ROIs for each tube associated with Fig. 2a, e, i.

Figure 2c, g, k show MIP images for the detection limit of ICG (36, 18, 9 μM, and DI water), MB (1.5, 0.75, 0.37 mM, and DI water), and DiR (136, 68, 34 μM, and DMSO). Figure 2d, h, l show the average photoacoustic intensity along the ROIs.

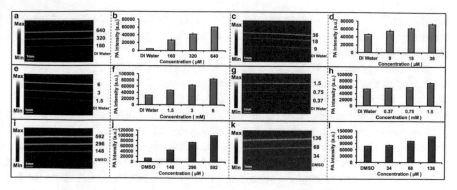

Fig. 2 Evaluation of LED-based photoacoustic imaging system for exogenous contrast agents. **a** MIP image of ICG solutions (640, 320, 160 μM, and DI water) with high concentration as positive control inside Teflon light wall tubes. **b** Statistical analysis of data in A. **c** MIP image detection limit experiment for ICG (36, 18, 9 μM, and DI water). **d** Statistical analysis of data in (**c**). **e** MIP images of MB solutions (6, 3, 1.5 mM, and DI water) with high concentration as positive control inside Teflon light wall tubes. **f** Statistical analysis of data in (**e**). **g** MIP image detection limit experiment for MB (1.5, 0.75, 0.37 mM, and DI water). **h** Statistical analysis of data in (**g**). **i** MIP images of DiR solutions (592, 320, 148 μM, and DMSO) with high concentration as positive control inside Teflon light wall tubes. **j** Statistical analysis of data in (**i**). **k** MIP image detection limit experiment for DiR (136, 68, 34 μM, and DMSO). **l** Statistical analysis of data in (**k**). All the error bars demonstrate standard deviation between different ROIs in each tube. Scan size is 10 mm [Reprinted with permission from 24]

The error bars show the standard deviation between ROIs in each tube. The limit of detection for ICG, MB, and DiR is 9 μM, 0.75 μM, and 68 μM, respectively when 850 nm is used for ICG and DiR and 690 nm is utilized for MB. The power for the LED-based system at 690 nm is almost three-fold lower than that at 850 nm. This might explain the lower detection limit for MB rather than ICG and DiR. This experiment also highlights how LED based systems are limited by the choice of wavelengths. While OPO based systems are capable of scanning wide-wavelength range, this system can only use two wavelengths at a time. Thus, it can be challenging to carefully match the absorption peak of the contrast agent with the excitation source. Nevertheless, many species absorb strongly at 690 or 850 nm and customized LED sets are available for ratiometric imaging.

2.3 Capability of LED PAUS System to Image Labeled Cells in Vivo

This work from Hariri et al. [24]. presents in vivo experiments to demonstrate the feasibility of LED-PAUS system for clinical applications. Several groups [37, 38] have previously used photoacoustic imaging for stem cell imaging. Here, labeled cells are used to understand the in vivo performance of this LED-PAUS system. This study used DiR which has been demonstrated as an effective contrast agent for cells checking [39, 40]. Figure 3a, e, i show photoacoustic images before injection of DiR, DiR @ HMSC, and HMSC, respectively.

The needle generates strong photoacoustic signal and overlaying the photoacoustic data with the ultrasound images offers more comprehensive structural information in addition to functional details from DiR-labeled cells. Figure 3b, f, J demonstrate B-mode photoacoustic/ultrasound images before injection. Figure 3c, d shows photoacoustic and photoacoustic/ultrasound images of injected DiR in the mice, respectively. These figures show strong photoacoustic signal in the presence of DiR. Figure 3g shows capability of LED-PAUS system to detect cells labeled with contrast agent. DiR was used as contrast agent for labeling the HMSCs. Unlabeled HMSCs

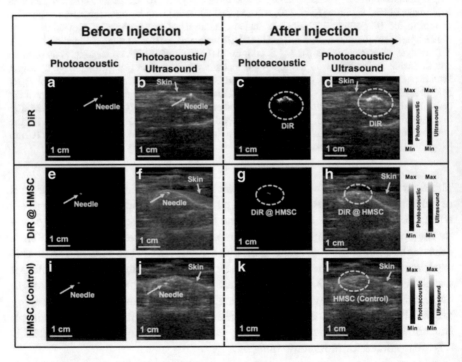

Fig. 3 In vivo evaluation of LED-PAUS system. **a** Photoacoustic image when needle is subcutaneously injected on spinal cord area before DiR injection. The needle has strong photoacoustic signal. **b** Photoacoustic/ultrasound image of (**a**). **c** Photoacoustic image after subcutaneously injection of DiR. **d** B-mode photoacoustic/ultrasound image of (**c**). **e** Photoacoustic image when needle is subcutaneously placed on the spinal cord area before HMSC labeled with DiR (DiR @ HMSC) injection. **f** B-mode photoacoustic/ultrasound image of (**e**). **g** Photoacoustic image after injection of HMSC labeled with DiR (DiR @ HMSC) on spinal cord. **h** B-mode photoacoustic/ultrasound image of (**g**).**i** Photoacoustic image in presence of needle before injection of unlabeled HMSC as control experiment. **j** B-mode photoacoustic/ultrasound image of (**i**). **k** Photoacoustic image of HMSC as control. This image shows no photoacoustic signal for HMSC. **l** B-mode photoacoustic/ultrasound image of (**k**). [Reprinted with permission from 24]

were also injected as control (Fig. 3k, l), but there was no increase in photoacoustic signal. Hence this study demonstrated the feasibility of LED-PAUS system for in vivo studies including photoacoustic cell imaging.

2.4 LED-PAUS System for Monitoring Angiogenesis in Fibrin Scaffolds

Laser speckle contrast analysis (LASCA), a widely used imaging technique within regenerative medicine, has high spatial resolution but offers limited imaging depth and is only sensitive to perfused blood vessels. As an emerging technology, PA imaging can provide centimeters of imaging depth and excellent sensitivity in vascular mapping. PA imaging in combination with conventional US imaging offers a potential solution to this challenge in regenerative medicine. This study from Zhu et al. [41] presented here used LED-PAUS dual system to image and monitor angiogenesis for 7 days in fibrin-based scaffolds subcutaneously implanted in mice. Scaffolds, with or without basic fibroblast growth factor (bFGF), were imaged on day 0 (i.e., post implantation), 1, 3, and 7 with both LASCA and LED-PAUS imaging systems. Quantified perfusion measured by LASCA and PA imaging were compared with histologically determined blood vessel density on day 7. Vessel density corroborated with changes in perfusion measured by both LASCA and PA. Unlike LASCA, PA imaging enabled delineation of differences in neovascularization in the upper and the lower regions of the scaffold. Overall, this study has demonstrated that PA imaging could be a noninvasive and highly sensitive method for monitoring vascularization at depth in regenerative applications.

2.4.1 In Situ Polymerization of Fibrin Scaffolds

This in vivo research [41] was conducted with the approval of the Institutional Animal Care and Use Committee at the University of Michigan. Female BALB/c mice (n = 5, 19.2 ± 1.0 g, 4–6 weeks old; Charles River Laboratories, Wilmington, MA) were anesthetized with isoflurane (5% for induction and 1.5% for maintenance). The lower dorsal hair was removed by shaving and applying depilatory cream (Nair, Church & Dwight Co, Ewing, NJ). The skin was disinfected with povidone-iodine (Betadine, Purdue Products L.P., Stamford, CT). The scaffold mixture (0.3 mL per implant) was injected subcutaneously using a 20-gauge needle (Becton Dickinson, Franklin Lakes, NJ) at two locations within the lower dorsal region and allowed to polymerize for 2 min before removal of the needle. The scaffold mixture consisted of the following: 10 mg/mL bovine fibrinogen (Sigma-Aldrich, St. Louis, MO) in Dulbecco's modified Eagle's medium (Life Technologies, Grand Island, NY), 0.05 U/mL bovine aprotinin (Sigma-Aldrich), 125 μg/mL Alexa Fluor 647-labeled human fibrinogen (Molecular Probes, Eugene, OR), 2 U/mL bovine thrombin (Thrombin-JMI, King Pharmaceuticals, Bristol, TN), 34 μg/mL bovine serum albumin (Sigma-Aldrich), and 6.6 mU/mL porcine heparin (EMD Millipore, Burlington, MA). Each mouse received one implant with 1 μg of bFGF (EMD Millipore) per scaffold, whereas the contralateral implant served as a negative control (i.e., 0 mg bFGF). The placement of the negative control (i.e., left or right side) was randomized in all mice.

2.4.2 LASCA and LED-PAUS Imaging of Scaffolds

The detailed experimental schedule, including scaffold implantation and imaging, is shown in Fig. 4. On days 2, 4, and 6, 50 μL of 20 μg/mL bFGF was subcutaneously injected into scaffolds initially containing bFGF on day 0 (i.e., +bFGF). Phosphate buffered saline (Life Technologies) was injected into the negative control scaffolds (i.e., −bFGF). LASCA and PA imaging procedures were done in no particular order on days 0, 1, 3, and 7. A brief schematic diagram of LASCA/PA imaging is also shown in Fig. 4. After the completion of imaging on day 7, the mice were euthanized, and the implants were retrieved for histology.

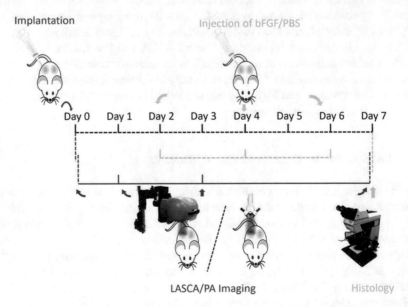

Fig. 4 The 7-day longitudinal experimental schedule. Scaffold implantation was done on day 0. LASCA/PA imaging was on day 0 (i.e., after implantation), 1, 3, and 7. bFGF or PBS was injected subcutaneously adjacent to each scaffold every 2 days. Scaffolds were retrieved on day 7 for H&E and CD31 staining. Each mouse received two scaffolds represented by the blue areas. bFGF, basic fibroblast growth factor; H&E, hematoxylin and eosin; LASCA, laser speckle contrast analysis; PA, photoacoustic; PBS, phosphate buffered saline [Reprinted with permission from 41]

LASCA imaging was employed to monitor the perfusion noninvasively and longitudinally in and around the subcutaneous implants placed in the lower dorsal region. To perform LASCA imaging, the mice were anesthetized with isoflurane and imaged with a PeriCam PSI HR (Perimed, Ardmore, PA) LASCA system. Figure 5a displays a macroscopic image of a mouse with implants along with longitudinal perfusion images from days 0 to 7. The ROIs are marked by black and red ellipses, indicating the −bFGF and +bFGF implants, respectively. The LASCA images qualitatively show more perfusion in the +bFGF scaf- folds than the −bFGF scaffolds in the days after implantation. Figure 5b shows a quantitative analysis of the ROIs, which was based on computing a relative change in average perfusion units for a given implant relative to day 0. Overall perfusion tended to increase over time, with the greatest increase observed for the +bFGF group relative to the −bFGF group, which approached statistical significance on day 7 (p = 0.055).

To perform LED-PAUS imaging, mice were anesthetized with isoflurane and secured to a platform in a prone position. The platform was partially submerged in a 37 °C water tank such that the implanted scaffolds were completely submerged for imaging. Two 850 nm LED arrays were used for imaging. A series of 2-D US and PA images of each scaffold were acquired at 10 Hz in the sagittal orientation while the probe was translated at 0.5 mm/s across the volume of the implant. Figure 6a

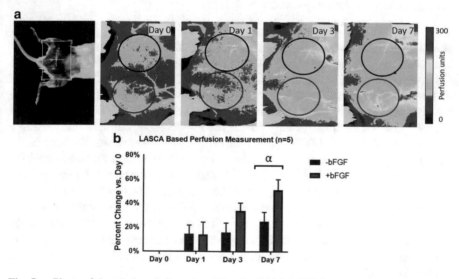

Fig. 5 **a** Photo of dorsal view (leftmost) and longitudinal LASCA images of a mouse with two subcutaneous implants. The ROIs were chosen based on the physical location of the implants, and are denoted by colored ellipses (red for +bFGF and black for −bFGF). For all LASCA images, the caudal direction is toward the left. ROI dimensions: 1.1 cm (major axis), 1.0 cm (minor axis). **b** Quantification of the change in perfusion relative to day 0, based on an ROI analysis of the LASCA images, shows an overall increase in perfusion over time. The greatest change in perfusion was observed on day 7, with +bFGF scaffolds trending toward greater perfusion than −bFGF scaffolds. α: −bFGF versus +bFGF on day 7 (p = 0.055). ROIs, regions of interest [Reprinted with permission from 41]

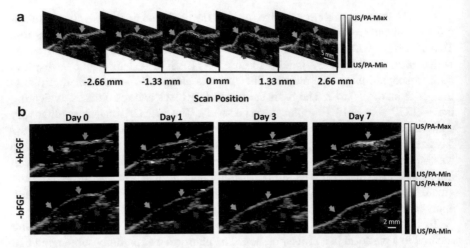

Fig. 6 Longitudinal LED-PAUS imaging of two subcutaneous implants. Green and red arrows indicate the upper and lower edges of the scaffold, as determined through the B-mode US. **a** A series of two-dimensional PAUS images from a +bFGF scaffold on day 7 at different scan positions. Note that only images within the range of −2.5 to 2.5 mm are used for MIP image. **b** A series of longitudinal MIP PAUS images of +bFGF and −bFGF scaffolds from the same mouse. PA intensity represented in red has the greatest difference on day 7. MIP, maximum intensity projected; US, ultrasound [Reprinted with permission from 41]

shows a series of 2-D PAUS images from a +bFGF scaffold on day 7 at different scan positions (i.e., sagittal planes). The PA signal, in red, is overlaid on the B-mode US image, in grayscale. Figure 6b shows a series of longitudinal MIP PAUS images of both scaffolds from the same mouse. Qualitatively, the +bFGF implant has a PA intensity that increased over time, especially adjacent to the skin. Some signal appears within the scaffold, likely due to the projection of multiple images into a single plane. With the −bFGF implant, only point-like contrast was detected in the scaffold, with no obvious trend during the 7-day experiment. The PA signal from the +bFGF and −bFGF scaffolds appeared similar on day 0. The results on different days of each mouse are normalized with day 0 and then compared the percentage change. The greatest difference was observed on day 7, which was also consistent with the LASCA results. On day 7, the +bFGF scaffold displayed a strong PA signal, especially between the skin and upper layer of the implant. Some PA signal was observed in the −bFGF scaffold, although at a lower level compared with the +bFGF scaffold.

Further, the perfusion in implants measured by both LASCA and LED-PAUS imaging techniques were validated by quantitative histology. Imaging was able to cover the entire scaffold volume, and enabled delineation of neo-vascularization in the upper and the lower regions of the scaffolds, respectively. The LED-PA imaging results well matched with the findings from histology, suggesting that PA imaging could be a non-invasive and highly sensitive method for monitoring angiogenesis at depth in regenerative applications.

2.5 High Speed Photoacoustic Imaging Using LED-PAUS System

Conventional laser-based PA imaging systems are not suitable for dynamic studies because of their low repetition rates and consequent low frame rates. LED's used in LED-PAUS system can be driven at high repetition rates of upto 4 kHz, offering the possibility of real-time PAUS imaging at frame rates close to 30 Hz. However, this frame rate is still not enough for applications involving dynamic tissue movements. This study from Sato et al. [42], presents a new high-speed (HSS) imaging mode in the LED-PAUS system. In this mode, instead of toggling between ultrasound and photoacoustic measurements, it is possible to continuously acquire only photoacoustic data for about 1.5 s with a time interval of 1 ms. With this improvement, photoacoustic signals can be recorded from the whole aperture (38 mm) at fast rate and can be reviewed later at different speeds for analyzing dynamic changes in the photoacoustic signals. This new high-speed feature opens up a feasible technical path for multiple dynamic studies that require high frame rates such as monitoring circulating tumor cells, voltage sensitive dye imaging, myocardial functional imaging etc.

This study validated the HSS mode by dynamically imaging the blood reperfusion in the finger of a human volunteer, thereby enabling the real-time measurement of the blood flow velocity. A rubber band was wrapped around the index finger, and the blood was pushed out as far as possible from the fingertip, causing temporary ischemia (Fig. 7a). Figure 7b shows the position of LED arrays on both sides of the US linear array probe and the finger placed between them. The finger and the LED-PAUS probe were positioned in a water bath as shown in Fig. 7c. After blocking the blood flow to the finger (by using rubber bands as shown in Fig. 7a), HSS mode was initiated and data acquisition was started. At the same instant, the rubber band wrapped around the finger was released in such a way that high-speed PA imaging can visualize the reperfusion of blood in the finger. To improve SNR, they pulsed the LED light source at 4 kHz and received the PA signal once every 0.25 ms. Four frames were averaged on board for improving SNR. To maintain high framerate, plane wave US imaging was used along with PA imaging. In addition, number of pulse-echo ultrasonic acquisitions was reduced to six frames to enable continuous

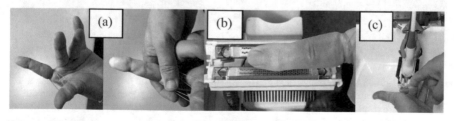

Fig. 7 **a** Rubber band used to block the blood flow to the finger, **b** positioning of finger and the LED-PAUS probe, **c** photograph of the probe and finger positioned inside the water bath [Reprinted with permission from 42]

high-speed PA signal acquisition for 1.5 s. Several measurements were done on the same finger of the volunteer to validate and evaluate the new high-speed imaging capability.

Figure 8 shows the results (first experiment) of imaging blood reperfusion in a human finger. Figure 8a shows the PA/US image at the start of acquisition, when there is almost no blood flow to the finger. Figure 2b–n shows the PA/US images at different time points (interval of 10 ms) during the high-speed acquisition. After the rubber band was released from the finger, PA signal intensity is clearly increasing in a blood vessel inside the imaging plane. It is visible that the blood vessel of interest is completely reperfused in about 130 ms.

Figure 9 shows the results of another example in which blood flow velocity is calculated from the reperfusion of blood in finger. The distance indicated by the dashed line was 6.4 mm, and the time taken for the blood to flow from the start point on the left side to the end on the right side was 35 ms. From this, blood flow was calculated to be about 18 cm/s.

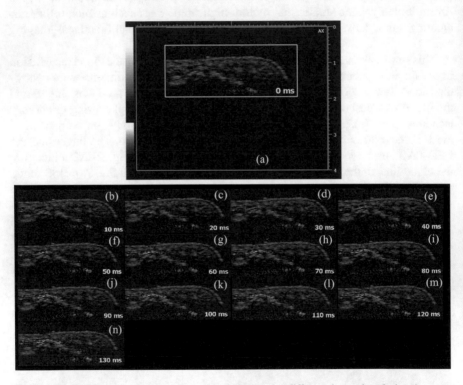

Fig. 8 LED-PAUS overlay images acquired and displayed at different time points during the reperfusion of blood vessel in a human finger. PA images are displayed in hot colormap and conventional US images are displayed in gray scale. It is clear from the images that blood is reperfused into one of the blood vessels as the time is increasing from 0–130 ms. By 130 ms, the blood vessel was completely reperfused [Reprinted with permission from 42]

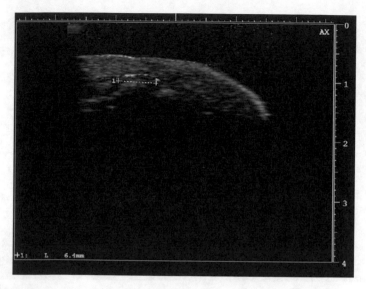

Fig. 9 LED-PAUS overlay image showing the reperfused blood vessel. Yellow line (indicated between '+' markers) shows the blood vessel from which the blood flow velocity was calculated [Reprinted with permission from 42]

2.6 Human Placental Vasculature Imaging Using LED-PAUS System

Minimally invasive fetal interventions, such as those used for therapy of twin-to-twin transfusion syndrome (TTTS), require accurate image guidance to optimize patient outcomes. Currently, TTTS can be treated fetoscopically by identifying anastomosing vessels on the chorionic (fetal) placental surface, and then performing photocoagulation. Incomplete photocoagulation increases the risk of procedure failure. Photoacoustic imaging can provide contrast for both haemoglobin concentration and oxygenation, and in this study, it was hypothesised that it can resolve chorionic placental vessels.

In this study, from Maneas et al. [43], to investigate the feasibility of the system to visualize superficial and subsurface placental vessels on the fetal chorionic placenta, a normal term placenta was collected with written informed consent after a caesarean section delivery at University College London Hospital. The umbilical cord was clamped immediately after the delivery to preserve the blood inside the vessels. The placenta was initially placed in a plastic container and subsequently it was coated with ultrasound gel for acoustic coupling and covered with cling film. The container was filled with water at room temperature for acoustic coupling and for free translation of the imaging probe. The experimental setup can be seen in Fig. 10.

LED-PAUS system with two 850 nm LED arrays was used to image several highly vascularized locations on the surface of the chronic fetal side. The linear stage was used to translate the imaging probe for 3-D image volume acquisition

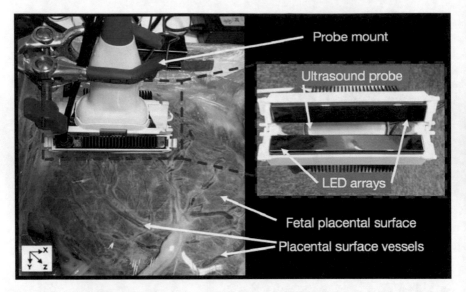

Fig. 10 Experimental setup to image a human placenta. A clamp was used to mount the ultrasound probe with the light emitting diode (LED) arrays to the linear motorized stage. The placenta was coated with ultrasound gel, covered with cling film, and placed inside a water-filled container [Reprinted with permission from 43]

along 40 mm scan region. Sample PA and US frames of a term human chorionic placental vasculature that were acquired in real-time are presented in Fig. 11. With PA imaging, superficial blood vessels and a subsurface structure were visible to a depth of approximately 5 mm from the placental chorionic fetal surface.

Some of these vascular structures were not apparent in US images. With US imaging, a large blood vessel located at a depth of approximately 7 mm could be identified, but this vessel was not apparent with PA imaging.

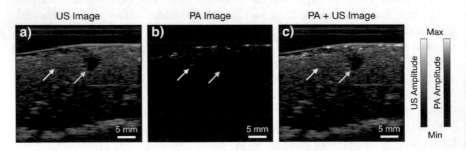

Fig. 11 Single frames of ultrasound (US), photoacoustic (PA), and merged US and PA images acquired from a human placenta, at one location (**a–c**). A large blood vessel (yellow arrow) that is visible in the US image was not visible in the PA image. A subsurface structure was visible with PA imaging (white arrow). All the images are displayed on logarithmic scales [Reprinted with permission from 43]

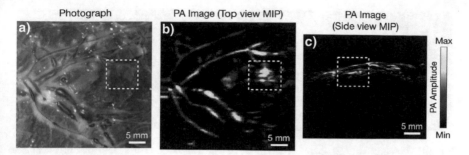

Fig. 12 Photograph (**a**) and photoacoustic (PA) images (**b, c**) of a portion of the human placenta. Superficial branching blood vessels are apparent in both the photograph and the PA images. High intensity PA signals (dashed squared box) that are not visible in the photograph might be attributable to subsurface vascular structures. The PA images are displayed on a logarithmic scale as maximum intensity projections (MIPs) of the reconstructed 3D photoacoustic image volume [Reprinted with permission from 43]

Figure 12 shows a photograph of the area that was imaged and the corresponding top and side maximum intensity projections (MIPs) of the reconstructed 3-D photoacoustic signals. Several superficial branching vessels were clearly resolved. In the top view MIP PA image, high intensity PA signals appeared to originate from vascular structures that were not visible in the photograph.

This feasibility study demonstrated that photoacoustic imaging can be used to visualize chorionic placental vasculature, and that it has strong potential to guide minimally invasive fetal interventions.

2.7 In Vivo Real-Time Oxygen Saturation Imaging Using LED-PAUS System

In this study, from Singh [44], potential of LED-based PAUS system in real-time oxygen saturation imaging is demonstrated using an in vivo measurement on a human volunteer. 2-D PA, US, and oxygen saturation imaging were performed on the index finger of a human volunteer. Results demonstrate that LED-based PAUS imaging system used in this study is promising for generating 2-D/3-D oxygen saturation maps along with PA and US images in real-time. Light illumination was provided by two combination LED arrays fixed on both sides of the US probe as shown in Fig. 13. Each LED arrays consists of 144 elements arranged in four rows. In this study, combination LED array were used in which first and third rows are embedded with 850 nm LED elements and second and fourth with 750 nm elements. For each LED array, energy per pulse is 50 μJ, and 100 μJ for 750 nm and 850 nm respectively. Light pulse duration can be varied from 30 to 100 ns and 70 ns pulse-width was used for the reported measurement.

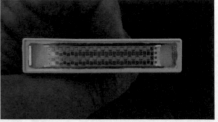

Fig. 13 Photograph of LED-based PAUS probe in which two LED arrays (750/850 nm) are placed on both sides of a linear array US probe (7 MHz) (left). Photograph of the LED array (right) with four rows of LED elements in which row 1 and 3 are 850 nm elements, row 2 and 4 are 750 nm elements. In this picture, 850 nm elements are activated and captured using an IR camera [Reprinted with permission from 44]

The system can pulse the LED arrays as well as transmit/acquire data parallel from all 128 elements of the US probe to generate interleaved PA/US (planewave or line-by-line) images at a frame rate of 30 Hz. System can drive the LED's at a rate of 4 kHz, providing the opportunity to average multiple frames without losing the frame rate. In the multispectral mode for oxygenation imaging, it can toggle between two wavelengths (750 and 850 nm in this case) at a rate of 4 kHz and provide oxygen saturation image overlaid with US or PA images at frame rates as high as 30 Hz.

Index finger of a human volunteer was immersed in water (imaging plane/location is marked in Fig. 14) and LED-PAUS probe was used to perform real-time (Frame rate: 10 Hz) interleaved oxygen saturation and US imaging. Real-time feedback allowed to align the probe with a pulsating radial artery (depth: 7 mm).

Figure 14 (right) shows the results of oxygenation imaging experiment using LED-PAUS system, in which tissue oxygenation map is overlaid on conventional US image. Alignment of probe was done in such a way that a pulsating radial artery was inside the imaging plane (this was verified using the pulsating nature of it while looking at one wavelength PA image). In the oxygenation image, arterial blood is visualized in red

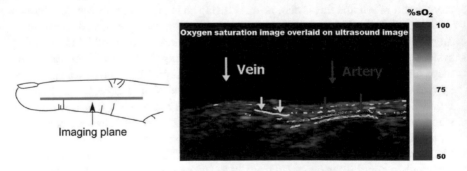

Fig. 14 Location on human finger where imaging was performed (left) and oxygen saturation image overlaid on conventional pulse echo image (right) [Reprinted with permission from 44]

color (oxygenation level close to 100%) and superficial venous blood in yellow shade (oxygenation level close to 70–75%) as expected. Also, it is interesting to see that skin melanin is visualized in blue color (above venous structure). It is commendable to mention that the system is capable to separate and visualize three different optical absorbers in tissue (arterial blood, venous blood, and melanin) using a simple 2 wavelength approach. It is important to note that, obtained oxygenation values are relative because of wavelength-dependent light fluence variations. However, form these results, it is clear that LED-PAUS system can differentiate arteries and veins in healthy human volunteers, where blood oxygenation changes in arterial and venous blood is expected to be less than around 20%.

2.8 *In Vivo Imaging of Human Lymphatic System Using LED-PAUS System*

Non-invasive in vivo imaging of lymphatic system is of paramount importance for analyzing the functions of lymphatic vessels, and for investigating their contribution to metastasis. This study from Singh [45] demonstrates the capabilities of LED-PAUS system to image human lymphatic system in real-time. Results demonstrate that the system is able to image vascular and lymphatic vessels simultaneously. This could potentially provide detailed information regarding the interconnected roles of lymphatic and vascular systems in various diseases, therefore fostering the growth of therapeutic interventions.

ICG has a peak spectral absorption at approximately 810 nm and has almost no optical absorption above the wavelength of 900 nm. Exploiting this, a combination LED array with 820 and 940 nm that can toggle between these wavelengths at a rate of 4 kHz was developed. This study hypothesized that ICG administered into the lymphatic vessels will generate PA signal only when the LED array emits light of 820 nm wavelength. On the other hand, other tissue optical absorbers like melanin and blood vessels possess absorption characteristic in both 820 and 940 nm wavelengths, which in turn generates PA signals in both wavelengths. For differentiating veins from lymphatic vessels with ICG, the PA images generated at 940 nm can be divided by images at 820 nm. After acquiring and reconstructing PA data for both wavelengths, images are normalized for variations in optical energy. Then the 940 nm PA image is divided by the 820 nm image and displayed real-time (10 Hz) along with the pulse echo US image. Following the division of images, the jet colormap is used in such a way that image intensity values close to or above 1 (veins) are color coded in red and those pixels with values below 0.5 (ICG) are coded in blue. Apart from this, the systems' user interface also enables visualization of the images acquired with the two wavelengths separately along with the US image for real-time validation of the experimental procedures.

To perform real-time imaging of human lymphatic vessels and blood vessels in vivo, measurements on the limb of a healthy volunteer were made. Under the guidance of a clinician, 0.1 ml of ICG (Diagnogreen 0.25%; Daiichi Sankyo Pharmaceutical, Tokyo, Japan) was injected subcutaneously into the first web space of the lower limb of a healthy person. ICG entering and flowing inside the lymph vessel was observed with a conventional fluorescent camera. After identifying the approximate position of a lymphatic vessel, real-time dual wavelength PAUS imaging was performed and the processed images were displayed (image generated by dividing 940 nm image with 820 nm image) along with conventional pulse echo US image.

Figure 15a and b shows PA images of 820 and 940 nm acquired at position 1 in which the probe was well aligned with a superficial vein. Beneath the melanin layer (marked with red arrow in 820 nm image), a superficial vein is clearly visualized in both PA images (double layered feature at a depth of ~2.3 cm, marked with pink arrow in 820 nm image). It is important to note that features visible in both the 820 and 940 nm images are identical at this position. Figure 15c and d show PA images of 820 and 940 nm acquired at position 2 in which the probe was aligned with a probable lymphatic vessel. At this position, common features evident in both wavelength images are likely to be veins (marked with pink arrows in 820 nm image). At a depth of ~2.6 cm, some bright features are visible only in the 820 nm image

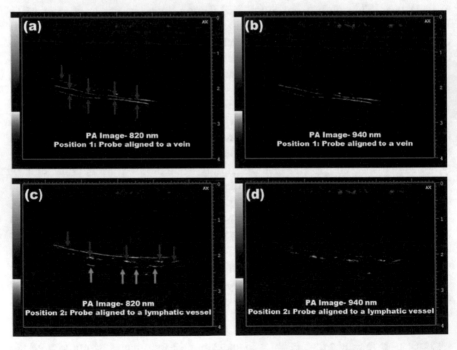

Fig. 15 **a** PA image—820 nm acquired when the probe was aligned to a superficial vein (position 1), **b** PA image—940 nm acquired when the probe was aligned to a superficial vein (position 1), **c** PA image—820 nm acquired when the probe was aligned to a lymphatic vessel (position 2), and **d** PA image—940 nm acquired when the probe was aligned to a lymphatic vessel (position 2). Red arrows—Melanin, Pink arrows—Veins, and Green arrows—Lymphatic vessel [Reprinted with permission from 45]

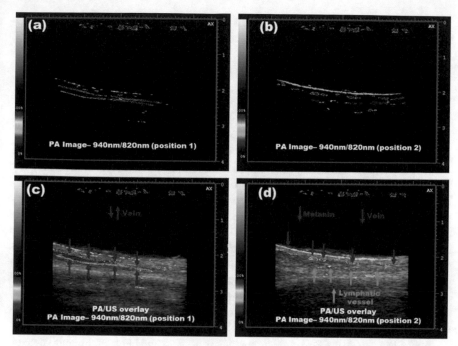

Fig. 16 **a** PA image—940/820 nm acquired when the probe was aligned to a superficial vein (position 1), **b** PA image—940/820 nm acquired when the probe was aligned to a lymphatic vessel (position 2), **c** 940/820 nm PA image overlaid on US image when the probe was aligned to a superficial vein (position 1), and **d** 940/820 nm PA image overlaid on US image when the probe was aligned to a lymphatic vessel (position 2). Red arrows—Melanin, Pink arrows—Veins, and Green arrows—Lymphatic vessel [Reprinted with permission from 45]

(Fig. 15c, marked with green arrows). These may be lymphatic vessels with ICG contrast.

Figure 16a shows the PA image obtained by dividing 940 and 820 nm images at position 1. As expected, the vein inside the imaging plane is visualized in red color since there is not much difference in absorption coefficient of venous blood in these two wavelengths. Figure 16b shows the PA image (940/820 nm) at position 2 where the probable lymphatic vessel is inside the imaging plane. In this case, we can see several features in blue color which confirms that these are lymphatic vessels. The ratio of 940 and 820 nm images resulted in low values in these areas since ICG inside lymphatic vessel is expected to absorb only 820 nm light. Figure 16c and d shows the same 940/820 nm PA images at two positions, but overlaid on conventional US images in gray scale.

From these results, it is evident that, by using a two-wavelength approach, we can simultaneously visualize and separate vein, lymphatic vessel, and melanin in vivo with high spatial and temporal resolution. It is worth mentioning that, these results were obtained at a frame rate of 10 Hz for two-wavelength PA imaging (along with processing for color-coded visualization) interleaved with pulse echo US imaging.

This is the first report on visualization of human lymphatic vessels using an LED-based PAUS system.

2.9 Multispectral Photoacoustic Characterization Using LED-PAUS System

The commercial LED-PAUS system has been so far used to perform either single or dual wavelength PA imaging. However, true advantages of photoacoustic imaging lies in being able to spatially un-mix multiple (more than two) tissue chromophores. This necessitates the use of more than two wavelengths. Towards this goal, this study from Shigeta [46] demonstrates the use of multiple wavelength LED arrays with the commercial LED-PAUS system. Here, the absorption spectra of ICG and porcine blood is photoacoustically measured using LED arrays with multiple wavelengths (405, 420, 470, 520, 620, 660, 690, 750, 810, 850, 925, 980 nm). Measurements were performed in a simple reflection mode configuration in which LED arrays where fixed on both sides of the linear array ultrasound probe. Phantom used consisted of micro-test tubes filled with ICG and porcine blood, which were placed in a tank filled with water, as shown in Fig. 17.

Figure 18a shows the PA signal intensities from ICG and hemoglobin with respect to different excitation wavelengths. Figure 18b shows PA intensities with respect to wavelengths normalized to optical output power of 660 nm LED arrays. It is evident that hemoglobin absorption is higher in lower wavelengths and ICG is highly absorbing in the range of 800–925 nm. For ICG, an absorption peak at the wavelength of 850 nm is visible. It is worth mentioning that the measured spectral behavior of

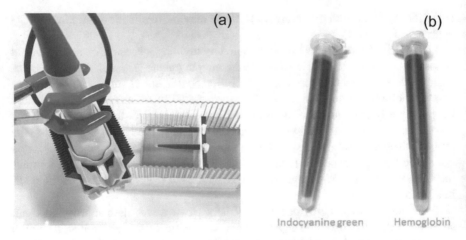

Fig. 17 **a** Experimental set up and, **b** photograph of micro-test tubes filled with ICG and porcine hemoglobin [Reprinted with permission from 46]

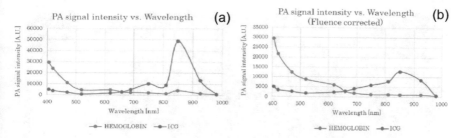

Fig. 18 **a** Measured PA signal intensities at different wavelengths, and **b** PA intensities with respect to wavelengths normalized to optical output power of 660 nm LED arrays [Reprinted with permission from 46]

ICG and hemoglobin (blood) is matching reasonably well with the reference values. These results demonstrate the potential capability of LED based PAUS system in performing clinical/pre-clinical multispectral photoacoustic imaging.

3 LED-Based PACT System

We recently developed a low-cost and portable PACT system [47, 48] using multi-wavelength LED arrays as optical sources and a linear ultrasound transducer array for the photoacoustic detection exploiting the commercial LED based B-mode PAUS system. In this section, we cover both the hardware implementation and the reconstruction of the LED-PACT system. We also present different validation studies comparing the capabilities of conventional LED-PAUS system with the LED based PACT system.

3.1 Design of LED-Based PACT System

A schematic of our experimental setup for the LED based photoacoustic computed tomography system is shown in Fig. 19d. A 3-D printed cylindrical tank, with an inner diameter of 38 mm, is used as the imaging cylinder. This imaging cylinder consists of five slots for housing the four LED arrays and one linear US probe. The cylinder with US probe and LEDs is attached to a rotational stage (PRMTZ8, ThorLabs Inc., Newton, NJ, USA). We have mounted this rotational stage in the inverted configuration as shown in Fig. 19d, f. The rotational axis of the stage is aligned with the vertical axes of the cylindrical imaging tank. The object to be imaged is embedded into a scattering phantom and this phantom is then attached to a phantom holder, as shown in Fig. 19d. The phantom is inserted into the imaging tank from the bottom (as shown in Fig. 19f) and is kept stationary during the rotational acquisition.

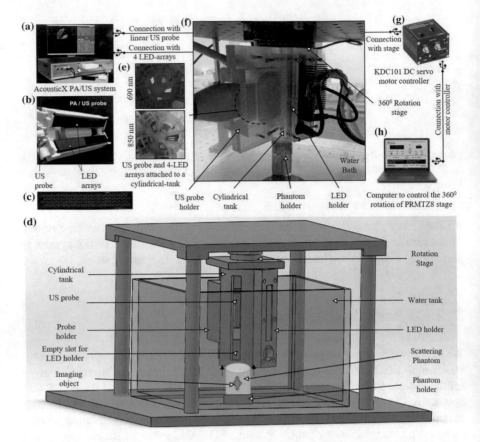

Fig. 19 Schematic representation of our LED based photoacoustic computed tomography (PAT) system. **a** Commercial LED-based combined photoacoustic/ultrasound (PA/US) system. **b** Typical arrangement of the two LED arrays and the US probe for B-mode PA/US imaging. **c** Optical image of an LED array consisting of four rows of 36 LEDs of dimension 1 mm × 1 mm. **d** Schematic showing the hardware implementation of our system consisting of a linear ultrasound probe and four LED arrays attached to an imaging cylinder mounted on rotation stage. Schematic also shows the placement of an imaging object. **e** Optical images showing the placement of four LED arrays (690/850 nm) and the US probe around a cylindrical tank of inner diameter 38 mm. **f** Photograph of our complete tomography setup with the LEDs, probe and the rotation stage controlled by the motor controller shown in (**g**). **h** Computer giving control signals to the controller for rotation of stage [Reprinted with permission from 47, 48]

Four LED holders (3-D printed, shown in Fig. 19d) housed in the cylindrical tank, hold the multi-wavelength LED arrays such that the stationary phantom is uniformly illuminated at all rotational positions. Figure 19e shows an example arrangement of four LED arrays (with each having 850/690 nm pair) and a linear US probe. The two sub-images in Fig. 19e shows the illumination at 690 and 850 nm wavelength achieved by selectively switching ON or OFF the respective wavelengths' LEDs from the LED array pairs, achieving uniform illumination at both the wavelengths.

The servo motor controller (KDC101 DC, ThorLabs Inc., Newton, NJ, USA, shown in Fig. 19g) controls the rotation stage with the help of a separate computer, shown in Fig. 19h. 2-D RF-Scan mode of commercial LED-PAUS system is used to acquire the B-mode PA images at all rotational positions and the acquired raw data is processed offline to reconstruct the volumetric 3-D photoacoustic computed tomography image at each wavelength.

While commercial B-mode LED-PAUS systems can only use two LED arrays at a time, the proposed LED-PACT system geometry is capable of employing more than four LED arrays used in this study. This not only increases the optical energy density inside the tissue medium, but also allows custom integration of multi-wavelength LEDs suitable for spectroscopic photoacoustic imaging.

3.2   LED-PACT Data Acquisition and Image Formation

The system acquires the PA data at a sampling rate of 40 MHz and the data is transferred to the graphical processing unit using USB 3.0 connection from the DAQ. One PA frame is acquired for each pulse of LED excitation. After the required PA frame averaging, the raw data is saved into the PC. During the 360° rotation, with a frame averaging of 2560 and pulse repetition frequency of 4 kHz, we have acquired a total of 90 frames, with 4° rotational steps. Total PA raw data corresponds to the 1024 time samples captured for each of the 128-transducer elements for each of these 90 frames, i.e. the size of the data matrix is $90 \times 1024 \times 128$.

A model-based time-reversal reconstruction algorithm [49] is applied, which numerically propagates the received photoacoustic pressure data back into the tissue medium from all the transducer elements. An Intel Xeon (2.1 GHz 32-core) based computer with 128 GB RAM and Nvidia Titan Xp GPU was used for the reconstruction. Since the computation time of model-based reconstruction methods increases exponentially with the size of the computational grid, two-dimensional computations can be orders of magnitude faster than three-dimensional computations. To be computationally efficient, we have applied the time-reversal algorithm in the 2-D plane formed by the rotation of a single transducer element. This is repeated for all 128-transducer elements individually, forming two-dimensional slices of the 3-D volume in 300 μm steps. The final 3-D image is formed by concatenating the 128 2-D slices into a single three-dimensional volume. With the above computational configuration, the total scan time for full tomography took around 102 s for each wavelength and about five minutes for the image reconstruction using the time-reversal algorithm.

3.3   Simulation and Experimental Studies

In this section, we first present a comparison study for the light fluence distribution along the imaging region in Sect. 3.3.1. Here, we have compared our proposed strategy for LED arrays' placement along the cylindrical tank for performing the LED based PACT, with the conventional approach of using two LED arrays in LED-PAUS. This section also presents several validation studies to compare the capabilities of our LED based PACT system with the conventional LED based PAUS system.

3.3.1 Comparison Study for Fluence Distribution

There are two distinct ways of numerically calculating the optical fluence distribution in a system. The first and most accurate method is to use a Monte Carlo simulation for light transport [50, 51]. While Monte Carlo simulations are able to more precisely model the light transport, there can be an exceptional computational burden for large spatial grids, especially in three dimensions. An alternative to Monte Carlo methods, when working in the photoacoustic regime, is to model the diffusion approximation to the radiative transport equation [52]. The diffusion approximation uses a partial differential equation that is computationally much faster than similar Monte Carlo methods, with acceptable accuracy when used in the photoacoustic regime [52]. In order to measure the difference in fluence distribution between LEDs and laser, as well as different geometries of light source, we have applied the finite difference method in MATLAB to solve the optical diffusion equation [52–54]:

$$\nabla \cdot D(x) \nabla \Phi(x) + \mu_a(x) \, \Phi(x) = 0, \; x \in X; \quad \Phi(y) = q(y), \; y \in \partial X \qquad (1)$$

In this equation, $D = [3(\mu_a + \mu_s')]^{-1}$ is the diffusion coefficient, where $\mu_a = 0.1 \, \text{cm}^{-1}$ and $\mu_s' = 10.0 \, \text{cm}^{-1}$ are the absorption and reduced scattering coefficients of the simulated tissue medium. $q(y)$ represents the optical source located at the boundary, either the LEDs or laser surrounding the region of interest.

As described in Sect. 2, the LED arrays used in this study consists of 1 mm x 1 mm LEDs arranged in a 2-D matrix form (4 rows and 36 columns). For a dual wavelength LED array (e.g. 850/690 nm), alternate two of the four rows are of same wavelength. Each of the element present in the array is separated by a 1 mm distance in all directions from the neighbor elements. We have defined these LED arrays in a three dimensional grid and have calculated the fluence distribution for a typical homogeneous tissue medium to study the effect of placing these arrays in the proposed approach.

Figure 20 presents a detailed study of the fluence distribution comparing our proposed approach (LED-PACT) with the conventional approach of placing the LED arrays for performing LED based PAUS imaging. Figure 20a shows an X–Y plane map with two LED arrays (shown in white, each array's two elements are shown corresponding to the 850 nm wavelength) placed at the left and right side of an ultrasound transducer (shown in blue) along the imaging circle (X–Y plane of the cylindrical imaging tank) shown in orange. Figure 20b shows a similar X–Y plane map with four LED arrays as proposed, placed along the orange circle and separated by 72° from each LED array and the ultrasound transducer (shown in blue). Figure 20c, d show the three-dimensional arrangement of the LED arrays along the cylindrical imaging

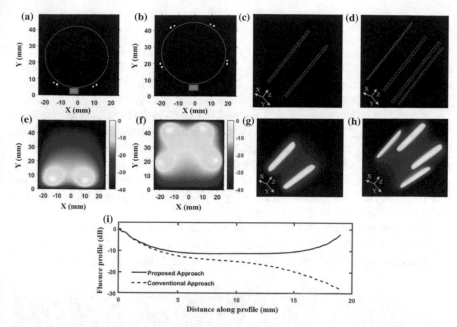

Fig. 20 Comparison of optical fluence distribution inside the tissue medium of 5 cm diameter for the proposed LED-based PACT and the conventional B-mode PAUS systems. **a** Schematic of LED-PAUS geometry consisting of two 850/690 nm LED arrays (white dots) and a linear ultrasound transducer array (blue rectangle) placed on the boundary (orange circle) of the tissue medium. **b** Similar schematic for the PACT geometry shows the arrangement four LED arrays with 72° separation and the ultrasound transducer. **c, d** show the corresponding 3-D schematic positions of the LED arrays in the PAUS and our PACT systems. Individual dots represent the positions of LED elements for a given wavelength in the dual-wavelength LED array. **e, f** show the 2-D optical fluence map inside the tissue medium for the two schematics shown in (**a**) and (**b**) respectively. **g, h** show the 3-D fluence distribution for the two schematics shown in (**c**) and (**d**) respectively. **i** Shows the fluence profile comparison along a diagonal in the imaging circles of the PAUS geometry shown in (**a**) and the PACT illumination shown in (**b**) [Reprinted with permission from 47]

tank for the conventional and the proposed approach, respectively. The X–Y cross-sections of the simulated fluence maps with the conventional and proposed approach are shown in Fig. 20e, f and the corresponding three dimension fluence are shown in Fig. 20g, h.

To study the changes in the fluence across the width of the imaging tank, we have plotted the magnitude (in dB) of the fluence in an X–Y plane (at the middle of the cylindrical tank) along the diagonal. Figure 20i shows the fluence profile plot for the conventional and the proposed approach clearly presenting the advantages of using the four LED arrays along the imaging tank.

3.3.2 Structural Imaging Studies

To compare the structural imaging capabilities of our developed PACT system with the commercial LED-PAUS system, we have imaged four pencil lead targets structurally placed in the imaging region. The four targets are embedded in to 1.5% agarose phantom cylinder with diameter of 35 mm, height of 80 mm.

To mimic the tissue scattering, intralipid (INTRALIPID 20% IV Fat Emulsion, VWR international, Radnor, PA, USA) was added to the agarose phantom to achieve reduced scattering coefficient of 10 cm^{-1}. Figure 21a–c shows the schematic side view, side view photograph and top view photograph of the phantom respectively. In this phantom, we have three 0.3 mm diameter pencil leads (marked as 2, 3 and 4 in Fig. 21a), and one bundle (group of 5 0.3 mm diameter pencil leads with total diameter of ~0.9 mm) marked as 1 in Fig. 21a.

To compare these results with the conventional B-mode LED-PAUS system 3-D scan, we have scanned the phantom in Fig. 21a using two LED arrays (850 nm) and the same ultrasound probe, with the arrangement shown in Fig. 19b. The 3-D

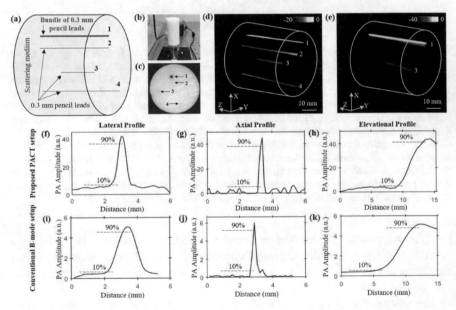

Fig. 21 Comparing the structural imaging capabilities of the LED-based PACT and PAUS systems using a pencil lead phantom. **a** Schematic showing side view of a tissue-mimicking intralipid phantom with four targets embedded. The depth of targets from the top surface of the phantom are as follows: 1 (bundle of five 0.3 mm pencil leads: at 10 mm), 2 (0.3 mm pencil lead: at 14 mm), 3 (0.3 mm pencil lead: at 23 mm), 4 (0.3 mm pencil lead: at 31 mm). **b, c** Photographs of the side and top views of the phantom. Reconstructed 3-D volume rendered photoacoustic image (**d**) using PACT, and **e** linear scanning of the conventional PAUS systems. Photoacoustic amplitude plots of the pencil lead target #3, located at 23 mm depth inside the medium, along the lateral, axial and elevational directions of the volume rendered (**f–h**) PACT image shown in (**d**) and **i–k** the linear scan image shown in (**e**) [Reprinted with permission from 47]

reconstructed volume rendered PA image for conventional b-mode 3-D scan is shown in Fig. 21e. Only the pencil lead "3", and the bundle "1" was seen in the reconstructed image. This is due to the following two limitations of using conventional B-mode LED-PAUS system. (1) With a lower LED light source, a small photoacoustic target located behind/below a thick/big photoacoustic target, is likely to be shadowed in conventional B-mode imaging. The same target when imaged with our proposed LED based PACT system, is detected with a decent SNR. (pencil lead "2" shown in Fig. 21). (2) While using conventional LED-PAUS system [24–29], the angular arrangement of LED arrays around the ultrasound probe as shown in Fig. 1, leaves about 8–10 mm of standoff region. The pencil lead "4" is about 31 mm from the top surface of the phantom. Hence, with the 8–10 mm of standoff, the conventional B-mode LED-PAUS system having a maximum data acquisition capability of 38 mm, fails to detect this target. However, due to negligible standoff with our PACT configuration, the same pencil lead target "4" is detected with a decent SNR.

We further studied the spatial resolution of our proposed LED based PACT system using the broken pencil lead "3" shown in Fig. 21a. A comparison study for LED-PACT versus LED-PAUS system is performed by calculating the lateral, axial and elevational resolutions with the plots shown in Fig. 21f–k. Peak photoacoustic amplitude for the pencil lead target "3" was plotted with respect to the lateral, axial and elevational distance as shown in Fig. 21f–h and i–k respectively for PACT and PAUS scans. The profiles shown in the plots were used to estimate the resolution of the system. Half of the distance between 90 and 10% of the peak photoacoustic amplitude was calculated as 300 μm in the lateral direction for PACT system as shown in Fig. 21f. Lateral resolution for PAUS system was calculated as 600 μm as shown in Fig. 21i. Similarly, the axial, elevational resolutions of PACT system and PAUS system were calculated as 120 μm, 2.1 mm and 130 μm, 3 mm respectively.

These experiments demonstrated that the PACT system can see through the shadow (blind spot) imaging regions of the conventional PAUS systems and visualize smaller targets hiding behind larger targets. These experiments also demonstrated that the spatial resolutions of the PACT system are better than the PAUS. These advantages can be attributed to the fact that the PACT system enables more uniform illumination of the imaging region during the 360° rotation. The conventional LED-PAUS systems require the imaging head to be 10 mm above the tissue surface to achieve uniform illumination of the phantom. This 10 mm standoff is usually filled with ultrasound coupling medium and leads to several complications such as (1) creation of bubbles and associated artifacts during the linear scan, (2) ultrasound and optical attenuation inside the thick coupling medium, (3) uncomfortable imaging of living subjects, and (4) extended imaging depth and computer memory which reduces imaging speed. The LED-PACT system demonstrated here required no such stand-off and therefore could image all 4 pencil lead targets inside the phantom, whereas the LED-PAUS system misses the target-4 at 31 mm depth.

3.3.3 Dual-Wavelength Imaging with Our PACT System

In this section, we have presented the capabilities of our system to perform dual wavelength photoacoustic computed tomography. To perform dual-wavelength imaging, we have employed the 850/690 nm LED array pairs. Each of the four LED arrays placed around the cylinder tank can be toggled to provide either 850 or 690 nm light, as shown in Fig. 19e. At each rotation step of 4°, two different frames are acquired corresponding to the two wavelengths of LEDs used. We have validated our approach by imaging two phantoms, with one having only endogenous photoacoustic targets (blood and melanin) and the other having a combination of blood with an exogenous contrast, i.e. Indocyanine-green (ICG).

Figure 22a shows the geometry of our first phantom. Here, highly oxygenated blood (Bovine Blood CITR, Carolina Biological Supply, Charlotte, NC, USA) and 0.1 mM Melanin solution (M8631, Sigma-Aldrich, St. Louis, MO, USA) were filled separately in 0.5 mm outer diameter tubes. These tubes were embedded in to 1.5% agarose phantom cylinder with diameter of 35 mm, height of 80 mm. To mimic the tissue scattering, intralipid (INTRALIPID 20% IV Fat Emulsion, VWR international, Radnor, PA, USA) was added to the agarose phantom to achieve reduced scattering

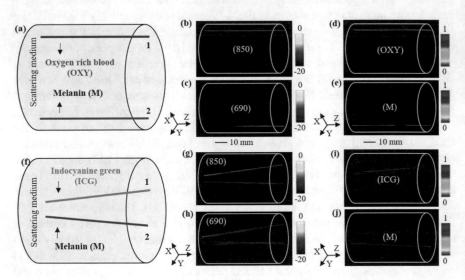

Fig. 22 Dual-wavelength imaging with LED based photoacoustic computed tomography system: validation over two phantoms. **a** First phantom: showing geometrical placement of the Oxygen rich blood (OXY) tube and a Melanin (M) tube inside scattering phantom. **b, c** 3-D PACT images of the phantom with 850 and 690 nm LED light illumination respectively. **d, e** Spectrally unmixed 3-D volumetric images of the phantom using the results in (**b, c**) highlighting the spatial distribution of (OXY) and (M) tubes respectively. **f** Second phantom: showing geometrical placement of the ICG tube and a Melanin (M) tube inside scattering phantom. **g, h** 3-D PACT images of the phantom at 850 nm and 690 nm illumination respectively. **i, j** Spectrally unmixed 3-D volumetric images of the phantom highlighting the spatial distribution of (ICG) and (M) tubes respectively [Reprinted with permission from 48]

coefficient of 10 cm^{-1}. Figure 22b, c shows the reconstructed volumetric 3-D PACT images for the above phantom with the 850 nm and 690 nm wavelengths respectively. With the dual wavelength tomography data acquired with our setup, we also applied a linear unmixing technique to separate the two types of chromophores imaged, viz. oxy rich blood and melanin. As the absorption of these chromophores differs significantly at the above two wavelengths, we could separate them spatially, as shown in Fig. 22d, e.

In our second study, we arranged ICG (with 1250 μM concentration) and Melanin (1 mM concentration) tubes inside similar scattering medium. Figure 22f shows the geometry of our second phantom. Figure 22g, h shows the reconstructed volumetric 3-D PACT images for our second phantom with the 850 nm and 690 nm wavelengths respectively. We further applied linear unmixing technique to separate these two chromophores, viz. ICG and melanin, and were able to separate them spatially, as shown in Fig. 22i, j.

3.3.4 Oxygen Saturation with Our PACT System

We further validated the vascular oxygen saturation imaging capabilities of the LED-PACT system by imaging a human finger mimicking phantom using four dual-wavelength 850/690 nm LED arrays. The human finger anatomy can be understood with the help of the schematic shown in Fig. 23a. For mimicking the finger, we used a high scattering intralipid phantom (reduced scattering coefficient of 15 cm^{-1}) and embedded an animal bone inside it. We further embedded the two types of blood tubes (Bovine Blood CITR, Carolina Biological Supply, Charlotte, NC, USA), namely, high-oxygenated (oxy-rich) tube and the low-oxygenated (oxy-poor) tube, both with outer diameter of 0.5 mm. The side view and top view optical images of the phantom are shown in Fig. 23b, c respectively.

The experimental setup used to image this phantom consisted of four 850/690 nm LED array pairs. Each of the four LED arrays placed around the cylinder tank can be toggled to provide either 850 or 690 nm light, as shown in Fig. 19e. At each rotation step of 4°, two different frames are acquired corresponding to the two wavelengths of LEDs used. Figure 23d–f represents the ultrasound, photoacoustic (at 850 nm) and co-registered (US + PA) frames at a single rotation step, captured during the full tomography acquisition. The structure of the finger is visible in the US frame whereas the presence of blood tubes can be seen using PA frame. With our proposed approach, the full 3-D volume rendered PACT images generated for the above phantom are shown in Fig. 23g, h, with 850 nm and 650 nm LED illumination respectively. Figure 23i shows the volumetric oxygen saturation map of the finger phantom obtained using the PACT images at the mentioned two wavelengths.

The LED-PACT configuration delivered higher (400 μJ) pulse energy for each 850 nm as well as 690 nm wavelengths than possible with the conventional LED-PAUS systems that can accommodate only two such LED arrays. This enabled mapping of vascular oxygen saturation from deeper regions.

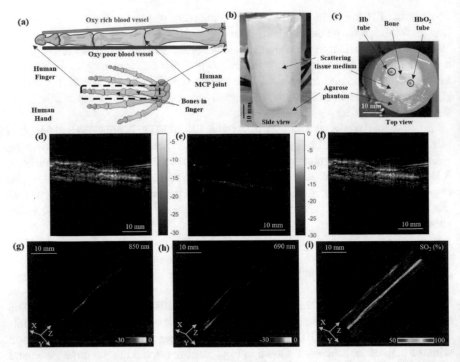

Fig. 23 LED-PACT-imaging of vascular oxygen saturation using a human finger-mimicking phantom. **a** Schematic sketch of a typical human finger with the location of bones and the blood vessels (oxy rich: HbO$_2$; and oxy poor: Hb). **b, c** Show the side and the top view photographs of the finger phantom consisting of bone and blood vessels. **d** The ultrasound (US), **e** photoacoustic (PA), and **f** co-registered US + PA frames acquired at one of the rotational steps during the 360° rotation around the phantom. **g, h** The reconstructed 3-D volume rendered PACT images of the finger phantom using 850 nm and 690 nm LED illuminations respectively. **i** Shows the spectrally unmixed volumetric oxygen saturation map for the finger phantom [Reprinted with permission from 47]

3.3.5 Multi-spectral Imaging with Our PACT System

One of the main advantages of our LED-PACT system, compared to the existing LED-based PAUS system, is that it can be easily adapted to allow custom designed multi-wavelength excitation using multiple LED arrays, to enable similar functional and molecular imaging capabilities of tunable lasers. This is demonstrated by imaging a tissue mimicking phantom embedded with three chromophores having different optical absorption spectra using two dual-wavelength 850/690 nm LED arrays and two 470 nm LED arrays. Each of these four LED arrays placed around the cylinder tank can be selectively switched "ON" or "OFF" to provide either 850, 690 or 470 nm light. At each rotation step of 4°, three different frames are acquired corresponding to the three wavelengths of LEDs used. In this study, we have imaged three biologically relevant chromophores, i.e. Indocyanine-green (ICG), Methylene blue (MB) and melanin, embedded in a high scattering phantom.

Figure 24a shows the geometry of our phantom. Here, 1 mM Melanin solution, 1 mM ICG solution and 1 mM MB solution were filled separately in 0.5 mm outer diameter tubes. These tubes were embedded in to 1.5% agarose phantom cylinder with diameter of 35 mm, height of 80 mm. To mimic the tissue scattering, intralipid (INTRALIPID 20% IV Fat Emulsion, VWR international, Radnor, PA, USA) was added to the agarose phantom to achieve reduced scattering coefficient of 10 cm^{-1}. Figure 24b shows the optical top view image of the phantom. Figure 24c–e shows the arrangement of US probe and the four LED arrays used for illuminating the phantom with 850 nm, 690 nm, and 470 nm light respectively. To maintain uniform distribution of light for the three wavelengths, out of the two 850/690 nm LED arrays, one is placed close to the transducer and the other one is placed at diagonally opposite

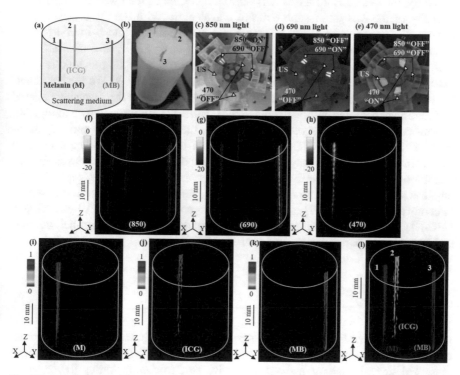

Fig. 24 LED based multispectral photoacoustic computed tomography. **a** Schematic view and **b** a photograph of the tissue mimicking cylindrical phantom of 35 mm diameter and 80 mm height. The phantom is embedded with 0.5 mm polyethylene tubes filled with 1 mM concertation solutions of melanin (M), indocyanine-green (ICG), and methylene blue (MB) **c–e** Show the arrangement of ultrasound (US) transducer array, two 850/690 nm LED arrays and two 470 nm LED arrays with sequentially switched 850 nm, 690 nm and 470 nm wavelength emissions from the arrays. **f–h** 3-D PACT images of the phantom acquired with 850 nm, 690 nm and 470 nm illumination respectively. **i–k** Spectrally unmixed volumetric images of the phantom, obtained from the multispectral PA images in (**f–h**), show the spatial distribution of M, ICG, and MB tubes respectively. **l** Superimposed 3-D unmixed image of the three chromophores. Scale bar is 10 mm in the images (**f–l**) [Reprinted with permission from 47]

corner. Similarly, one 470 nm LED array is placed closed to the transducer and the other one is placed at diagonally opposite corner. Figure 24c shows the configuration where we only switch ON the 850 nm light from the 850/690 nm pairs. The total output energy in this case is 200 µJ. Similarly, in Fig. 24d, we switch ON the 690 nm light from the 850/690 pairs giving the same output energy of 200 µJ. To maintain the same energy and distribution for 470 nm light, we switch ON only half of the LEDs in each of the 470 nm arrays as shown in Fig. 24e.

Figure 24f–h shows the reconstructed volumetric PAT images for the above phantom with the 850 nm, 690 nm, and 470 nm wavelengths respectively. Since the optical absorption of melanin decreases with increase in the optical wavelength, the PA contrast of the melanin is higher at 470 nm and lower in the 870 nm PACT images. Similarly, the peak absorption of MB ~680 nm correlates well with the highest PA intensity of the MB tube at the 690 nm wavelength. With the three wavelengths tomography data acquired with our setup, we also applied a linear unmixing technique to separate the three types of chromophores imaged, viz. Melanin, ICG, and MB. Based on these spectral trends, the linear spectral unmixing technique could easily separate the three chromophores, as shown in Fig. 24i–k. We further superimposed the three unmixed images to visualize and confirm the respective spatial distribution of the three chromophores in the given 3-D volume, as shown in Fig. 24l.

This current study is designed to demonstrate the above described various advantages of our novel LED-PACT system using proof-of-concept experiments on tissue mimicking phantoms. The imaging performance of the LED-PACT system can be further improved from multiple directions. This includes a better imaging geometry that employs more than 4 LED arrays, faster data acquisition, model-based image reconstruction algorithms [55–57] and deep learning approaches [58–60]. Fully developed LED-PACT will can be validated on living subjects such as imaging small animals and human body parts such as finger, wrist and breast.

4 Conclusion

In this chapter, we first discussed the commercial LED-based dual mode Photoacoustic and Ultrasound imaging system, referred to as LED-PAUS system. We have presented several phantom as well as in vivo studies demonstrating the capabilities as well as limitations of the commercial LED-PAUS system. This chapter also presents our recently developed LED based photoacoustic computed tomography system, referred to as LED-PACT system that integrates four LED arrays and a linear ultrasound transducer array in a cylindrical housing. The LED-PACT system has several benefits compared to the existing LED-based PAUS systems, such as multispectral photoacoustic imaging, better spatial resolution, uniform illumination, and improved imaging depth. Validation experiments on different tissue mimicking phantoms demonstrated the structural, functional and molecular photoacoustic imaging capabilities of the system. With further optimization, such as increase in the number of LED arrays and model-based image reconstruction, LED based PACT imaging systems herald a promising biomedical imaging tool of living subjects.

Acknowledgements LED-PACT project was partially funded by the NIH-NIBIB R00EB017729-04 (SRK) and Penn State Cancer Institute (SRK). We also acknowledge the support of NVIDIA Corporation with the donation of the Titan X Pascal GPU used for the reconstruction of LED-PACT images. We further thank CYBERDYNE Inc. for their technical support.

References

1. P. Beard, Biomedical photoacoustic imaging. Interface Focus **1**, 602–631 (2011)
2. X. Wang, Y. Pang, G. Ku, X. Xie, G. Stoica, L.V. Wang, Noninvasive laser-induced photoacoustic tomography for structural and functional in vivo imaging of the brain. Nat. Biotechnol. **21**, 803 (2003)
3. L. Lin, P. Hu, J. Shi, C.M. Appleton, K. Maslov, L. Li, R. Zhang, L.V. Wang, Single-breath-hold photoacoustic computed tomography of the breast. Nat. Commun. **9**, 2352 (2018)
4. R.A. Kruger, C.M. Kuzmiak, R.B. Lam, D.R. Reinecke, S.P. Del Rio, D. Steed, Dedicated 3D photoacoustic breast imaging. Med. Phys. **40**, 113301 (2013)
5. L. Li, L. Zhu, C. Ma, L. Lin, J. Yao, L. Wang, K. Maslov, R. Zhang, W. Chen, J. Shi et al., Single-impulse panoramic photoacoustic computed tomography of small-animal whole-body dynamics at high spatiotemporal resolution. Nat. Biomed. Eng. **1**, 0071 (2017)
6. H.P.F. Brecht, R. Su, M.P. Fronheiser, S.A. Ermilov, A. Conjusteau, A.A. Oraevsky, Whole-body three-dimensional optoacoustic tomography system for small animals. J. Biomed. Opt. **14**, 064007 (2009)
7. R. Ma, A. Taruttis, V. Ntziachristos, D. Razansky, Multispectral optoacoustic tomography (MSOT) scanner for whole-body small animal imaging. Opt. Express **17**, 21414–21426 (2009)
8. H. Chen, S. Agrawal, A. Dangi, C. Wible, M. Osman, L. Abune, H. Jia, R. Rossi, Y. Wang, S.R. Kothapalli, Optical-resolution photoacoustic microscopy using transparent ultrasound transducer. Sensors **19**(24), 5470 (2019)
9. K. Maslov, H.F. Zhang, S. Hu, L.V. Wang, Optical-resolution photoacoustic microscopy for in vivo imaging of single capillaries. Opt. Lett. **33**, 929–931 (2008)
10. T.T. Wong, R. Zhang, C. Zhang, H.C. Hsu, K.I. Maslov, L. Wang, J. Shi, R. Chen, K.K. Shung, Q. Zhou et al., Label-free automated three-dimensional imaging of whole organs by microtomy-assisted photoacoustic microscopy. Nat. Commun. **8**, 1386 (2017)
11. L. Xi, H. Jiang, High resolution three-dimensional photoacoustic imaging of human finger joints in vivo. Appl. Phys. Lett. **107**, 063701 (2015)
12. K. Fukutani, Y. Someda, M. Taku, Y. Asao, S. Kobayashi, T. Yagi, M. Yamakawa, T. Shiina, T. Sugie, M. Toi, Characterization of photoacoustic tomography system with dual illumination, in *Photons Plus Ultrasound: Imaging and Sensing 2011*, vol. 7899 (International Society for Optics and Photonics, 2011), p. 78992J
13. X. Wang, D.L. Chamberland, D.A. Jamadar, Noninvasive photoacoustic tomography of human peripheral joints toward diagnosis of inflammatory arthritis. Opt. Lett. **32**, 3002–3004 (2007)
14. P. van Es, S.K. Biswas, H.J.B. Moens, W. Steenbergen, S. Manohar, Initial results of finger imaging using photoacoustic computed tomography. J Biomed Opt **19**, 060501 (2014)
15. Z. Deng, C. Li, Noninvasively measuring oxygen saturation of human finger-joint vessels by multi-transducer functional photoacoustic tomography. J Biomed Opt **21**, 061009 (2016)
16. M. Zafar, K. Kratkiewicz, R. Manwar, M. Avanaki, Development of low-cost fast photoacoustic computed tomography: system characterization and phantom study. Appl Sci **9**, 374 (2019)
17. A. Dangi, S. Agrawal, J. Lieberknecht, J. Zhang, S.R. Kothapalli, Ring Ultrasound transducer based miniaturized photoacoustic imaging system. IEEE, 1–4 (2018 IEEE SENSORS, 2018)

18. A. Dangi, S. Agrawal, S. Tiwari, S. Jadhav, C. Cheng, G.R. Datta, S. Trolier-McKinstry, R. Pratap, S.R. Kothapalli, Ring PMUT array based miniaturized photoacoustic endoscopy device, in *Photons Plus Ultrasound: Imaging and Sensing*, vol. 10878 (International Society for Optics and Photonics, February 2019), p. 1087811
19. J. Zhang, S. Agrawal, A. Dangi, N. Frings, S.R. Kothapalli, Computer assisted photoacoustic imaging guided device for safer percutaneous needle operations, in *Photons Plus Ultrasound: Imaging and Sensing*, vol. 10878 (International Society for Optics and Photonics, Feb 2019), p. 1087866
20. A. Dangi, S. Agrawal, G.R. Datta, V. Srinivasan, S.R. Kothapalli, Towards a low-cost and portable photoacoustic microscope for point-of-care and wearable applications. IEEE Sens. J. (2019)
21. K. Daoudi, P. Van Den Berg, O. Rabot, A. Kohl, S. Tisserand, P. Brands, W. Steenbergen, Handheld probe integrating laser diode and ultrasound transducer array for ultrasound/photoacoustic dual modality imaging. Opt. Express **22**, 26365–26374 (2014)
22. H. Zhong, T. Duan, H. Lan, M. Zhou, F. Gao, Review of low-cost photoacoustic sensing and imaging based on laser diode and light-emitting diode. Sensors **18**, 2264 (2018)
23. T.J. Allen, P.C. Beard, High power visible light emitting diodes as pulsed excitation sources for biomedical photoacoustics. Biomed. Opt. Express **7**, 1260–1270 (2016)
24. A. Hariri, J. Lemaster, J. Wang, A.S. Jeevarathinam, D.L. Chao, J.V. Jokerst, The characterization of an economic and portable LED-based photoacoustic imaging system to facilitate molecular imaging. Photoacoustics **9**, 10–20 (2018)
25. Y. Zhu, G. Xu, J. Yuan, J. Jo, G. Gandikota, H. Demirci, T. Agano, N. Sato, Y. Shigeta, X. Wang, Light emitting diodes based photoacoustic imaging and potential clinical applications. Sci. Rep. **8**, 9885 (2018)
26. W. Xia, M. Kuniyil Ajith Singh, E. Maneas, N. Sato, Y. Shigeta, T. Agano, S. Ourselin, S.J. West, A.E. Desjardins, Handheld real-time LED-based photoacoustic and ultrasound imaging system for accurate visualization of clinical metal needles and superficial vasculature to guide minimally invasive procedures. Sensors **18**, 1394 (2018)
27. T. Agano, N. Sato, H. Nakatsuka, K. Kitagawa, T. Hanaoka, K. Morisono, Y. Shigeta, Attempts to increase penetration of photoacoustic system using LED array light source, in *Photons Plus Ultrasound: Imaging and Sensing 2015*, vol. 9323 (International Society for Optics and Photonics, 2015), p. 93233Z.
28. Y. Adachi, T. Hoshimiya, Photoacoustic imaging with multiple-wavelength light-emitting diodes. Jp. J. Appl. Phys. **52**, 07HB06 (2013)
29. J. Jo, G. Xu, Y. Zhu, M. Burton, J. Sarazin, E. Schiopu, G. Gandikota, X. Wang, Detecting joint inflammation by an LED-based photoacoustic imaging system: a feasibility study. J. Biomed. Opt. **23**, 110501 (2018)
30. G.P. Luke, D. Yeager, S.Y. Emelianov, Biomedical applications of photoacoustic imaging with exogenous contrast agents. Ann. Biomed. Eng. **40**, 422–437 (2012)
31. S.R. Kothapalli, G.A. Sonn, J.W. Choe, A. Nikoozadeh, A. Bhuyan, K.K. Park, P. Cristman, R. Fan, A. Moini, B.C. Lee, J. Wu, T.E. Carver, D. Trivedi, L. Shiiba, I. Steinberg, D.M. Huland, M.F. Rasmussen, J.C. Liao, J.D. Brooks, P.T. Khuri-Yakub, S.S. Gambhir, Simultaneous transrectal ultrasound and photoacoustic human prostate imaging. Sci. Transl. Med. **11**, eaav2169 (2019)
32. M. Jeon, W. Song, E. Huynh, J. Kim, J. Kim, B.L. Helfield, B.Y. Leung, D.E. Geortz, G. Zheng, J. Oh et al., Methylene blue microbubbles as a model dual-modality contrast agent for ultrasound and activatable photoacoustic imaging. J. Biomed. Opt. **19**, 016005 (2014)
33. J. Wang, C.Y. Lin, C. Moore, A. Jhunjhunwala, J.V. Jokerst, Switchable photoacoustic intensity of methylene blue via sodium dodecyl sulfate micellization. Langmuir **34**, 359–365 (2017)
34. K. Cheng, S.R. Kothapalli, H. Liu, A.L. Koh, J.V. Jokerst, H. Jiang, M. Yang, J. Li, J. Levi, J.C. Wu et al., Construction and validation of nano gold tripods for molecular imaging of living subjects. J. Am. Chem. Soc. **136**, 3560–3571 (2014)
35. A. Dragulescu-Andrasi, S.R. Kothapalli, G.A. Tikhomirov, J. Rao, S.S. Gambhir, Activatable oligomerizable imaging agents for photoacoustic imaging of furin-like activity in living subjects. J. Am. Chem. Soc. **135**, 11015–11022 (2013)

36. M.T. Berninger, P. Mohajerani, M. Wildgruber, N. Beziere, M.A. Kimm, X. Ma, B. Haller, M.J. Fleming, S. Vogt, M. Anton et al., Detection of intramyocardially injected DiR-labeled mesenchymal stem cells by optical and optoacoustic tomography. Photoacoustics 6, 37–47 (2017)

37. J.V. Jokerst, M. Thangaraj, P.J. Kempen, R. Sinclair, S.S. Gambhir, Photoacoustic imaging of mesenchymal stem cells in living mice via silica-coated gold nanorods. ACS Nano **6**, 5920–5930 (2012)

38. S.Y. Nam, L.M. Ricles, L.J. Suggs, S.Y. Emelianov, In vivo ultrasound and photoacoustic monitoring of mesenchymal stem cells labeled with gold nanotracers. PLoS ONE **7**, e37267 (2012)

39. B. Zhang, X. Sun, H. Mei, Y. Wang, Z. Liao, J. Chen, Q. Zhang, Y. Hu, Z. Pang, X. Jiang, LDLR-mediated peptide-22-conjugated nanoparticles for dual-targeting therapy of brain glioma. Biomaterials **34**, 9171–9182 (2013)

40. J. Huang, H. Zhang, Y. Yu, Y. Chen, D. Wang, G. Zhang, G. Zhou, J. Liu, Z. Sun, D. Sun et al., Biodegradable self-assembled nanoparticles of poly (D, L-lactide-co-glycolide)/hyaluronic acid block copolymers for target delivery of docetaxel to breast cancer. Biomaterials **35**, 550–566 (2014)

41. Y. Zhu, X. Lu, X. Dong, J. Yuan, M.L. Fabiilli, X. Wang, LED-Based Photoacoustic imaging for monitoring angiogenesis in fibrin scaffolds. Tissue Eng. Part C: Methods (2019)

42. N. Sato, M.K.A. Singh, Y. Shigeta, T. Hanaoka, T. Agano, High-speed photoacoustic imaging using an LED-based photoacoustic imaging system, in *Photons Plus Ultrasound: Imaging and Sensing 2018*, vol. 10494 (International Society for Optics and Photonics, 2018), p. 104943N.

43. E. Maneas, W. Xia, M.K.A. Singh, N. Sato, T. Agano, S. Ourselin, S.J. West, A.L. David, T. Vercauteren, A.E. Desjardins, Human placental vasculature imaging using an LED-based photoacoustic/ultrasound imaging system, in *Photons Plus Ultrasound: Imaging and Sensing 2018*, vol. 10494 (International Society for Optics and Photonics, 2018), p. 104940Y

44. M.K.A. Singh, N. Sato, F. Ichihashi, Y. Sankai, In vivo demonstration of real-time oxygen saturation imaging using a portable and affordable LED-based multispectral photoacoustic and ultrasound imaging system, in *Photons Plus Ultrasound: Imaging and Sensing 2019*, vol. 10878 (International Society for Optics and Photonics, 2019), p. 108785N

45. M.K.A. Singh, T. Agano, N. Sato, Y. Shigeta, T. Uemura, Real-time in vivo imaging of human lymphatic system using an LED-based photoacoustic/ultrasound imaging system, in *Photons Plus Ultrasound: Imaging and Sensing 2018*, vol. 10494 (International Society for Optics and Photonics, 2018), p. 1049404

46. Y. Shigeta, N. Sato, M.K.A. Singh, T. Agano, Multispectral photoacoustic characterization of ICG and porcine blood using an LED-based photoacoustic imaging system, in *Photons Plus Ultrasound: Imaging and Sensing 2018*, vol. 10494 (International Society for Optics and Photonics, 2018), p. 104943O

47. S. Agrawal, C. Fadden, A. Dangi, X. Yang, H. Albahrani, N. Frings, S. Heidari Zadi, S.R. Kothapalli, Light-emitting-diode-based multispectral photoacoustic computed tomography system. Sensors **19**(22), 4861 (2019)

48. S. Agrawal, X. Yang, H. Albahrani, C. Fadden, A. Dangi, M.K.A. Singh, S.R. Kothapalli, Low-cost photoacoustic computed tomography system using light-emitting-diodes, in *Photons Plus Ultrasound: Imaging and Sensing 2020*, vol. 11240 (International Society for Optics and Photonics, 2020), p. 1124058

49. B.E. Treeby, E.Z. Zhang, B.T. Cox, Photoacoustic tomography in absorbing acoustic media using time reversal. Inverse Prob. **26**, 115003 (2010)

50. B. Wilson, G. Adam, A Monte Carlo model for the absorption and flux distributions of light in tissue. Med. Phys. **10**, 824–830 (1983)

51. L. Wang, S.L. Jacques, L. Zheng, MCML—Monte Carlo modeling of light transport in multi-layered tissues. Comput. Methods Programs Biomed. **47**, 131–146 (1995)

52. W.M. Star, Diffusion theory of light transport, in *Optical-thermal response of laser-irradiated tissue* (Springer, 1995), pp. 131–206

53. S. Agrawal, A. Dangi, N. Frings, H. Albahrani, S.B. Ghouth, S.R. Kothapalli, Optimal design of combined ultrasound and multispectral photoacoustic deep tissue imaging devices using hybrid simulation platform, in *Photons Plus Ultrasound: Imaging and Sensing 2019*, vol. 10878 (International Society for Optics and Photonics, 2019), p. 108782L
54. C. Fadden, S.R. Kothapalli, A single simulation platform for hybrid photoacoustic and RF-acoustic computed tomography. Appl. Sci. **8**, 1568 (2018)
55. G. Bal, K. Ren, On multi-spectral quantitative photoacoustic tomography in diffusive regime. Inverse Prob. **28**, 025010 (2012)
56. A. Javaherian, S. Holman, Direct quantitative photoacoustic tomography for realistic acoustic media. Inverse Prob. (2019)
57. J. Poudel, Y. Lou, M.A. Anastasio, A survey of computational frameworks for solving the acoustic inverse problem in three-dimensional photoacoustic computed tomography. Phys. Med. Biol. (2019)
58. E.M.A. Anas, H.K. Zhang, J. Kang, E. Boctor, Enabling fast and high quality LED photoacoustic imaging: a recurrent neural networks based approach. Biomed. Opt. Express **9**, 3852–3866 (2018)
59. K. Johnstonbaugh, S. Agrawal, D. Abhishek, M. Homewood, S.P.K. Karri, S.R. Kothapalli, Novel deep learning architecture for optical fluence dependent photoacoustic target localization, in *Photons Plus Ultrasound: Imaging and Sensing 2019*, vol. 10878 (International Society for Optics and Photonics, 2019), p. 449108781L
60. K. Johnstonbaugh, S. Agrawal, D.A. Durairaj, C. Fadden, A. Dangi, S.P.K. Karri, S.R. Kothapalli, A deep learning approach to photoacoustic wavefront localization in deep-tissue medium. IEEE Trans. Ultrason. Ferroelectr. Freq. Contr. (2020)

LED-Based Functional Photoacoustics—Portable and Affordable Solution for Preclinical Cancer Imaging

Marvin Xavierselvan and Srivalleesha Mallidi

Abstract Photoacoustic imaging (PAI) is an imaging modality with promising results in cancer theranostics, both in preclinical and clinical applications. Its applicability in image-guided drug delivery and monitoring therapeutic response holds great promise for clinical translation. Current PAI techniques rely on using bulky lasers to provide the nanosecond pulsed light for photoacoustic signal generation. Tremendous growth in semiconductor industry within the last decade has led to creation of low-cost powerful LEDs that can be used as an alternate light source in lieu of laser to generate photoacoustic signal. In this chapter, we provide an overview of PAI usage in preclinical cancer research and provide examples of the LED based PAI performance in similar settings. LEDs will play a major role in catapulting PAI into clinics at an earlier pace and low cost than expected.

1 Introduction

Today cancer is one of the leading causes of death worldwide. In the United States approximately 1.8 million people will be diagnosed with cancer and around 600,000 deaths due to cancer are projected to occur in 2019 [1]. Preclinical cancer research in small and large animals have been central to comprehend many physiological and biomolecular processes involved in tumorigenesis, not only at the organ level but also systemically in the whole body. These studies were vital in development of new treatments or in designing better targeting and dosimetry of current treatments. Gauging the heterogeneity in tumors especially with vasculature being unevenly

M. Xavierselvan · S. Mallidi (✉)
Department of Biomedical Engineering, Tufts University, Medford, MA 02155, USA
e-mail: Srivalleesha.Mallidi@tufts.edu

M. Xavierselvan
e-mail: Marvin.Xavierselvan@tufts.edu

S. Mallidi
Wellman Center for Photomedicine, Massachusetts General Hospital, Harvard Medical School, Boston, MA 02114, USA

© Springer Nature Singapore Pte Ltd. 2020
M. Kuniyil Ajith Singh (ed.), *LED-Based Photoacoustic Imaging*,
Progress in Optical Science and Photonics 7,
https://doi.org/10.1007/978-981-15-3984-8_12

303

distributed [2] leading to hypoxia, rapid proliferation of tumor cells, deteriorating microenvironments [2, 3] is critical in evaluating therapeutic efficacy. Hypoxia is a major factor for the resistance to radiotherapy, chemotherapy and photodynamic therapy. Hence it could be used as a potential biomarker for treatment efficacy [4, 5]. Blood vessels are usually disrupted in the core of tumor while the periphery vessels are deformed and hyperpermeable, causing the drugs to be distributed unevenly with higher concentration in the periphery and very low concentration in the tumor core affecting treatment efficacy [6, 7]. Understanding the intratumor drug accumulation and vascular heterogeneity is of clinical relevance, as it would provide important information for designing personalized cancer therapeutic strategies.

Imaging can play a huge role in early detection of cancer, monitoring the response after a treatment and providing information on the tumor microenvironment non-invasively in real time. Commonly employed imaging techniques that are used in medical settings for the diagnostic and treatment procedures are magnetic resonance imaging (MRI), ultrasonography, X-ray based examinations (computed tomography), nuclear medicine tomography like positron emission tomography (PET), single photon emission computed tomography (SPECT). Each imaging technique has its benefits for specific applications. However, some of these techniques are very expensive and involve the use of ionizing radiation or radioisotopes and external contrast agents to obtain image with better contrast. Currently there are no standard techniques available to measure oxygen levels in tumors in clinical use. Oxygen sensing pO_2 electrodes (polarographic needles, phosphorescence, fluorescence based) can be used to acquire the information of oxygen at the site of electrode, but they are invasive which limits the longitudinal measurements at multiple site of the tumor. Additionally it only provides point measurement and cannot provide the spatial information of oxygen levels in the tumor [8].

Non-invasive imaging techniques such as MRI, PET and CT scan can be used for studying oxygen levels, but they are expensive and involves ionizing radiation or radioisotopes or external contrast agents for the measurement. Blood oxygenation level dependent (BOLD) and tumor oxygenation level dependent (TOLD) MRI can be used to assess tumor oxygenation as the contrast generated by oxygenated and deoxygenated hemoglobin and eventually predict radiation response [5, 9]. However, MRI is a "macroscopic" realm imaging modality and cannot match the resolution of an ultrasound imaging system. Among the newer imaging modalities that do not use ionizing radiation and can detect blood oxygen saturation at ultrasonic resolution is photoacoustic imaging (PAI) [10–12]. PAI has been in development for several years, gained lot of attention recently and is currently in clinical trials for breast cancer and other malignancies [13–15]. PAI is a hybrid imaging technique based on the photoacoustic effect which is the generation of acoustic waves when the sample is irradiated with pulsed light. The intensity of light is varied either periodically or as a single pulse to get this effect. Optical imaging techniques like stimulated emission depletion microscopy and photoactivation localization microscopy have super spatial resolution of 10–100 nm, while confocal microscopy has 0.5–1 μm but these techniques have a poor imaging depth typically 10–100 μm [16]. High frequency ultrasound can achieve spatial resolution of 10–100 μm but clinical systems only reaches about 0.1–1 mm resolution with a penetration depth of 5–50 mm. The spatial resolution of MRI ranges from 10 to 100 μm with an imaging depth of 10–100 cm [17]. PAI fits in the middle range (mesoscale) providing high optical contrast image with a resolution of 0.5–100 μm and at a reasonable penetration depth of 0.5–10 mm [16]. Medical ultrasound is widely used in the clinics to visualize the structure, size and pathological lesions of muscles, various internal organs. and are employed for both diagnostic and therapeutic applications. PAI can be easily coupled with clinical ultrasound systems as the ultrasound transducer can be used to receive the acoustic waves generated as a result of the photoacoustic effect. Studies have shown that PA can detect cancer while providing tumor vasculature information complementing the structural and site information of the tumor deep inside the body by making use of the endogenous contrast agents or by using targeted external agents towards the tumor. The endogenous molecule such as hemoglobin, myoglobin or melanin are optical absorbers in the biological tissues which can give photoacoustic signals. Since PA can provide information about blood vasculature network in the tumor, it can be used to monitor and evaluate the treatment response [18]. Information on angiogenesis, oxygen saturation and total hemoglobin content in the tumor can be obtained using PAI without the use of external contrast agents [12]. Oxygenated hemoglobin and deoxygenated hemoglobin have different optical absorption at different wavelengths. Deoxy Hb has peak absorption at 750 nm, while oxy Hb has a peak absorption at 850 nm and they both have similar absorption at 805 nm. The photoacoustic signal at these wavelengths gives information about oxygen saturation while their sum gives the total hemoglobin content.

2 LED Based Photoacoustic Imaging

PAI systems used in preclinical applications or clinical applications use bulky and expensive Optical Parametric Oscillator (OPO) laser or Q-switched Nd:YAG laser and have large footprints. They have low repetition rate leading to low image frame rate and acquisition speed. These issues bear huge hinderance to the translation of PAI for the clinical applications [19].

Recent advances in the Light emitting diode (LED) technology has led to the development of LED based PAI systems that are of low cost, compact, portable and can be readily translated into various clinical applications given high frame rates. AcousticX, an LED based PAI system from Cyberdyne Inc. (Tsukuba, Japan) performs alternating US and PA imaging at a frame rate of up to 30 Hz and displays the image individually or on overlay [20, 21]. The LEDs in the AcousticX can be driven at various repetition rate of up to 4 kHz [21]. LED's have low output power when compared with class IV lasers, when used in groups of 2D arrays, the delivered output power which is two orders of magnitude smaller than the power produced by lasers [22]. With high repetition rate and signal averaging the low output power from LEDs can be compensated and be used to improve the signal to noise ratio (SNR) comparable to that of laser-based PA systems. The radio frequency (RF) data generated during the acquisition can be read through MATLAB for further analysis and visualization. So far, the tumor vasculature and oxygen saturation imaging has been performed with laser based PAI system. There are very few reports on the use of LED based PAI for monitoring tumor response to therapies and oxygen saturation. Various studies have been performed using the LED based PAI for its sensitivity by imaging exogeneous contrast agents and superficial vascular network in the body. In this book chapter, we present results from the recent papers on utilizing laser based PAI for monitoring oxygen saturation in pre-clinical cancer research. We demonstrate the LED based PAI system's potential to monitor the change in oxygen saturation and perform contrast enhancement using nanoparticles in an in vivo tumor model. We also review the current status of PAI in monitoring vasculature and blood oxygen saturation for preclinical cancer research and elucidate the role of LED based PAI in preclinical cancer research prior to such technologies being translated to clinic.

2.1 PAI to Monitor the Tumor Microenvironment

The tumor microenvironment, particularly the physiological aspects such as hypoxia and vascular density, is known to have to have major impact on cancer treatment outcomes. Hence it is critical to understand these aspects and monitor the dynamic treatment dependent changes in the tumor vascular and oxygenation function as has been demonstrated by several studies [5, 9, 18, 23–25]. An example is the study conducted by Rich and Seshadri [8] where tumor oxygenation was studied in a mouse with patient derived head and neck squamous cell carcinoma (HNSCC) tumor xenografts and the photoacoustic images were compared with MRI pre and post-radiotherapy. The tumor response to hyperoxia was monitored with PAI by measuring oxygen saturation (SO_2) and longitudinal relaxation rate (R_1) using MRI. The viable region in the tumor showed an increase in SO_2 and R_1 while the necrotic region did not show any change. Furthermore, good correlation was observed between MRI and PAI as well as histology and PAI (Fig. 1). Changing the breathing gas to oxygen from the room air resulted in significant increase in the tumor SO_2 and R_1. Reverting to normal air from oxygen showed a decrease in both values. The median value of the PAI signal at 850 nm post oxygen breathing showed a positive correlation with the median R_1 values. They used PAI to predict the outcome of the tumor response to radiation and chemoradiation by measuring the SO_2 before and 24 h post treatment. The change in SO_2 24 h post treatment with the change in tumor volume at 2 weeks following the treatment, showed that an increase in SO_2 following the treatment had a good outcome (growth of tumor was inhibited) compared to the ones that did not change much or had a reduction in SO_2.

A study led by Wilson et al. [26] used spectroscopic PAI to differentiate among different breast tissues (normal, hyperplasia, ductal carcinoma in situ (DCIS), and invasive breast carcinoma) in a mouse model by measuring oxygen saturation, total hemoglobin, and lipid content. Normal and hyperplasia tissue had significantly higher total hemoglobin content than the DCIS and invasive breast carcinoma, while the

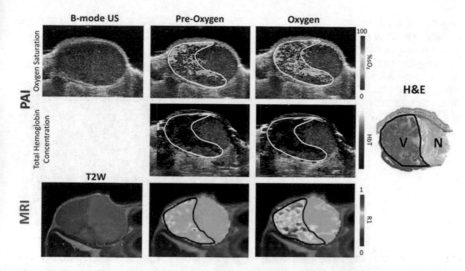

Fig. 1 Top row: ultrasound image overlaid with oxygen saturation map of tumor during exposure to air (Pre-Oxygen) and followed by 100% oxygen inhalation, Middle row: ultrasound image overlaid with hemoglobin concentration map of same tumor following the oxygen challenge. Bottom row: T2 weighted image overlaid with color map of longitudinal relaxation rate (R1). Histology section of tumor is shown on the right, (V) is the viable region in tumor where response to the oxygen challenge is found and (N) is necrotic region in tumor where no responses were detected in PAI and MRI. Figure reprinted with permissions from Rich and Seshadri [8]

oxygen saturation for all the tissues was higher than the normal tissue. The lipid content was significantly decreased in DCIS and invasive breast carcinoma than the normal or hyperplasia. Combining these parameters and performing spectral analysis of photoacoustic images allows for differentiating normal and hyperplasia from DCIS and invasive breast carcinoma with decent accuracy. Using clinical grade PAI systems, spectroscopic PAI can be assessed with focal breast lesions and this technique when used along with ultrasound can improve the diagnostic accuracy of breast cancer [27, 28].

We tested the LED based PAI system for its ability to image the vasculature in the tumor and track the changes in oxygen saturation when breathing is challenged with normal air and 100% pure oxygen. For our study, we used subcutaneous head and neck tumor (FaDu) xenografts in nude mice. When the tumor size reached about 100 mm^3, the tumors were imaged using the AcousticX, LED based PAI system. The animal was anesthetized and immersed in a warm water bath for the acoustic coupling with the ultrasound transducer. First, the B-mode image of the tumor is acquired using a 7 MHz ultrasound transducer. Two sets of multiple arrays of 850 nm LED light source on a 2D block was used to irradiate the tumor from both sides of the US probe. The LED were fired at a repetition rate of 4 kHz and the signal was averaged 384 times to image at a frame rate of 10 Hz. Photoacoustic images of the blood vessel in the tumor was acquired and simultaneously raw RF data was saved. Using custom written MATLAB codes based on previously reported algorithms [29–32],

Fig. 2 Overlay of PA and US image (red and gray colormap respectively) showing the vasculature in subcutaneous FaDu tumor xenografts in mice. PA images were scanned using the LED based PAI system (850 nm wavelength)

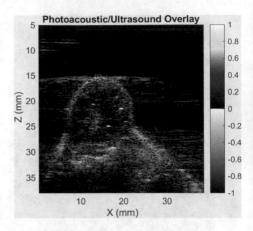

the RF data was read and processed to get the US/PA images of the tumor blood vessels as shown in Fig. 2. The PA image of the vasculature is shown in orange-red pseudo colormap in Fig. 2. Clearly the image shows heterogenous vascular density in the tumor which demonstrates the ability of the LED based PAI system to obtain vascular information from tissues that are more than 1 cm deep.

2.2 Oxygen Enhanced PAI

Taking cues from the oxygen enhanced MRI, oxygen enhanced PA imaging provides high spatially resolved images with hemoglobin concentration and blood oxygenation by measuring the change in oxygenation level in blood preceded by inhaling 100% oxygen. Oxygen enhanced imaging can differentiate hypoxic regions from well oxygenated regions. The biomarkers derived from the measurement of hemoglobin change in oxygen enhanced PAI performs better than the biomarkers derived from the static PAI in terms of repeatability and robustness and also correlate well with the histopathologic analysis of tumor vasculature function [24]. Dynamic contrast enhanced PAI makes use of external contrast agents to boost the photoacoustic image contrast. Multiple wavelength PAI is performed to separate the contrast agent signal from oxygenated and deoxygenated hemoglobin which provides information on the oxygenation and vasculature perfusion in the tumor [33]. With PAI making its headway in clinical testing with endogenous contrast, we anticipate that this technology will soon be applied with FDA approved contrast agents to obtain molecular maps of the lesions under consideration.

A study conducted by Tomaszewski et al. [23], compared oxygen enhanced (OE) PAI and dynamic contrast enhanced (DCE) PAI as biomarkers for tumor vasculature, hypoxia, and necrosis. They co-registered OE-PAI and DCE-PAI and showed a quantitative spatial per pixel correlation between the biomarkers derived from OE and DCE PAI for the assessment of tumor hypoxia and vascular maturity with the immunohistochemistry. The change in oxygen saturation (ΔSO_2) post switching of breathing gas to 100% oxygen from air and the increase in ICG (ΔICG) signal following its administration is shown for two tumor types. A strong per pixel correlation was observed between the change in oxygen saturation and change in ICG signal. No correlation was observed between change in ICG signal and oxygen saturation (SO_2) measured at baseline for air or 100% oxygen, showing that just measuring the oxygen saturation does not provide information about the tumor perfusions. Further analysis of DCE-PAI revealed areas in the tumor with distinct ICG kinetics. The first group was dubbed as 'clearing' which showed an increase in ICG signal followed by an exponential decay of the signal. The second group referred to as 'retaining' showed a slow increase in signal with no clearance. The signal either increased gradually or remained high and stable over the course of the experiment. The fraction of the clearing regions was significantly higher in the rim of the tumor than the tumor core.

OE responding fraction was calculated for each tumor and scan as a ratio of number of tumor pixels responding to the total number of pixels in the tumor. OE-PAI response for these two regions display significant differences. The retaining regions showed lower OE responding fraction than the clearing region as well as weaker correlation between ΔSO_2 and ΔICG. Immunohistochemistry of the tumor sections for vascular maturity (CD31 positive cells), hypoxia (CAIX), and tumor necrosis (H&E) showed a positive correlation between CD31, DCE, and OE PAI, while showing a negative correlation between CAIX, tumor necrosis, and OE-PAI. The results show that DCE-PAI signals are markers for vascular maturity while OE PAI signals are for tumor hypoxia and necrosis.

For oxygen saturation measurements, a combined array of 750 nm and 850 nm LED light source is used. The animal was set to breathe normal air initially and a B-mode image of the hind leg muscle of mice is imaged. Using the 750/850 nm LED array, the oxygen saturation of the muscle was acquired (cycle 1). The breathing gas was then switched to 100% oxygen and breathing was allowed to stabilize for two minutes followed by acquisition of an oxygen saturation image (cycle 2). The cycle was repeated multiple times and oxygen saturation of the muscle at each cycle was acquired. At each imaging timepoint, corresponding RF data was also saved. An ROI was used on the oxygen saturation image acquired at different timepoints to calculate the relative change (Fig. 3a). We can also infer from Fig. 3b that as the breathing gas was set to 100% oxygen, the relative change in SO_2 increased and when the breathing gas was set to normal air, the change in SO_2 decreased. This experiment showed the ability of the LED based PAI system to track the changes in the oxygen saturation and image it in real time.

2.3 PAI to Predict the Tumor Response to Treatment

Tumor microenvironment, particularly the vasculature and its function (delivering oxygen), is a central theme for many drug development and cancer therapeutic

a)   b)

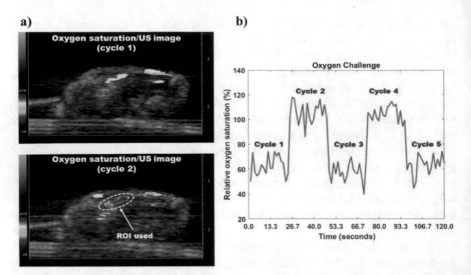

Fig. 3 **a** Oxygen saturation change in hind leg muscle of mice when the animal breathing was challenged with cyclic changes of air and 100% oxygen, imaged using 750/850 nm LED array with red being oxygenated and blue being hypoxic regions. **b** Relative change in oxygen saturation in the muscle calculated using the ROI

research groups. PAI plays a major role in this realm as it measures the vascular density and blood oxygen saturation longitudinally for diagnosis as well as to monitor therapeutic response. Many cancer treatments rely on the oxygen availability in the tumor and by acquiring the vascular information using PAI and monitoring the oxygen content in the tumor before and after treatment, tumor response to the therapy can be gauged and appropriate subsequent therapies can be designed.

Hysi et al. [34] used oxygen saturation measurement and radiofrequency signals from PAI to estimate the response of the tumor in a mouse for a heat activated cytotoxic (HaT-DOX) liposome containing the drug doxorubicin (DOX) and saline as control. PAI imaging was performed for 30 min pre and post treatment as well as 2, 5, 24 h, and 7 days post treatment. The responders for the HaT-DOX treatment showed a 50% decrease in the tumor volume while the non-responders showed an increase. The ROI from B-mode is used to segment PA images at two wavelengths, 750 nm and 850 nm, and compute the oxygen saturation of the tumor (Fig. 4a). A reference phantom was used to obtain the spectral parameters using the normalized power spectra. A SO_2 histogram was computed and used to quantify the changes in blood vessel oxygenation throughout the tumor. The responders had a drop in the oxygenation which was evident in early as 30 min and was observed till 24 h timepoint. The early change in the blood vessel oxygenation correlated with the treatment response. The mode of oxygen saturation was plotted against the PA spectral slope (SS) (average value of each treatment group at that timepoint). The responders showed a decrease in oxygenation and PA SS at 750 nm and 850 nm (Fig. 4b). SS further decreased after the 24-h timepoint but the SO_2 mode increased at 7 days post treatment. HaT-DOX

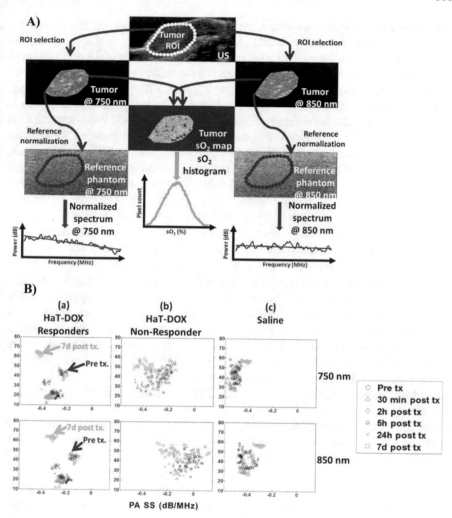

Fig. 4 **A** Schematic illustration for generating tumor oxygenation maps and PA spectral parameters. Using the B-mode ultrasound image, an ROI is selected to generate oxygen saturation map of tumor by segmenting PA images at 750 and 850 nm. Average mode of SO_2 is calculated from the histogram of oxygen saturation values from all the slices of tumor. Reference phantom is imaged at 750 nm and 850 nm and the collected frequency information and normalized power spectra is used to obtain the RF spectra and PA spectral parameters of the tumor. **B** Mode SO_2 plotted against the spectral slope (PA SS) calculated from 750 nm (top row) and 850 nm (bottom row) and for HaT-DOX **a** responders (n = 5), **b** non-responder (n = 1), and **c** saline (n = 7). Each dot represents the average SS across at least 100 normalized power spectra within 21 tumor slices at each timepoint for every mouse. Figure reprinted with permissions from Elsevier and Hysi et al. [34]

non-responder's oxygenation did not drop significantly 24 h post treatment, and there was high variation and no clear trend in SS distribution. The control saline group did not exhibit large change in SS. This shows that the difference in SS post treatment can be useful to identify the non-responders for the treatment.

Photodynamic therapy (PDT) is a forthcoming treatment option for cancer therapy. PDT is a photochemical process where a molecule called photosensitizer (PS) is excited by a particular wavelength of light to generate cytotoxic species which kills tumor cells [35, 36]. PDT can be of type I process where electron transfer happens with excited PS to generate radicals and radical anion species while in type II process, the excited PS transfers energy to molecular oxygen producing singlet oxygen which upon contact with cells are toxic [37]. Photosensitizers get distributed all through the cells of the body but stays longer in the cancerous cells because of enhanced permeability and retention effect (EPR) [38]. After a specific time delay (Drug Light Interval, DLI) depending on the photosensitizer, PS gets specifically accumulated in the cancer cells, following that tumor is exposed to light delivered through a device suitable for the site location of the tumor [39]. PDT response depends on tumor oxygenation and DLI. With high DLI, most of the PS gets accumulated in the tumor cellular compartment and PDT causes inflammation in the tumor region while in a low DLI case, PS is mostly in the blood vessel and PDT causes vascular shutdown within the irradiate region. Furthermore, in addition to DLI, the oxygenation status of the tumor also determines PDT efficacy. A study conducted by [25] in oral cancers shows highly oxygenated tumors respond better to PDT than the tumors with low oxygen content. Therefore, it is important to monitor for tumor oxygenation as it provides valuable information to calculate PDT light dose and PS concentration effectively.

Mallidi et al. [18] used PAI to find the changes in oxygen saturation (SO_2) of the tumor and using it as a surrogate biomarker for the predicting the response to PDT and identify recurrence of tumor post treatment. BPD was used as PS and 1-h DLI and 3-h DLI was chosen for PDT treatment. 65% of the mice in 1-h DLI group had complete suppression of tumor 30 days post treatment dubbed as responders, while 35% had regrowth after two weeks. 3-h DLI group had six-fold increase in tumor growth post treatment and was not significantly different from no treatment group. 3-h DLI group was referred to as non-responders. PAI was performed for 1-h and 3-h DLI groups at pre-PDT, immediately post-PDT, 6-h and 24-h post-PDT and SO_2 and total hemoglobin content (HbT) in the tumor was measured. Figure 5a describes the average SO_2 and HbT values at various time points in both 1-h DLI and 3-h DLI groups. While the SO_2 did not change significantly for 3-h DLI immediately post-PDT due to less disruption to the vasculature, 1-h DLI showed a significant increase immediately post-PDT followed by decrease to 3% and 8% at 6-h and 24-h post-PDT respectively in tumors that responded to the treatment (Fig. 5a). An algorithm was devised based on the SO_2 values at various timepoints post treatment. Average SO_2 value for every frame of the PAI scan was calculated for 6-h and 24-h post-PDT. If the average SO_2 at 6-h post-PDT and 24-h post-PDT for a particular B-scan frame was less than 6.2 and 16.3%, the region is considered treated (colored as green) or non-responsive (colored as red) as shown in Fig. 5b. The algorithm was repeated

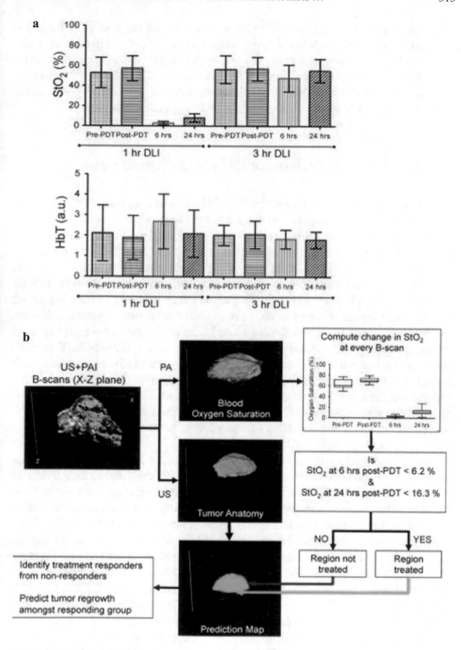

Fig. 5 **a** Mean SO$_2$ and HbT values at various time points (Pre-PDT, Post-PDT, 6-h and 24-h post-PDT) in the 1-h DLI and 3-h DLI groups. **b** Graphical outline of the algorithm used to classify the PDT treatment responders from non-responders and to generate the prediction map for predicting treatment response and tumor regrowth using the ultrasound (US) and photoacoustic (PA) images. Figure adapted with permissions from Mallidi et al. [18]

for all the B-scan frames in the 3-D tumor volume to get the prediction map. The pre-PDT or immediately post-PDT SO_2 values alone did not predict the treatment outcome, but the combined average SO_2 value at 6-h and 24-h at each frame made the prediction. The timepoint for SO_2 measurement and threshold levels need to be optimized for different tumor models along with the algorithm to get a better prediction for patient specific treatment design and response outcome.

2.4   Contrast Enhancement in PAI Using Nanoparticles

Advancement in the field of nanotechnology led to the development of various types of nanoconstructs and nanoparticles that act as both drug carriers (theranostic agents) and contrast agents (diagnostic agents). These nanoparticles can also be loaded with dyes that have high optical absorption such as methylene blue, IRdye800, Indocyanine green. In addition to contrast dye, chemotherapeutic drugs like doxorubicin which are toxic to normal tissues are also encapsulated in nanoconstructs to minimize the toxic effects on healthy cells [40, 41]. The presence of imaging agents in these nanosystems will enable PAI to monitor the nanoparticle uptake, retention, and interaction with the tumor tissue [42–45]. The nanosystems are generally non-targeted or specifically targeted to biomarkers expressed in malignancies. Especially in tumors, non-targeted nanoparticles are selectively retained in the cancerous cells because of enhanced permeability and retention effect while the targeted nanoparticles get localized in the target cell through receptor mediated endocytosis or other biological mechanisms [46–48]. Furthermore, the drugs and contrast agents from nanoparticles can be released using external energy sources such as light, heat, etc. and these processes can be monitored with PAI.

This provides PAI with the opportunity to image for contrast agents and give information on the tumor profile, pharmacokinetics, and drug distribution inside the tumor [44, 45, 49–51]. The nanoparticles-based contrast agents are categorized into metallic and non-metallic nanoparticles [14]. Plasmonic metal nanoparticles utilize surface plasmon resonance effect to increase the optical absorption to provide greater acoustic signal than the normal dye. For example, gold nanoparticles in the form of nanorods or nanospheres are employed in PAI as exogeneous contrast agents because of the surface plasmon resonance (SPR) effect which gives rise to their tunable optical absorption properties [12, 52, 53]. Non-metallic nanoparticles like single walled carbon nanotubes (CNT) are strong light absorbers by nature and can serve as high contrast molecular agents for photoacoustic imaging [54]. Carbon nanotubes and its variant single referred to as CNTs are widely used as carrier vehicles for chemo drugs and molecular agents [55–57]. PAI is increasingly used in image guided drug delivery of contrast agents and nanoparticles into tumors and also used in image guided cancer treatment like PDT, phototherapy, and chemoradiation [48, 53, 57, 58].

Given the rise of PAI in monitoring nanoparticle uptake, we believe LED based photoacoustic imaging is a more affordable option available for various research groups to avail the opportunity of utilizing the technology to understand nanoparticle

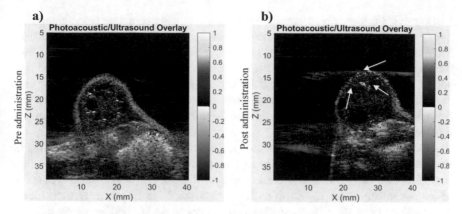

Fig. 6 Contrast enhancement in LED based PAI using exogeneous contrast agents. Combined overlay of PA and US image of subcutaneous FaDu tumor in mice before (**a**) and after (**b**) the NC dye administration. PA images were acquired using 850 nm LED light source (areas with greater contrast is shown with white arrows)

or drug uptake non-invasively at high resolution. Figure 6 demonstrates the capability of the LED based PAI system in imaging a naphthalocyanine (NC) dye which has a strong absorption in the NIR region (~absorption peak at 860 nm). Figure 6a shows the photoacoustic image overlaid with ultrasound of FaDu tumor xenografts in nude mice. An 850 nm LED light source was used for the photoacoustic scan of tumors. A 100 μL NC dye was injected intratumorally into the FaDu tumor interstitium using a 32-gauge needle and the distribution of dye inside the tumor was imaged. Figure 6b shows the distribution profile of the NC dye in the tumor. The NC dye has distributed into almost all parts of the tumor, but strong PA signal was received from the top of tumor which is potentially due to these areas receiving stronger light energy. Several studies demonstrate unmixing of the dye signal from the blood using spectroscopic photoacoustic imaging using tunable lasers [42]. Such studies will be possible with the availability of multi-wavelength LEDs for PAI.

3 Future Directions

The AcousticX system currently has only a few LEDs arrays operating at 470, 520, 620, 660, 690, 750, 820, 850, 940 or 980 nm. The dual wavelength combination LED arrays are available at 690/850, 750/850 and 820/940 nm. The possibility of having only fewer wavelengths in LED array significantly limits the potential of PAI in terms of multiwavelength spectroscopic imaging and unmixing signals from multiple photoacoustic contrast generating chromophores. The dual combination LED are developed based on optical signature of hemoglobin in blood and widely used photoacoustic contrast agent Indocyanine green. Any other contrast agent with different peak optical absorption properties probably cannot be imaged with the current

LED arrays. Furthermore, to unmix blood photoacoustic signature (oxygenated and deoxygenated hemoglobin) from the contrast agent, there should be at least three illumination wavelengths available. The choice of wavelength depends on the chromophores used in the study. Cyberdyne Inc. the pioneers in fabricating LED light source for PAI understand the need of various light source and are in pursuit of developing other wavelengths in both single source and multiwavelength combination LED arrays in order to further advance the utility of LED based PAI in preclinical research.

Functional imaging of tumors is one of major application of PAI. Depending on the tumor model and focus of study in the preclinical research, tumor size ranges from few mm to 1–2 cm. The current shape of LED arrays is a rectangular box of approximately 8 cm in length to conform around the linear array transducers. The LED array is large in comparison to tumors being studied by many research groups and most of the light from LED is delivered outside the tumor region of interest. Studies to progress towards designing custom shaped LED are currently being undertaken based on niche applications where such a technology could make significant impact. Moreover, making use of flexible electronics in fabricating LED arrays would be useful for myriad preclinical or clinical applications.

4 Conclusion

The examples and in vivo demonstrations presented in the chapter showcase the utility of LED based PAI for imaging vasculature, monitoring changes in oxygen saturation, and gauging the distribution of nanoparticles within the tumor. The low power light delivered by the LEDs may cause poor signal to noise ratio, however the high frame rate LED-based PAI produces images on par with slower frame rate laser-based PAI due to the possibility of averaging many frames obtained within a short time range. With the advances in semiconductor industry leading to production of powerful LEDs, exciting opportunities await for the portable, low-cost LED-based PAI systems in the pre-clinical and clinical application realm.

Acknowledgements The authors gratefully acknowledge funds from NIH R41CA221420, NIH EY028839 and School of Engineering, Tufts University. The authors would also like to thank Christopher D. Nguyen for editing our book chapter.

References

1. R.L. Siegel, K.D. Miller, A. Jemal, Cancer statistics, 2019. CA Cancer J. Clin. **69**(1), 7–34 (2019)
2. A. Eberhard, S. Kahlert, V. Goede, B. Hemmerlein, K.H. Plate, H.G. Augustin, Heterogeneity of angiogenesis and blood vessel maturation in human tumors: implications for antiangiogenic tumor therapies. Cancer Res. **60**(5), 1388–1393 (2000)
3. M. Hockel, P. Vaupel, Tumor hypoxia: definitions and current clinical, biologic, and molecular aspects. J. Natl. Cancer Inst. **93**(4), 266–276 (2001)
4. J.L. Tatum, G.J. Kelloff, R.J. Gillies, J.M. Arbeit, J.M. Brown, K.S. Chao, J.D. Chapman, W.C. Eckelman, A.W. Fyles, A.J. Giaccia, R.P. Hill, C.J. Koch, M.C. Krishna, K.A. Krohn, J.S. Lewis, R.P. Mason, G. Melillo, A.R. Padhani, G. Powis, J.G. Rajendran, R. Reba, S.P. Robinson, G.L. Semenza, H.M. Swartz, P. Vaupel, D. Yang, B. Croft, J. Hoffman, G. Liu, H. Stone, D. Sullivan, Hypoxia: importance in tumor biology, noninvasive measurement by imaging, and value of its measurement in the management of cancer therapy. Int. J. Radiat. Biol. **82**(10), 699–757 (2006)
5. R.R. Hallac, H. Zhou, R. Pidikiti, K. Song, S. Stojadinovic, D. Zhao, T. Solberg, P. Peschke, R.P. Mason, Correlations of noninvasive BOLD and TOLD MRI with pO2 and relevance to tumor radiation response. Magn. Reson. Med. **71**(5), 1863–1873 (2014)
6. T. Stylianopoulos, J.D. Martin, M. Snuderl, F. Mpekris, S.R. Jain, R.K. Jain, Coevolution of solid stress and interstitial fluid pressure in tumors during progression: implications for vascular collapse. Cancer Res. **73**(13), 3833–3841 (2013)
7. S. Garattini, I. Fuso Nerini, M. D'Incalci, Not only tumor but also therapy heterogeneity. Ann. Oncol. **29**(1), 13–19 (2018)
8. L.J. Rich, M. Seshadri, Photoacoustic monitoring of tumor and normal tissue response to radiation. Sci. Rep. **6**, 21237 (2016)
9. H. Zhou, S. Chiguru, R.R. Hallac, D. Yang, G. Hao, P. Peschke, R.P. Mason, Examining correlations of oxygen sensitive MRI (BOLD/TOLD) with [^{18}F]FMISO PET in rat prostate tumors. Am. J. Nucl. Med. Mol. Imaging **9**(2), 156–167 (2019)
10. M. Xu, L.V. Wang, Photoacoustic imaging in biomedicine. Rev. Sci. Instrum. **77**(4) (2006)
11. V. Ntziachristos, D. Razansky, Molecular imaging by means of multispectral optoacoustic tomography (MSOT). Chem. Rev. **110**(5), 2783–2794 (2010)
12. S. Mallidi, G.P. Luke, S. Emelianov, Photoacoustic imaging in cancer detection, diagnosis, and treatment guidance. Trends Biotechnol. **29**(5), 213–221 (2011)
13. K.S. Valluru, J.K. Willmann, Clinical photoacoustic imaging of cancer. Ultrasonography **35**(4), 267–280 (2016)
14. S. Zackrisson, S.M.W.Y. van de Ven, S.S. Gambhir, Light in and sound out: emerging translational strategies for photoacoustic imaging. Can. Res. **74**(4), 979–1004 (2014)
15. J.A. Guggenheim, T.J. Allen, A. Plumb, E.Z. Zhang, M. Rodriguez-Justo, S. Punwani, P.C. Beard, Photoacoustic imaging of human lymph nodes with endogenous lipid and hemoglobin contrast. J. Biomed. Opt. **20**(5), 50504 (2015)
16. S. Gigan, Optical microscopy aims deep. Nat. Photonics **11**(1), 14–16 (2017)
17. S. Wang, I.V. Larina, High-resolution imaging techniques in tissue engineering, in *Monitoring and Evaluation of Biomaterials and Their Performance In Vivo* (2017), pp. 151–180
18. S. Mallidi, K. Watanabe, D. Timerman, D. Schoenfeld, T. Hasan, Prediction of tumor recurrence and therapy monitoring using ultrasound-guided photoacoustic imaging. Theranostics **5**(3), 289–301 (2015)
19. M. Kuniyil Ajith Singh, W. Steenbergen, S. Manohar, Handheld probe-based dual mode ultrasound/photoacoustics for biomedical imaging, in *Frontiers in Biophotonics for Translational Medicine. Progress in Optical Science and Photonics* (2016), pp. 209–247
20. A. Hariri, J. Lemaster, J. Wang, A.S. Jeevarathinam, D.L. Chao, J.V. Jokerst, The characterization of an economic and portable LED-based photoacoustic imaging system to facilitate molecular imaging. Photoacoustics **9**, 10–20 (2018)

21. N. Sato, T. Agano, T. Hanaoka, Y. Shigeta, M. Kuniyil Ajith Singh, High-speed photoacoustic imaging using an LED-based photoacoustic imaging system. Paper presented at the photons plus ultrasound: imaging and sensing 2018

22. Y. Zhu, G. Xu, J. Yuan, J. Jo, G. Gandikota, H. Demirci, T. Agano, N. Sato, Y. Shigeta, X. Wang, Light emitting diodes based photoacoustic imaging and potential clinical applications. Sci. Rep. **8**(1), 9885 (2018)

23. M.R. Tomaszewski, M. Gehrung, J. Joseph, I. Quiros-Gonzalez, J.A. Disselhorst, S.E. Bohndiek, Oxygen-enhanced and dynamic contrast-enhanced optoacoustic tomography provide surrogate biomarkers of tumor vascular function, hypoxia, and necrosis. Cancer Res. **78**(20), 5980–5991

24. M.R. Tomaszewski, I.Q. Gonzalez, J.P. O'Connor, O. Abeyakoon, G.J. Parker, K.J. Williams, F.J. Gilbert, S.E. Bohndiek, Oxygen Enhanced Optoacoustic tomography (OE-OT) reveals vascular dynamics in Murine models of prostate cancer. Theranostics **7**(11), 2900–2913 (2017)

25. A.L. Maas, S.L. Carter, E.P. Wileyto, J. Miller, M. Yuan, G. Yu, A.C. Durham, T.M. Busch, Tumor vascular microenvironment determines responsiveness to photodynamic therapy. Can. Res. **72**(8), 2079–2088 (2012)

26. K.E. Wilson, S.V. Bachawal, L. Tian, J.K. Willmann, Multiparametric spectroscopic photoacoustic imaging of breast cancer development in a transgenic mouse model. Theranostics **4**(11), 1062–1071 (2014)

27. L. Lin, P. Hu, J. Shi, C.M. Appleton, K. Maslov, L. Li, R. Zhang, L.V. Wang, Single-breath-hold photoacoustic computed tomography of the breast. Nat. Commun. **9**(1), 2352 (2018)

28. T.T.W. Wong, R. Zhang, P. Hai, C. Zhang, M.A. Pleitez, R.L. Aft, D.V. Novack, L.V. Wang, Fast label-free multilayered histology-like imaging of human breast cancer by photoacoustic microscopy. Sci. Adv. **3**(5), e1602168 (2017)

29. M. Jaeger, S. Schüpbach, A. Gertsch, M. Kitz, M. Frenz, Fourier reconstruction in optoacoustic imaging using truncated regularized inversek-space interpolation. Inverse Prob. **23**(6), S51–S63 (2007)

30. A. Hussain, W. Petersen, J. Staley, E. Hondebrink, W. Steenbergen, Quantitative blood oxygen saturation imaging using combined photoacoustics and acousto-optics. Opt. Lett. **41**(8), 1720–1723 (2016)

31. W. Xia, E. Maneas, N. Trung Huynh, M. Kuniyil Ajith Singh, N. Montaña Brown, S. Ourselin, E. Gilbert-Kawai, S.J. West, A.E. Desjardins, A.A. Oraevsky, L.V. Wang, Imaging of human peripheral blood vessels during cuff occlusion with a compact LED-based photoacoustic and ultrasound system. Paper presented at the photons plus ultrasound: imaging and sensing 2019

32. M. Kuniyil Ajith Singh, N. Sato, F. Ichihashi, Y. Sankai, In vivo demonstration of real-time oxygen saturation imaging using a portable and affordable LED-based multispectral photoacoustic and ultrasound imaging system. Paper presented at the photons plus ultrasound: imaging and sensing 2019

33. J. Weber, P.C. Beard, S.E. Bohndiek, Contrast agents for molecular photoacoustic imaging. Nat. Methods **13**(8), 639–650 (2016)

34. E. Hysi, L.A. Wirtzfeld, J.P. May, E. Undzys, S.D. Li, M.C. Kolios, Photoacoustic signal characterization of cancer treatment response: correlation with changes in tumor oxygenation. Photoacoustics **5**, 25–35 (2017)

35. J.C. Chen, L. Keltner, J. Christophersen, F. Zheng, M.R. Krouse, A. Singhal, S. Wang, New technology for deep light distribution in tissue for phototherapy. Cancer J. **8**(2), 154–163 (2002)

36. S. Mallidi, S. Anbil, A.L. Bulin, G. Obaid, M. Ichikawa, T. Hasan, Beyond the barriers of light penetration: strategies, perspectives and possibilities for photodynamic therapy. Theranostics **6**(13), 2458–2487 (2016)

37. L.B. Josefsen, R.W. Boyle, Photodynamic therapy and the development of metal-based photosensitisers. Met. Based Drugs **2008**, 276109 (2008)

38. H. Kobayashi, R. Watanabe, P.L. Choyke, Improving conventional enhanced permeability and retention (EPR) effects; what is the appropriate target? Theranostics **4**(1), 81–89 (2013)

39. D.E.J.G.J. Dolmans, D. Fukumura, R.K. Jain, Photodynamic therapy for cancer. Nat. Rev. Cancer **3**(5), 380–387 (2003)

40. J. You, R. Zhang, G. Zhang, M. Zhong, Y. Liu, C.S. Van Pelt, D. Liang, W. Wei, A.K. Sood, C. Li, Photothermal-chemotherapy with doxorubicin-loaded hollow gold nanospheres: a platform for near-infrared light-trigged drug release. J. Controlled Release: Official Journal of the Controlled Release Society **158**(2), 319–328 (2012)

41. H. Wang, J. Yu, X. Lu, X. He, Nanoparticle systems reduce systemic toxicity in cancer treatment. Nanomedicine (Lond.) **11**(2), 103–106 (2016)

42. G. Obaid, S. Bano, S. Mallidi, M. Broekgaarden, J. Kuriakose, Z. Silber, A.-L. Bulin, Y. Wang, Z. Mai, W. Jin, D. Simeone, T. Hasan, Impacting pancreatic cancer therapy in heterotypic in vitro organoids and in vivo tumors with specificity-tuned, NIR-activable photoimmunonanoconjugates: towards conquering desmoplasia? Nano Lett. (2019)

43. C. Moore, F. Chen, J. Wang, J.V. Jokerst, Listening for the therapeutic window: advances in drug delivery utilizing photoacoustic imaging. Adv. Drug Deliv. Rev. **144**, 78–89 (2019)

44. L. Cui, J. Rao, Semiconducting polymer nanoparticles as photoacoustic molecular imaging probes. Wiley Interdisc. Rev. Nanomed. Nanobiotechnol. **9**(2), e1418 (2017)

45. D. Cui, C. Xie, K. Pu, Development of semiconducting polymer nanoparticles for photoacoustic imaging. Macromol. Rapid Commun. **38**(12) (2017)

46. B. Yameen, W.I. Choi, C. Vilos, A. Swami, J. Shi, O.C. Farokhzad, Insight into nanoparticle cellular uptake and intracellular targeting. J. Controlled Release: Official Journal of the Controlled Release Society **190**, 485–499 (2014)

47. P. Foroozandeh, A.A. Aziz, Insight into cellular uptake and intracellular trafficking of nanoparticles. Nanoscale Res. Lett. **13**(1), 339 (2018)

48. Y. Liu, S. Chen, J. Sun, S. Zhu, C. Chen, W. Xie, J. Zheng, Y. Zhu, L. Xiao, L. Hao, Z. Wang, S. Chang, Folate-targeted and oxygen/indocyanine green-loaded lipid nanoparticles for dual-mode imaging and photo-sonodynamic/photothermal therapy of ovarian cancer in vitro and in vivo. Mol. Pharm. (2019)

49. C. Moore, J.V. Jokerst, Strategies for image-guided therapy, surgery, and drug delivery using photoacoustic imaging. Theranostics **9**(6), 1550–1571 (2019)

50. J. Xia, C. Kim, J.F. Lovell, Opportunities for photoacoustic-guided drug delivery. Curr. Drug. Targets **16**(6), 571–581 (2015)

51. Y. Zhang, J. Yu, A.R. Kahkoska, Z. Gu, Photoacoustic drug delivery. Sensors (Basel) **17**(6) (2017)

52. W. Li, X. Chen, Gold nanoparticles for photoacoustic imaging. Nanomedicine (Lond.) **10**(2), 299–320 (2015)

53. K. Yang, Y. Liu, Y. Wang, Q. Ren, H. Guo, J.B. Matson, X. Chen, Z. Nie, Enzyme-induced in vivo assembly of gold nanoparticles for imaging-guided synergistic chemo-photothermal therapy of tumor. Biomaterials **223**, 119460 (2019)

54. J.-W. Kim, E.I. Galanzha, E.V. Shashkov, H.-M. Moon, V.P. Zharov, Golden carbon nanotubes as multimodal photoacoustic and photothermal high-contrast molecular agents. Nat. Nanotechnol. **4**(10), 688–694 (2009)

55. L. Meng, X. Zhang, Q. Lu, Z. Fei, P.J. Dyson, Single walled carbon nanotubes as drug delivery vehicles: targeting doxorubicin to tumors. Biomaterials **33**(6), 1689–1698 (2012)

56. B.S. Wong, S.L. Yoong, A. Jagusiak, T. Panczyk, H.K. Ho, W.H. Ang, G. Pastorin, Carbon nanotubes for delivery of small molecule drugs. Adv. Drug Deliv. Rev. **65**(15), 1964–2015 (2013)

57. J. Liu, C. Wang, X. Wang, X. Wang, L. Cheng, Y. Li, Z. Liu, Mesoporous silica coated single-walled carbon nanotubes as a multifunctional light-responsive platform for cancer combination therapy. Adv. Func. Mater. **25**(3), 384–392 (2015)

58. L. Xie, G. Wang, H. Zhou, F. Zhang, Z. Guo, C. Liu, X. Zhang, L. Zhu, Functional long circulating single walled carbon nanotubes for fluorescent/photoacoustic imaging-guided enhanced phototherapy. Biomaterials **103**, 219–228 (2016)

LED-Based Photoacoustic Imaging for Guiding Peripheral Minimally Invasive Procedures

Eleanor Mackle, Efthymios Maneas, Wenfeng Xia, Simeon West, and Adrien Desjardins

Abstract Photoacoustic imaging (PAI) could be useful for improving guidance of peripheral minimally invasive procedures. In this clinical application space, light emitting diodes (LEDs) have several potential advantages as excitation sources. B-mode ultrasound (US) imaging is often used for guiding invasive medical devices; however, most anatomical structures do not have unique US appearances, so the misidentification of tissues is common. There are several potential uses for LED-based PAI, including identifying procedural targets, avoiding critical tissue structures, and localising invasive medical devices relative to external imaging probes. In this chapter, we discuss the state-of-the-art of visualising tissue structures and

E. Mackle (✉) · E. Maneas · W. Xia · S. West · A. Desjardins
Department of Medical Physics and Bioengineering, University College London, Gower Street, London WC1E 6BT, UK
e-mail: eleanor.mackle.14@ucl.ac.uk

E. Maneas
e-mail: efthymios.maneas@ucl.ac.uk

W. Xia
e-mail: wenfeng.xia@kcl.ac.uk

S. West
e-mail: simeon.west@nhs.net

A. Desjardins
e-mail: a.desjardins@ucl.ac.uk

E. Mackle · E. Maneas · A. Desjardins
Wellcome-EPSRC Centre for Interventional & Surgical Sciences, Charles Bell House, Foley Street, London W1W 7TY, UK

W. Xia
School of Biomedical Engineering & Imaging Sciences, King's College London, 4th Floor, Lambeth Wing St. Thomas' Hospital, Westminster Bridge Road, London SE1 7EH, UK

S. West
Department of Anaesthesia, University College Hospital, Main Theatres, Maple Bridge Link Corridor, Podium 3, 235 Euston Road, London NW1 2BU, UK

© Springer Nature Singapore Pte Ltd. 2020
M. Kuniyil Ajith Singh (ed.), *LED-Based Photoacoustic Imaging*,
Progress in Optical Science and Photonics 7,
https://doi.org/10.1007/978-981-15-3984-8_13

medical devices relevant to minimally invasive procedures with PAI, key clinical considerations, and challenges involved with translating LED-based PAI systems from the benchtop to clinical use.

1 Introduction

1.1 Photoacoustic Imaging of Peripheral Vasculature

Photoacoustic imaging (PAI) is an emerging modality that provides information complementary to conventional B-mode ultrasound (US) imaging. With PAI, pulsed excitation light is delivered to tissue, where it is absorbed by specific tissue constituents and the corresponding temperature rise generates US waves via the photoacoustic effect. Received US signals are processed to generate an image [1]. Whereas B-mode US imaging yields information about variations in the mechanical properties of tissue, image contrast in PAI stems from optical absorption by endogenous or exogenous chromophores. In the visible and near-infrared (NIR) wavelength ranges, haemoglobin in blood is a prominent optical absorber, so that vasculature can be visualised with high contrast [1] and blood oxygen saturation can be estimated [2]. Lipids can also be prominent optical absorbers in the NIR, so that nerves can be directly visualised [3–5]. Studies performed to date indicate that PAI has strong potential for clinical translation in broad range of applications. This chapter is focused on the use of PAI for guiding minimally invasive procedures.

Several studies have demonstrated that PAI can visualise human peripheral vasculature in vivo, including vessels in the human palm [6–8] and arm [9]. In a study by Fronheiser et al., photoacoustic imaging was used to map and monitor vasculature for haemodialysis treatment [9]. It has also been shown to be promising for assessing skin microvasculature pathology, or for investigating soft tissue damage such as burns [10]. Favazza et al. were interested in visualising vasculature to obtain indicators of cardiovascular health [11]. More recently, Plumb et al. [12] obtained highly detailed 3D images of the human microvasculature (Fig. 1) and investigated the dynamic potential of their system by assessing the response of the microvasculature to thermal stimuli. Taken together, these studies indicate that PAI can provide clinically compelling images of vasculature, which motivates its application to guiding minimally invasive procedures.

Developing PAI systems so that they can be used routinely to guide minimally invasive devices such as needles and catheters involves multiple challenges. From a translational standpoint, both the size and the cost of these systems are important factors. This consideration has led to interest in light emitting diodes (LEDs) as PA excitation sources [13, 14]. However, a prominent challenge associated with using LEDs is that they tend to have much lower pulse energies than conventional excitation sources such as Nd:YAG lasers and optical parametric oscillators (OPOs), so that US signals are correspondingly weaker. Pioneering work by Hansen [15] demonstrated

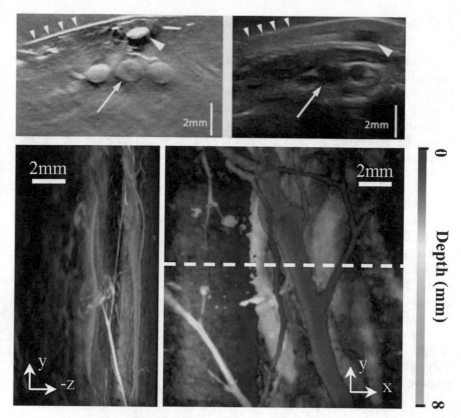

Fig. 1 Photoacoustic imaging (PAI) of human peripheral vasculature, acquired with an optical parametric oscillator (OPO) as the excitation light source. The top panel shows a photoacoustic (PA) image (top left) alongside the corresponding ultrasound image (top right), obtained simultaneously from a volunteer; the bottom panel shows corresponding volumetric PAI data sets, presented as coronal and sagittal maximum intensity projections. The dashed line in the bottom right image indicates plane through which axial image (top left) was reconstructed. Adapted with permission from Plumb et al. [12]

the feasibility of LED-based PAI, and Allen and Beard [16] investigated the use of LEDs for biomedical applications. Recently, LEDs have been integrated alongside a clinical, handheld imaging probe as part of a commercial system (CYBERDYNE INC., Tsukuba, Japan) [17]. In this system, an LED array is positioned on each side of the imaging probe, angled so that their axes of illumination intersect the US imaging plane. B-mode US images and photoacoustic images can be acquired sequentially, with photoacoustic information overlaid onto the US images in real-time.

1.2   Applications to Minimally Invasive Procedures

LED-based PAI may be well suited to guiding minimally invasive procedures that are targeted at peripheral blood vessels or nerves. In current clinical practice, B-mode US guidance is used to visualise both anatomical structures and invasive medical devices. Procedures that use US guidance include peripheral venous access (e.g. for central line placement and vascular access in varicose vein procedures), peripheral arterial access (e.g. for arterial pressure monitoring and percutaneous coronary interventions), biopsies, nerve blocks, fascial plane blocks and interventional pain procedures. In expert hands, B-mode US imaging can provide reliable identification of structures such as nerves, arterial and venous blood vessels, muscle, solid organs and tumours. However, none of these structures give rise to unique US appearances with B-mode imaging; the reflected US waves are dependent upon the mechanical properties of the tissue and the angle of insonation, so that misidentification of tissues is common. In these contexts, LED-based PAI could be useful in a variety of ways. Firstly, it could help to identify procedural targets, such as blood vessels or nerves. Blood vessels can be visualised directly with PAI due to the presence of haemoglobin in red blood cells, with excitation wavelengths spanning visible and NIR wavelengths. Likewise, direct image contrast for nerves and surrounding adipose tissues can be obtained with specific NIR wavelengths where optical absorption by lipids is prominent [3–5]. Secondly, LED-based PAI could be used to avoid damaging critical structures or puncturing arteries. Finally, it could be used to localise invasive devices relative to external imaging probes.

2   LED-Based Photoacoustic Imaging of Vasculature

Several studies have explored the use of LED-based PAI to image superficial human vasculature. In the study of Xia et al. [18], the human finger and wrist were imaged, and strong PA signals from subsurface vascular structures in both imaging locations were observed. This study showed relatively strong visual correspondence between US and PA modalities, although the authors noticed distinct differences between the features visible with US and those visible with PAI (Fig. 2). In another study by the same group, it was shown that LED-based PAI could provide sufficient depth and resolution to image superficial vasculature, such as the digital vessels [19]. These results suggest that LED-based PAI might be useful clinically for identifying, avoiding or targeting superficial vasculature.

Maneas et al. used an LED-based system (AcousticX, CYBERDYNE INC., Tsukuba, Japan) to image placental vasculature, in the context of fetal medicine [20]. They imaged post-partum human placentas ex vivo and were able to detect superficial blood vessels with PAI that were not apparent with US alone (Fig. 3). They compared images acquired from this system to those acquired from a Fabry-Pérot based system and concluded that the two systems were complementary: the

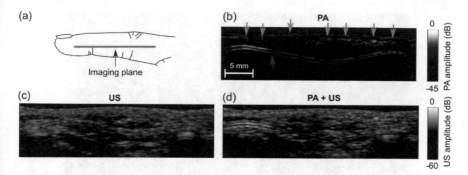

Fig. 2 Photoacoustic (PA) imaging of human finger with a light-emitting diode (LED)-based system. **a** Schematic showing the location of the imaging plane; **b** PA imaging; **c** ultrasound (US) B-mode imaging; **d** PA overlaid on US. Interpretations: digital artery (red arrow), veins (blue arrows) and skin surface (yellow arrow). Adapted with permission from Xia et al. [18]

former allowed for rapid 2D PA and B-mode US imaging whilst the latter yielded finer detail.

In order to be useful for guiding vascular access procedures, a PAI device would need to visualise vessels to depths of 40 mm or more, thereby ensuring its suitability for patients with a high body mass index (BMI). For vascular access applications, imaging vessels with diameters greater than 2 mm would be necessary. However, it would also be useful to identify vessels with much smaller diameters and to ensure that collapsed vessels and those parallel to the imaging plane can be seen. Visualising these vessels could improve the safety of minimally invasive procedures by enabling clinicians to avoid vascular structures where necessary.

Beyond identifying human vasculature, unequivocally distinguishing between arteries and veins is a prominent clinical objective, particularly in the context of avoiding puncturing arterial structures when targeting veins. In current clinical practice, this can be challenging. US-guided clinical procedures that involve percutaneous access to vessels would benefit from enhanced visualisation of arterial and venous structures. For instance, with central venous access, misidentification of arterial structures can lead to catastrophic bleeding; the risk of arterial puncture has been estimated to be as high as 6% [21]. The addition of imaging modalities such as colour US Doppler imaging can help to discriminate pulsatile blood flows, which are indicative of arterial blood. Nonetheless, arterial puncture is still a risk [22], with many underlying factors. Higher risk procedures include those where the vasculature is too small to identify on US, and also ones in low-flow or no-flow states where Doppler imaging has limited utility, which can occur when patients are in shock or cardiac arrest.

LED-based PAI is promising for differentiating between arterial and venous structures, based on differences in the optical absorption spectra of oxy- and deoxy-haemoglobin. Zhu et al. explored the use of LED-based PAI to quantify blood oxygenation levels in a human volunteer [2], using two dual wavelength LED bars emitting alternatively at 690 and 850 nm. Their high imaging frame rates, which

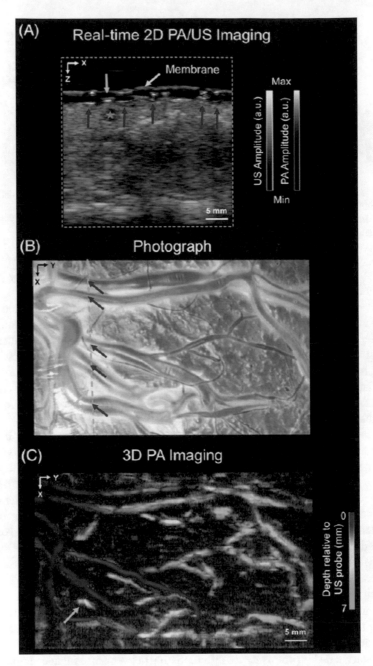

Fig. 3 Photoacoustic (PA) imaging of human placenta with a light-emitting diode (LED)-based system. **a** Combined PA and ultrasound (US) imaging; **b** photograph of the placenta (fetal side); **c** 3D PA image, displayed as a maximum intensity projection of the reconstructed image volume. Adapted with permission from Maneas et al. [20]

reached 500 Hz, could be well suited to visualising rapid changes in oxygenation levels. Additionally, LED-based PAI provided dynamic measurements of vasculature resulting from cuff occlusion [19]. The authors of that study discussed how LED-based PAI imaging of superficial vasculature could potentially be used in clinical settings to measure diagnostic parameters such as the heart rate and recovery time from cuff occlusion.

3 LED-Based Imaging of Invasive Medical Devices

Many types of needle-based procedures could benefit enhanced visualisation of nerves. As an example, nerve blocks involve injections of anaesthetics around a nerve or group of nerves, to treat pain or to prepare a patient for surgery. In these procedures, accurate and efficient identification of nerves is essential. In current practice, US guidance is used to ensure that the needle reaches the target nerve and avoids other structures. Whilst the use of colour Doppler can be beneficial in these contexts, smaller vessels and those oblique to the imaging plane can often be overlooked. Although US has enabled deeper and more challenging blocks to be performed by a greater number of practitioners, inadvertent vascular injury remains a risk [23, 24]. These complications can be even greater in anticoagulated patients. Imaging nerves is challenging, as their appearance can mimic other structures such as tendons [25]. Moreover, nerves exhibit anisotropy, so their appearance is strongly dependent on the angle of insonation. Despite its advantages for visualising neural structures, the resolution of B-mode US can be insufficient to consistently recognise smaller branches, which are increasingly of clinical interest. For example, advanced practitioners may wish to selectively block small sensory branches (less than 1 mm in diameter) in the forearm or ankle, which are extremely challenging to visualise with US. If PAI could provide enhanced visualisation of small superficial neural structures, there would be strong potential to improve procedural outcomes.

4 Prospects for LED-Based Photoacoustic Imaging of Peripheral Nerves

To the authors' knowledge, the use of LED-based PAI for visualising nerves during minimally invasive procedures has remained elusive. The use of PAI with conventional excitation light sources to image nerves has been explored to a limited extent. Ex vivo pilot studies indicate that PAI may provide higher contrast for nerves than that obtained with B-mode US, and that it could be useful for differentiating nerves from tendons [26] (Fig. 4). Even with high energy sources such as OPOs, when excitation light is delivered from the tissue surface, obtaining sufficient signal from

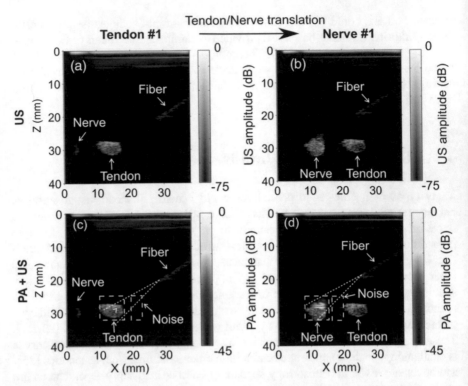

Fig. 4 Photoacoustic (PA) imaging of a nerve/tendon pair, using an interventional multispectral system with an optical parametric oscillator as an excitation source. **a, b** Ultrasound (US) images of nerve and tendon samples; **c, d** the PA images superimposed onto the corresponding US images. Adapted with permission from Mari et al. [26]

lipids at clinically relevant depths is challenging. One solution, which may be relevant to future implementations of LED-based PAI, can be found with interventional PAI, where excitation light is delivered through a needle to reach targets several cm beneath the surface [27–30]. With an LED-based PAI system, either identifying nerves as procedural targets or as structures to avoid could be useful, depending on the clinical context. Ideally, nerves could be visualised at depths up to 60 mm with these systems. As this could be very challenging, given the relatively low pulse energies of LEDs, visualisation of nerves to depths of 30 mm would still be useful.

Imaging of percutaneous devices such as needles and catheters using US can be challenging. As a result, identification of the needle tip is vital to prevent damage to underlying structures. With steep insertion angles, US waves are reflected away from the transducer, so that the needle is not visible. In addition, catheters can be very poorly visible when positioned in soft tissues. Visualising needles and catheters is another area where LED-based PAI could be very useful, and the combination of PAI with current US techniques could provide better real-time guidance. In a study by Xia et al. [18], the performance of LED-based PAI for guiding needle insertions was

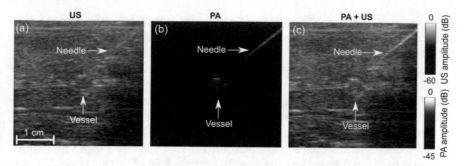

Fig. 5 Light-emitting diode (LED)-based photoacoustic imaging (PAI) of needle insertions. **a** Ultrasound (US) imaging; **b** photoacoustic (PA) imaging at 850 nm; **c** US image with PA overlay. Adapted with permission from Xia et al. [18]

evaluated for the first time. Using an ex vivo blood mimicking phantom, needles were visualised to depths of 38 mm. At insertion angles of 26°–51°, the signal-to-noise ratio (SNR) achieved was 1.2–2.2 times higher than that measured with B-mode US alone; the SNR decreased as the needle insertion angle increased. Although the spatial resolution was similar for both US and PA imaging, the inserted needle was visible down to 2 cm with PAI but it was barely visible with US (Fig. 5). In a second study by Xia et al. [31] medical devices were coated with a carbon nanotube polydimethylsiloxane composite to enhance visibility for photoacoustic imaging. In this study, two experiments were performed: first, a metal needle was inserted into chicken breast and in the second, a catheter dipped in the composite coating was put into the chicken breast. In both cases, the devices were barely visible with US but were visible with LED-based PAI. The uncoated and coated needles were visible to depths greater than 20 mm and 30 mm, respectively.

5 Challenges for Clinical Translation

A significant challenge for the clinical translation of LED-based PAI for minimally invasive procedures is to overcome poor SNR arising from low pulse energies and long pulse durations. Poor SNR, which limits the imaging depth, has also been encountered with the use of laser diodes for PAI [32–34]. One solution, which was suggested by Allen and Beard [16, 35] and implemented by Dai et al. [36], is to overdrive LEDs when they are driven at low duty cycles [37]. Another solution is to perform signal averaging, to which LEDs can be well suited due to their high repetition rate. However this solution comes that expense of decreasing the frame rate [14]. Allen and Beard [35] used signal averaging to demonstrate that LEDs can be used as an excitation source for imaging superficial vascular anatomy. Coded excitation sequences such as Golay code pairs [16] can also be used, as a type of averaging, to improve the SNR. In practice, motion can limit the lengths of these code pairs. In 2013, Allen and Beard [16] demonstrated that Golay code pairs could be used to simultaneously acquire signals from a tissue-mimicking phantom at multiple wavelengths.

A second challenge with the clinical translation of LED-based PAI is to manage the heat that LEDs produce, which can be transferred to the patient and can also result in shifts in the emission wavelengths of the LEDs [38]. These thermal considerations will be important when considering integration of LEDs directly into US imaging probes, as an evolution from bulky, side-mounted arrays (Fig. 6) [39].

Exogenous contrast agents could also improve the SNR and depth penetration achievable with LED-based PAI systems. To increase SNR of vessels, contrast agents such as gold or silver nanoparticles have been used to generate larger signals [40–45]. However, adding contrast agents is usually sub-optimal or not possible for clinical translation; many of the ones used in pre-clinical studies are unapproved for human use and are known to be toxic. In this respect, injections of indocyanine green (ICG) are promising, as this contrast agent is approved for use in human patients and has optical absorption spectra that can be matched to LEDs. Singh et al. [39] used ICG as a contrast agent to show simultaneous imaging of both vascular and lymphatic structures in vivo. In combination with contrast agents such as ICG, LED-based PAI could potentially be used to increase contrast for nerves, with injections around the nerves during hydrodissection.

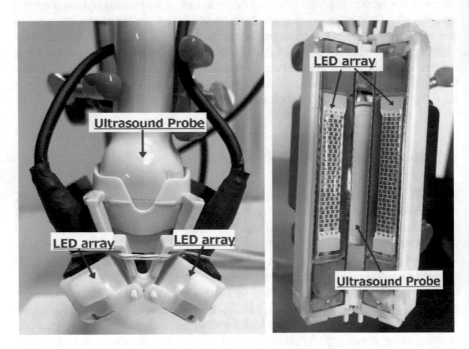

Fig. 6 The imaging probe of a light-emitting diode (LED)-based PAI system (AcousticX, CYBERDYNE INC., Tsukuba, Japan). Images adapted with permission from Singh et al. [39]

Tissue-mimicking phantoms will be important for training clinicians with PAI systems. Anatomically accurate phantoms, with optical and acoustic properties similar to those of human tissue, can be challenging to develop. To this end, there has recently been work to develop anatomically realistic vascular [46] and photoacoustic phantoms [47]. Another key step to facilitate clinical training with LED-based PAI systems is the use of computer assisted segmentation of photoacoustic images. Recent studies have applied deep learning algorithms to photoacoustic images, for instance to segment vessel structures [48] and to assist with breast cancer diagnostics [49].

6 Conclusion

In summary, LED-based PAI systems have strong potential for guiding a wide range of minimally invasive clinical procedures with peripheral targets. As photoacoustic excitation sources, LEDs have the advantage of being compact, so that they can potentially be tightly integrated with clinical US imaging probes. However, they come with the significant challenge of overcoming low SNR that results from smaller pulse energies and longer pulse durations than many conventional photoacoustic excitation sources. Recent demonstrations with LED-based systems to visualise vasculature and minimally invasive devices are promising indications of how PAI could be used in clinical practice.

Acknowledgements The authors gratefully acknowledge the support of the Wellcome/EPSRC Centre for Interventional and Surgical Sciences at the University College London, and Dr Mithun Singh for helpful feedback on this Chapter.

References

1. P. Beard, Biomedical photoacoustic imaging. Interface Focus **1**, 602–631 (2011). https://doi.org/10.1098/rsfs.2011.0028
2. Y. Zhu, G. Xu, J. Yuan et al., Light emitting diodes based photoacoustic imaging and potential clinical applications. Sci. Rep. **8**, 9885 (2018). https://doi.org/10.1038/s41598-018-28131-4
3. A. Balthasar, A.E. Desjardins, M. van der Voort et al., Optical detection of peripheral nerves: an in vivo human study. Reg. Anesth. Pain Med. **37**, 277–282 (2012). https://doi.org/10.1097/AAP.0b013e31824a57c2
4. R. Nachabé, B.H.W. Hendriks, A.E. Desjardins et al., Estimation of lipid and water concentrations in scattering media with diffuse optical spectroscopy from 900 to 1600 nm. J. Biomed. Opt. **15**, 037015 (2010). https://doi.org/10.1117/1.3454392
5. M.K. Dasa, C. Markos, M. Maria et al., High-pulse energy supercontinuum laser for high-resolution spectroscopic photoacoustic imaging of lipids in the 1650–1850 nm region. Biomed. Opt. Express **9**, 1762 (2018). https://doi.org/10.1364/BOE.9.001762
6. S. Hu, L. Wang, Photoacoustic imaging and characterization of the microvasculature, in *Proceedings of SPIE* (2010), p. 011101

7. H.F. Zhang, K. Maslov, G. Stoica, L.V. Wang, Functional photoacoustic microscopy for high-resolution and noninvasive in vivo imaging. Nat. Biotechnol. **24**, 848–851 (2006). https://doi.org/10.1038/nbt1220

8. E.Z. Zhang, J.G. Laufer, R.B. Pedley, P.C. Beard, In vivo high-resolution 3D photoacoustic imaging of superficial vascular anatomy. Phys. Med. Biol. **54**, 1035–1046 (2009). https://doi.org/10.1088/0031-9155/54/4/014

9. M.P. Fronheiser, S.A. Ermilov, H.-P. Brecht et al., Real-time optoacoustic monitoring and three-dimensional mapping of a human arm vasculature. J. Biomed. Opt. **15**, 021305 (2010). https://doi.org/10.1117/1.3370336

10. B. Zabihian, J. Weingast, M. Liu et al., In vivo dual-modality photoacoustic and optical coherence tomography imaging of human dermatological pathologies. Biomed Opt Express **6**, 3163 (2015). https://doi.org/10.1364/BOE.6.003163

11. C.P. Favazza, L.A. Cornelius, L.V. Wang, In vivo functional photoacoustic microscopy of cutaneous microvasculature in human skin. J. Biomed. Opt. **16**, 026004 (2011). https://doi.org/10.1117/1.3536522

12. A.A. Plumb, N.T. Huynh, J. Guggenheim et al., Rapid volumetric photoacoustic tomographic imaging with a Fabry-Perot ultrasound sensor depicts peripheral arteries and microvascular vasomotor responses to thermal stimuli. Eur. Radiol. **28**, 1037–1045 (2018). https://doi.org/10.1007/s00330-017-5080-9

13. Q. Yao, Y. Ding, G. Liu, L. Zeng, Low-cost photoacoustic imaging systems based on laser diode and light-emitting diode excitation. J. Innov. Opt. Health Sci. **10**, 1730003 (2017). https://doi.org/10.1142/S1793545817300038

14. H. Zhong, T. Duan, H. Lan et al., Review of low-cost photoacoustic sensing and imaging based on laser diode and light-emitting diode. Sensors **18**, 2264 (2018). https://doi.org/10.3390/s18072264

15. R.S. Hansen, Using high-power light emitting diodes for photoacoustic imaging, in *Proceedings of SPIE. International Society for Optics and Photonics*, eds. by J. D'hooge, M.M. Doyley, p. 79680A

16. T.J. Allen, P.C. Beard, Light emitting diodes as an excitation source for biomedical photoacoustics, in *Proceedings of SPIE*, eds. by A.A. Oraevsky, L.V. Wang (International Society for Optics and Photonics, 2013), p. 85811F

17. A. Hariri, J. Lemaster, J. Wang et al., The characterization of an economic and portable LED-based photoacoustic imaging system to facilitate molecular imaging. Photoacoustics **9**, 10–20 (2018). https://doi.org/10.1016/J.PACS.2017.11.001

18. W. Xia, M. Kuniyil Ajith Singh, E. Maneas et al., Handheld real-time LED-based photoacoustic and ultrasound imaging system for accurate visualization of clinical metal needles and superficial vasculature to guide minimally invasive procedures. Sensors **18**, 1394 (2018). https://doi.org/10.3390/s18051394

19. W. Xia, E. Maneas, N. Trung Huynh et al., Imaging of human peripheral blood vessels during cuff occlusion with a compact LED-based photoacoustic and ultrasound system, in *Proceedings of SPIE*, eds. by A.A. Oraevsky, L.V. Wang (SPIE, 2019), p. 3

20. E. Maneas, R. Aughwane, N. Huynh et al., Photoacoustic imaging of the human placental vasculature. J. Biophotonics (2019). https://doi.org/10.1002/jbio.201900167

21. C. Thompson, T. Barrows, Carotid arterial cannulation: removing the risk with ultrasound? Can. J. Anesth. **56**, 471–472 (2009). https://doi.org/10.1007/s12630-009-9082-1

22. A. Balthasar, A.E. Desjardins, M. van der Voort et al., Optical detection of vascular penetration during nerve blocks: an in vivo human study. Reg. Anesth. Pain Med. **37**, 3–7 (2012). https://doi.org/10.1097/AAP.0b013e3182377ff1

23. L. Song, Y. Zhou, D. Huang, Inadvertent posterior intercostal artery puncture and haemorrhage after ultrasound-guided thoracic paravertebral block: a case report. BMC Anesthesiol. **18**, 196 (2018). https://doi.org/10.1186/s12871-018-0667-5

24. K. Shirozu, S. Kuramoto, S. Kido et al., Hematoma after transversus abdominis plane block in a patient with HELLP syndrome. A A Case Rep. **8**, 257–260 (2017). https://doi.org/10.1213/XAA.0000000000000487

25. L. Helen, B.D. O'Donnell, E. Moore, Nerve localization techniques for peripheral nerve block and possible future directions. Acta Anaesthesiol. Scand. **59**, 962–974 (2015). https://doi.org/10.1111/aas.12544

26. J.M. Mari, W. Xia, S.J. West, A.E. Desjardins, Interventional multispectral photoacoustic imaging with a clinical ultrasound probe for discriminating nerves and tendons: an ex vivo pilot study. J. Biomed. Opt. **20**, 110503 (2015). https://doi.org/10.1117/1.JBO.20.11.110503

27. D. Piras, C. Grijsen, P. Schütte et al., Photoacoustic needle: minimally invasive guidance to biopsy. J. Biomed. Opt. **18**, 070502 (2013). https://doi.org/10.1117/1.JBO.18.7.070502

28. W. Xia, D.I. Nikitichev, J.M. Mari et al., Performance characteristics of an interventional multispectral photoacoustic imaging system for guiding minimally invasive procedures. J. Biomed. Opt. **20**, 086005 (2015). https://doi.org/10.1117/1.JBO.20.8.086005

29. W. Xia, D.I. Nikitichev, J.M. Mari et al., An interventional multispectral photoacoustic imaging platform for the guidance of minimally invasive procedures, in *Opto-Acoustic Methods and Applications in Biophotonics II* (OSA, Washington, D.C., 2015), p. 95390D

30. W. Xia, E. Maneas, D.I. Nikitichev et al., Interventional photoacoustic imaging of the human placenta with ultrasonic tracking for minimally invasive fetal surgeries, *International Conference on Medical Image Computing and Computer-Assisted Intervention* (Springer, Cham, 2015), pp. 371–378

31. W. Xia, S. Noimark, E. Maneas et al., led-based photoacoustic imaging of medical devices with carbon nanotube-polydimethylsiloxane composite coatings, in *Biophotonics Congress: Biomedical Optics Congress 2018 (Microscopy/Translational/Brain/OTS)* (OSA, Washington, D.C., 2018), p. JW3A.4

32. T. Allen, B. Cox, P.C. Beard, Generating photoacoustic signals using high-peak power pulsed laser diodes, in *Photons Plus Ultrasound: Imaging and Sensing 2005: The Sixth Conference on Biomedical Thermoacoustics, Optoacoustics, and Acousto-Optics* (2005), pp. 233–242

33. T. Allen, 2006 undefined pulsed near-infrared laser diode excitation system for biomedical photoacoustic imaging. Opt. Lett. http://osapublishing.org

34. R.G.M. Kolkman, W. Steenbergen, T.G. van Leeuwen, In vivo photoacoustic imaging of blood vessels with a pulsed laser diode. Lasers Med. Sci. **21**, 134–139 (2006). https://doi.org/10.1007/s10103-006-0384-z

35. T.J. Allen, P.C. Beard, High power visible light emitting diodes as pulsed excitation sources for biomedical photoacoustics. Biomed. Opt. Express **7**, 1260 (2016). https://doi.org/10.1364/BOE.7.001260

36. X. Dai, H. Yang, H. Jiang, In vivo photoacoustic imaging of vasculature with a low-cost miniature light emitting diode excitation. Opt. Lett. **42**, 1456 (2017). https://doi.org/10.1364/OL.42.001456

37. T.J. Allen, P.C. Beard, Pulsed near-infrared laser diode excitation system for biomedical photoacoustic imaging. Opt. Lett. **31**, 3462 (2006). https://doi.org/10.1364/ol.31.003462

38. C. Willert, B. Stasicki, J. Klinner, S. Moessner, Pulsed operation of high-power light emitting diodes for imaging flow velocimetry. Meas. Sci. Technol. **21**, 075402 (2010). https://doi.org/10.1088/0957-0233/21/7/075402

39. M. Singh, T. Agano, N. Sato et al., Real-time in vivo imaging of human lymphatic system using an LED-based photoacoustic/ultrasound imaging system, in *Photons Plus Ultrasound: Imaging and Sensing 2018*, eds. by A.A. Oraevsky, L.V. Wang (SPIE, 2018), p. 3

40. L. Leggio, S. Gawali, D. Gallego et al., Optoacoustic response of gold nanorods in soft phantoms using high-power diode laser assemblies at 870 and 905 nm. Biomed. Opt. Express **8**, 1430 (2017). https://doi.org/10.1364/BOE.8.001430

41. F. Gao, L. Bai, X. Feng et al., Remarkable in vivo nonlinear photoacoustic imaging based on near-infrared organic dyes. Small **12**, 5239–5244 (2016). https://doi.org/10.1002/smll.201602121

42. F. Gao, L. Bai, S. Liu et al., Rationally encapsulated gold nanorods improving both linear and nonlinear photoacoustic imaging contrast in vivo. Nanoscale **9**, 79–86 (2017)

43. M. Beckmann, B.S. Gutrath, Size dependent photoacoustic signal response of gold nanoparticles using a multispectral laser diode system, in *IEEE International Ultrasonics Symposium, IUS* (2012), pp. 2336–2339

44. B.S. Gutrath, M.F. Beckmann, A. Buchkremer et al., Size-dependent multispectral photoacoustic response of solid and hollow gold nanoparticles. Nanotechnology **23**, 225707 (2012). https://doi.org/10.1088/0957-4484/23/22/225707
45. V. Cunningham, H. Lamela, Laser optoacoustic spectroscopy of gold nanorods within a highly scattering medium. Opt. Lett. **35**, 3387–3389 (2010)
46. E.C. Mackle, E. Maneas, C. Little et al., Wall-less vascular poly(vinyl) alcohol gel ultrasound imaging phantoms using 3D printed vessels, in *Design and Quality for Biomedical Technologies XII*, eds. by R. Liang, T.J. Pfefer, J. Hwang (SPIE, 2019), p. 25
47. E. Maneas, W. Xia, O. Ogunlade et al., Gel wax-based tissue-mimicking phantoms for multispectral photoacoustic imaging. Biomed. Opt. Express **9**, 1151 (2018). https://doi.org/10.1364/BOE.9.001151
48. P. Raumonen, T. Tarvainen, Segmentation of vessel structures from photoacoustic images with reliability assessment. Biomed. Opt. Express **9**, 2887–2904 (2018). https://doi.org/10.1364/BOE.9.002887
49. J. Zhang, B.I.N. Chen, M. Zhou, H. Lan, Photoacoustic image classification and segmentation of breast cancer: A feasibility study. IEEE Access **7**, 5457–5466 (2019). https://doi.org/10.1109/ACCESS.2018.2888910

Application of LED-Based Photoacoustic Imaging in Diagnosis of Human Inflammatory Arthritis

Yunhao Zhu, Janggun Jo, Guan Xu, Gandikota Girish, Elena Schiopu, and Xueding Wang

Abstract Using low cost and small size light emitting diodes (LED) as the alternative illumination source for photoacoustic (PA) imaging has many advantages, and can largely benefit the clinical translation of the emerging PA imaging (PAI) technology. To overcome the challenge of achieving sufficient signal-to-noise ratio by the LED light that is orders of magnitude weaker than lasers, extensive signal averaging over hundreds of pulses is performed. According to our research, the LED-based PAI could be a promising tool for several clinical applications, such as assessment of peripheral microvascular function and dynamic changes, and diagnosis of inflammatory arthritis. In this chapter, we will first introduce a commercially available LED-based PAI system, and then show the ability of this system in identifying inflammatory arthritis in human hand joints. B-mode ultrasound (US), Doppler, and PA images were obtained from 12 joints with clinically active arthritis, five joints with subclinically active arthritis, and 12 normal joints. The quantitative assessment of hyperemia in joints by PAI demonstrated statistically significant differences among the three conditions. The imaging results from the subclinically active arthritis joints also suggested that the LED-based PAI has a higher sensitivity to

Y. Zhu · J. Jo · X. Wang (✉)
Department of Biomedical Engineering, University of Michigan, Ann Arbor, MI, USA
e-mail: xdwang@umich.edu

Y. Zhu
e-mail: yunhaoz@umich.edu

J. Jo
e-mail: janggunj@umich.edu

G. Xu · G. Girish · X. Wang
Department of Radiology, University of Michigan, Ann Arbor, MI, USA
e-mail: guanx@med.umich.edu

G. Girish
e-mail: ggirish@med.umich.edu

E. Schiopu
Division of Rheumatology, Department of Internal Medicine, University of Michigan, Ann Arbor, MI, USA
e-mail: eschiopu@med.umich.edu

© Springer Nature Singapore Pte Ltd. 2020
M. Kuniyil Ajith Singh (ed.), *LED-Based Photoacoustic Imaging*,
Progress in Optical Science and Photonics 7,
https://doi.org/10.1007/978-981-15-3984-8_14

angiogenic microvasculature compared to US Doppler imaging. This initial clinical study on arthritis patients validates that PAI can be a potential imaging modality for the diagnosis of inflammatory arthritis.

1 Introduction

Inflammatory arthritis caused by autoimmune disorders is a chronic, progressive set of diseases with worldwide prevalence [1]. Rheumatoid arthritis (RA), one such type of inflammatory arthritis, has symptoms of stiffness, pain, and swelling of the joints. In addition, RA synovium shows hypoxia, neoangiogenesis and synovial proliferation within the peripheral joints [2–4]. Synovial angiogenesis is an important feature in the early stage of development and perpetuation of inflammatory arthritis. Magnetic resonance imaging (MRI) and ultrasound (US) Doppler imaging has been employed as the main modalities in identifying inflammatory arthritis [5, 6]. The use of MRI is not widespread for general clinical screening and diagnosis due to the high cost and limited accessibility. B-mode US Doppler can offer high-resolution images of joint structures and has proven sensitive in detecting blood flow. US Doppler imaging, however, is more sensitive to the fast blood flowing in relatively larger vessels. Slow blood flowing in smaller capillaries, which are more clinically and pathologically relevant to early active synovitis [7], could be missed by Doppler imaging.

Laser-induced photoacoustic imaging (PAI) has shown the capability of identifying active synovitis in human finger joints [8, 9]. With the unique capability of mapping highly sensitive optical information in deep tissue with excellent spatial resolution [10], this emerging imaging technique has been developed and investigated for various preclinical and clinical applications [11–13]. Presenting endogenous optical absorption contrast in tissues, PAI, when combined with B-mode US, can provide additional functional and molecular information such as blood volume and blood oxygen saturation which are highly valuable in diagnosis of many pathological conditions [14–17].

The advancement of light emitting diode (LED) technology offers a unique opportunity to solve the challenges in clinical applications of PA imaging [18, 19]. Compared to expensive class-IV laser systems, the LED-based light source is much lower in both owning and operating costs. The lower costs together with the significantly reduced footprints of the LED light source can make PA imaging a practical option for point-of-care screening or diagnosis of a variety of diseases. Furthermore, unlike class-IV laser systems which need to be placed in dedicated spaces securing safety and operation requirements (e.g., light shielding, high electric power, and air or water cooling), LED light source can operate in almost any place, including resource deficient settings such as on battlefields or in ambulances. At the light fluence we are working with, there is no need for wearing laser safety glasses for anyone within the operation area.

However, the biggest challenge for LED-based PA imaging is to produce sufficient signal-to-noise ratio (SNR) in biological samples. During limited pulse duration of 100 ns or less, LED, even working as a group such as a 2D panel, can only deliver light energy that is about two orders of magnitude lower than that from a class-IV laser. For example, 2D LED array panels along with Acoustic X system (Cyberdyne, Inc.) can deliver up to 200 μJ of pulsed light at 850-nm wavelength. Without performing signal averaging, however, this low pulse energy will still not be able to produce detectable PA signal from biological samples even at the surfaces. Fortunately, taking advantage from the high pulse repetition rate from 1 kHz to 16 kHz, extensive signal averaging from dozens to hundreds of pulses can be performed. In this way, the LED-based PA imaging system is able to achieve SNR comparable to those in laser-based PA imaging systems without sacrificing the capability of performing real-time imaging [20, 21]. If there is no special description, all results below were acquired with a 7-MHz linear array (128 elements), with two 850-nm LED bars driven at 4 kHz repetition rate, and 384 times averaging.

2 LED-Based PA Imaging of Subsurface Microvasculature In Vivo: A Feasibility Study

Via the experiments on human finger and other parts, the performance of LED-based PA imaging for 2D and 3D mapping of subsurface microvasculature in vivo are discussed first, including sensitivity, spatial resolution, penetration depth, and imaging speed, as well as its feasibility in sensing and quantifying the hemodynamic properties and changes in motion such as arterial pulsation, blood reperfusion, and blood oxygen saturation.

2.1 Imaging Microvasculature in 2D and 3D

Figure 1 shows an example result of PA and US combined 2D and 3D imaging of a human finger. The 3D PA and US combined image acquired via the linear scan of the probe along the digit is presented along different views including coronal, axial, and sagittal in Fig. 1a–c. In each 2D rendering, the PA image of microvessels is presented in pseudo-color, and superimposed on the background gray-scale US image. Figure 1d shows perspective view of the spatially distributed vessels in the finger acquired by PA imaging.

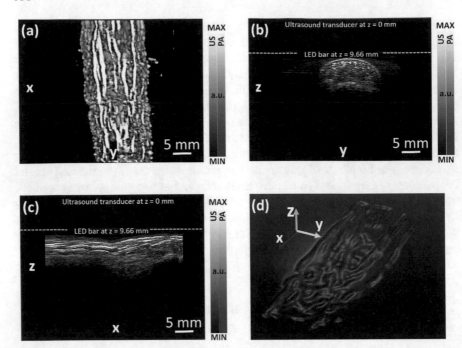

Fig. 1 **a** Coronal, **b** axial, and **c** sagittal views of the 3D PA and US combined image of a human finger from a volunteer. **d** Perspective view of the 3D PA image showing the microvessels in the finger. Reprinted with permission from [22]

2.2 Arterial Pulsation and Blood Reperfusion

Figure 2a shows a real-time acquired PA and US B-scan results of a human finger along the sagittal section with a frame rate of 10 Hz. To present the arterial pulsation in motion (as marked by the arrows), four frames from the cine loop are presented. In the pseudo-color PA images, besides the marked artery, spatially distributed microvessels in the finger up to 5-mm deep can also be recognized. A higher central frequency of 10-MHz linear array (128 elements) was employed.

Working at a higher frame rate of 500 Hz, Fig. 2b shows another results of real-time PA and US B-scan of a finger demonstrating the capability of this system in imaging and quantifying fast hemodynamic changes in vivo. Note that the two 850-nm LED bars were driven at 16 kHz pulse repetition rate, and 32 times averaging was conducted and a 7-MHz linear array was employed. Both the two frames in Fig. 2b are from a cine loop scan showing the blood reperfusion into the finger after releasing of a rubber band which tied around the root of the finger. The left and the right images were acquired right after and at 46 ms after the rubber band releasing respectively. As we can see, during a time period of 46 ms (i.e., 23 frames), the blood signal expanded along the vessel for a total length of 4.6 mm (from right to

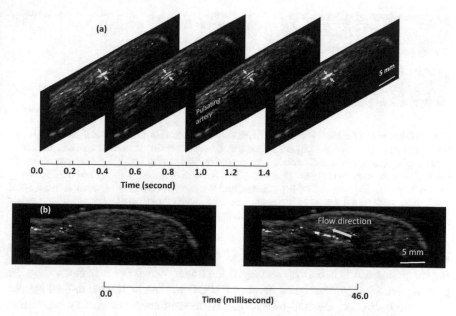

Fig. 2 **a** Four frames from a video showing the cine loop of B-scan PA imaging of a human finger (index finger, sagittal section) presenting the pulsation of an artery marked by the arrows. Imaging frame rate: 10 Hz. **b** Two frames from a video showing the cine loop of B-scan PA imaging of a human finger (index finger, sagittal section) showing the blood reperfusion into the finger after releasing of the rubber band which tied around root of the finger. (Left) The image acquired right after the rubber band releasing; (Right) The image acquired at 46 ms after the rubber band releasing. The flow rate Imaging frame rate: 500 Hz. The distance between two asterisks is measured to be 4.6 mm. Blood flow direction is marked by the arrow. Reprinted with permission from [22]

left, as marked by the two asterisks). The speed of blood reperfusion in this vessel is quantified to be 100 mm/s.

2.3 Blood Oxygen Saturation

Hypoxia is an important biomarker reflecting the onset and progression of many diseases such as cancer. It has been validated that multispectral PA imaging, by probing the spectroscopic difference between oxygenated and deoxygenated hemoglobin, can evaluate relative hemoglobin oxygen saturation and hypoxia in biological samples in vivo, in a non-invasive manner [23–25]. By design, the LED-based PA imaging for measurement of blood oxygenation could be realized by using a pair of dual-wavelength LED bars which can emit 690 and 850-nm wavelength light alternatively as shown in Fig. 3a. Due to the fast switch between the two wavelengths (lagging time <0.5 ms), the accuracy in quantitative imaging of blood oxygenation does not suffer from body motion. PA functional imaging of blood oxygenation in the vessels

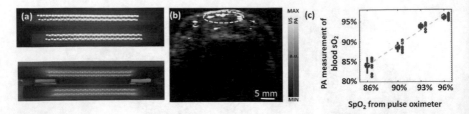

Fig. 3 **a** Photo of a pair of dual-wavelength LED bars that emits 690 and 850-nm light alternatively. **b** PA (pseudo color) and US (gray scale) combined image showing the microvessels in the cross-section of a human finger. **c** Correlation between the dual-wavelength PA measurements of the blood sO2 in the finger and the SpO2 readouts from a pulse oximeter. A fitting line (dashed line, y = ax + b, a = 1.251, b = –0.234) is presented, and R-square of 0.9838 is achieved. At each SpO2 level, the asterisk on red line shows the mean, and the distance above or below it shows the standard deviation of the PA measurements. Reprinted with permission from [22]

in an index finger of a volunteer was shown. Figure 3b is an example PA image showing the microvessels in a cross-section of the finger, which is superimposed on the gray-scale US image. The region of interest (ROI) was marked by the dashed yellow. As shown in Fig. 3c, the dual-wavelength PA measurements of blood oxygenation and the readouts from the pulse oximeter achieved a good correlation (R-square = 0.9838).

2.4 Imaging of Peripheral Vasculature and Response to Cold Exposure

Disturbance of peripheral microvascular function can be associated with many pathological conditions such as diabetes mellitus [26–31], heart failure [32, 33], and Raynaud's phenomenon (RP) [34, 35]. RP, seen as the first manifestation in 70% of the patients with systemic sclerosis, refers to constriction of the microvessels of the hands or feet in response to cold exposure. Objective evaluation of peripheral microcirculatory flow plays a key role in the characterization and treatment assessment of RP [35]. For example, it has been demonstrated that, for patients with RP, the blood flow in skin measured by laser Doppler decreases significantly after local cold exposure [36].

Here we show the LED-based PA imaging in mapping peripheral microvessels in foot, and in sensing the decrease in microvascular flow in response to cold temperature. The decrease in PA signal intensity should be caused mainly by the vasocontraction instead of the temperature change in blood which may also affect PA signal intensity via the temperature-dependent Grüneisen parameter. This is due to the fact that the temperature change in blood should be small, since the experimental duration was short and the blood in the vessels was also continuously flowing.

PA imaging of peripheral vasculature acquired from the dorsal surface of the left foot of a volunteer is shown in Fig. 4. The dashed rectangle in Fig. 4a indicates

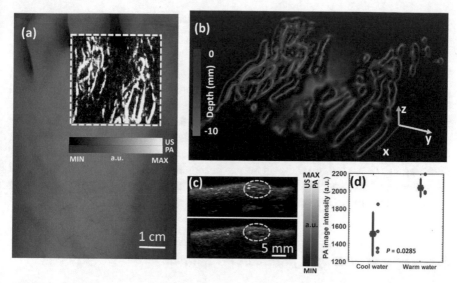

Fig. 4 **a** Maximum intensity projection PA and US combined image showing the vasculature in the dorsal surface of a human foot. The dashed rectangle indicates the imaged area. **b** Perspective view image showing the 3D vasculature. The depth (i.e., the position along the z axis) is color encoded. **c** B-scan PA and US combined images of the vasculature in foot surface acquired at a temperature of 40.8 and 34.2 °C. The dashed circles indicate the regions of interest (ROI) for quantifying the change of PA image intensity in response to the local cold exposure. **d** Data distribution showing the averaged PA image intensities of vessels within the ROI at the two different temperatures (i.e., 40.8 vs. 34.2 °C). For the PA measurements at each temperature, the asterisk on red line shows the mean, and the distance above or below it shows the standard deviation. A *p*-value of 0.0285 was achieved for a two-tailed *t*-test (n = 4), demonstrating the capability of PA imaging in detecting the change in peripheral microvascular flow in response to the local cold exposure. Reprinted with permission from [22]

the scanned area on the foot surface. With the 3D image acquired, a maximum intensity projection (MIP) PA and US combined image and a perspective view PA image with depth color-encoded are presented in Fig. 4a and Fig. 4b, respectively. Spatially distributed microvessels within the depth up to 10 mm can be recognized. Figure 4c shows 2D B-scan images of the foot surface acquired at 40.8 and 34.2 °C respectively. Pseudo-color PA image showing the vessels is superimposed on the gray-scale US image. The PA intensities of the blood vessels within the ROI marked by the dashed circle were averaged. The quantified PA measurements at the two different temperatures (40.8 vs. 34.2 °C) are compared, as shown in Fig. 4d. With four independent measurements at each temperature (n = 4), a two-tailed *t*-test was conducted with a hypothesis that there is no difference between the PA measurements at the two temperatures. A *p*-value of 0.0285 was achieved, suggesting that the decrease in peripheral microvascular flow in response to the local cold exposure (i.e., temperature drop from 40.8 to 34.2 °C) can be detected by the LED-based PA imaging.

3   LED-Based PA Imaging of Inflammatory Arthritis: A Clinical Study

We have demonstrated the capability of PA technique in imaging human peripheral joints [37–41]. These relatively smaller joints are usually among the first to be affected by inflammatory arthritis. The early research findings from animal models and human subjects suggest that PA imaging holds promise for rheumatology clinic, and can provide a low-cost and non-invasive tool for early diagnosis and treatment monitoring of inflammatory arthritis [8, 42–45]. In this section, we explored the feasibility of detecting soft tissue inflammation in human peripheral joints by using the LED-based PA system, and its capability in differentiating arthritic joints from normal joints by evaluating the enhanced microvascular flow in the synovial tissue.

3.1   Introduction of Inflammatory Arthritis

Synovial angiogenesis is an important early feature in the development and perpetuation of RA [46]. Angiogenesis from a combination of hypoxia and high metabolic demand increases the number of synovial vessels [47], which drives synovial infiltration and hyperplasia. Neoangiogenesis, hyperemia and hypoxia in pathological synovium are therefore essential hallmarks of inflammatory arthritis and regarded as the key features for the early diagnosis of the disease [48–60].

During the progression of inflammatory arthritis, at the very early stage we can see inflammatory markers in tissues, then we will notice functional changes such as hyperemia, hypoxia and metabolic activity as shown in Fig. 5. Later in the progress, anatomical changes will happen including synovium thickening and effusion.

3.2   LED-Based PA Imaging in Three Groups of Joints: Clinically Active Arthritis, Subclinically Active Arthritis and Healthy Joints

The LED-based PA imaging can potentially be a low-cost and non-invasive tool for early diagnosis and treatment monitoring of inflammatory arthritis by evaluating the enhanced microvascular flow in the synovial tissue. US Doppler (left) and PA (right) images as shown in Fig. 6 of the three groups are compared in this section, including clinically active arthritis joints, subclinically active arthritis joints, and normal healthy joints.

As an example of clinically active RA, Fig. 6a showed Doppler US B-scan and the corresponding PAI of a human metacarpophalangeal (MCP) joint with inflammation. Doppler US image (left) was obtained by a sonographer right before the PAI scan. Right image was the superimposed PA and US image acquired immediately after US

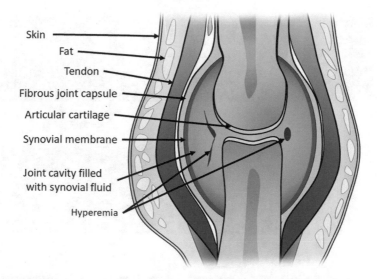

Skin
Fat
Tendon
Fibrous joint capsule
Articular cartilage
Synovial membrane

Joint cavity filled
with synovial fluid

Hyperemia

Fig. 5 Sketch of an inflamed human MCP joint showing hyperemia in synovium. Reprinted with permission from [61]

Doppler scan. The US images in both showed highly correlated MCP joint structures. The hyperemia displayed by colored Doppler images can also be found at the same position, where the pseudo-color red pixels were blood signals detected by PAI.

As an example of subclinically active, PA and US Doppler images are shown in Fig. 6b. In this case, we found hyperemia can be recognized in PA images but not by Doppler US imaging using the ZONARE system right before the PAI scan. Same as the clinically active arthritis patients, this group of patients also had swelling and pain in affected finger joints, as confirmed by the board-certified rheumatologists following the American College of Rheumatology (ACR) criteria. However, the activity in the affected joints was not strong enough to be detected by the US Doppler imaging systems used. These patients, with hyperemia seen only on PAI images, were categorized as a separate group which was defined as sub-clinically active arthritis.

Following a similar procedure, we had healthy volunteer performed an US Doppler scan, followed by a subsequent scan of the same finger joints using the LED PAI as shown in Fig. 6c. Unlike the results from the arthritic joints, no prominent hyperemia can be identified in the synovium of the normal joints, which was confirmed by both US Doppler imaging and PAI.

3.3 Statistic Results Based on Imaging

To characterize the capability of PAI utilizing LED light source in differentiating the three groups studied [i.e., clinically active arthritis group (n = 12), subclinically

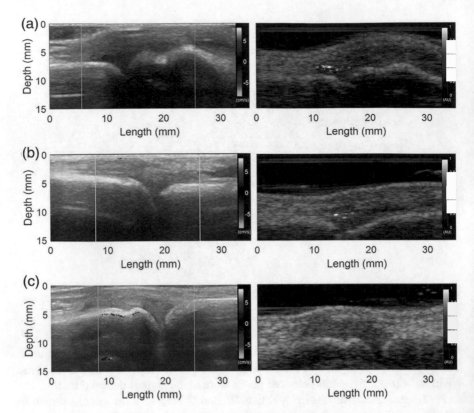

Fig. 6 (Left column) US Doppler images and (right column) PA images of human MCP joints. **a** The images of a clinically active inflammatory arthritis joint showing hyperemia in both US Doppler and PA images. **b** The images of a subclinically active inflammatory arthritis joint showing hyperemia in PA image only but not in US Doppler image. **c** The images of a normal joint showing no hyperemia in either US Doppler image or PA image. Reprinted with permission from [61]

active arthritis group (n = 5), and normal group (n = 12)], the imaging quantified results from the three groups were compared. With the pseudo-color PA images of each joint acquired, we evaluated the hyperemia as a biomarker of joint inflammation by quantifying two parameters, including (1) the density of colored pixels and (2) the average intensity of colored pixels in the joint area. For each pseudo-color PA image of a joint, the density of colored pixels was calculated by dividing the number of colored pixels by the number of total pixels in the joint area; the average intensity of colored pixels was calculated by the sum of the intensities of all colored pixels divided by the number of colored pixels in the joint area.

The quantified parameters of the three groups are compared in Fig. 7. Figure 7a shows the box plots of the density of colored pixels in the joint area. The averages and the standard deviations of the three groups are 14.0 ± 5.80 (%), 7.0 ± 3.20 (%), and 0.4 ± 0.51 (%), respectively. To examine whether there is statistically significant difference in this first parameter between any of the two groups, two tailed t-test was

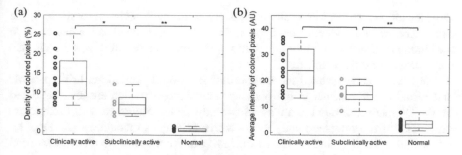

Fig. 7 Statistical studies comparing the hyperemia in the three groups of joints (i.e., clinically active arthritis, n = 12; subclinically active arthritis, n = 5; and normal, n = 12) as quantified by LED PAI. **a** The quantified results showing the density of colored pixels in pseudo-color PA images of the three groups. **b** The quantified results showing the average intensity of colored pixels in pseudo-color PA images of the three groups. *Note* * stands for $p < 0.05$, and ** stands for $p < 0.005$. Reprinted with permission from [61]

performed using the built-in functions of the MATLAB (R2016b, Mathworks). The statistical analyses show that any of the two groups can be differentiated by PAI based on the quantified density of colored pixels in the joint area. The p-values from the two-tailed t-tests were 0.024 for differentiating the clinically active group and the subclinically negative group, 5.6×10^{-8} for differentiating the clinically active group and the normal group, and 3.1×10^{-6} for differentiating the subclinically active group and the normal group.

Figure 7b shows the box plots of the average intensity of colored pixels in the joint area for the three groups. The averages and the standard deviations of the three groups are 24.53 ± 8.28, 15.03 ± 4.54, and 3.56 ± 1.97, respectively. Similarly, the statistical analyses show that any of the two groups can be differentiated by PAI based on the quantified average intensity of colored pixels in the joint area. The p values were 0.030 for differentiating the clinically active group and the subclinically negative group, 2.0×10^{-8} for differentiating the clinically active group and the normal group, and 2.0×10^{-6} for differentiating the subclinically active group and the normal group.

All above results have demonstrated that LED PAI is capable of differentiating arthritic joints from the normal joints.

4 Future Perspective

In the future, we will focus on imaging the Achilles tendon area. Because the psoriatic arthritis effects not only the joints, but also the tendon. In clinic, Doppler ultrasound imaging has been used as a standard procedure to detect the flow in tendon. Clinicians conventionally scan the Achilles tendon of the patients in two ways—scanning along the tendon, and across the tendon to find the hyperemia in the tendon area using

Doppler ultrasound, and PAI can do the same with much higher resolution. In the same way, we could identify the hyperemia in the joint as well as the enthesitis in the tendon area with PAI. Enthesitis is an inflammation in the enthesis, which is also a symptom of psoriatic arthritis.

We have already scanned the tendon area of several psoriatic arthritis patients and visualized hyperemia in the photoacoustic images, which are also validated by the Doppler ultrasound in a feasibility study. In the future, we will accumulate more results from volunteers and patients with two imaging modalities—the Doppler ultrasound and PAI, to find the statistical significance between the patient and normal groups.

5 Conclusion

The imaging results from human subjects have demonstrated that the LED-based PA imaging can be a potential clinical tool for assessment of a variety of disease conditions associated with peripheral microvascular function and the diagnosis of inflammatory arthritis. In a clinical study, the results from arthritis patients and normal volunteers demonstrated that the LED-based PAI can detect the early functional changes of inflammation in human peripheral joints. In addition to the structural details and blood flow detected by the pulse-echo and Doppler US, the PAI provides unique information regarding subtle changes in blood content independent of flow. The quantitative PA measurements have also demonstrated narrower and sharper criteria for identifying neovascularity in the synovium.

Acknowledgements We thank Cyberdyne, Inc for technical support and long-term collaboration with our lab. This research was funded by the National Institute of Health (R01AR060350, 5R21AI122098-021, R21AI12209801A1) and Michigan Institute for Clinical and Health Research (MICHR) (UL1TR000433).

Ethical Approval All procedures for human subjects in this study were approved by the Institutional Review Board (IRB) of the University of Michigan Medical School (HUM00003693).

References

1. J.L. Hoving et al., Non-pharmacological interventions for preventing job loss in workers with inflammatory arthritis. Cochrane Database Syst. Rev. **11** (2014)
2. S.Y. Park et al., HMGB1 induces angiogenesis in rheumatoid arthritis via HIF-1α activation. Eur. J. Immunol. **45**(4), 1216–1227 (2015)
3. M. Biniecka et al., Dysregulated bioenergetics: a key regulator of joint inflammation. Ann. Rheum. Dis. **75**(12), 2192–2200 (2016)
4. C.M. Quiñonez-Flores, S.A. González-Chávez, C. Pacheco-Tena, Hypoxia and its implications in rheumatoid arthritis. J. Biomed. Sci. **23**(1), 62 (2016)

5. B.N. Weissman, Imaging of arthritis and metabolic bone disease. (Elsevier Health Sciences, 2009)
6. W.A. Schmidt, Technology insight: the role of color and power Doppler ultrasonography in rheumatology. Nat. Rev. Rheumatol. **3**(1), 35 (2007)
7. I. Goldie, The synovial microvascular derangement in rheumatoid arthritis and osteoarthritis. Acta Orthop. Scand. **40**(6), 751–764 (1969)
8. J. Jo et al., A functional study of human inflammatory arthritis using photoacoustic imaging. Sci. Rep. **7**(1), 15026 (2017)
9. P.J. van den Berg et al., Feasibility of photoacoustic/ultrasound imaging of synovitis in finger joints using a point-of-care system. Photoacoustics **8**, 8–14 (2017)
10. M.W. Schellenberg, H.K. Hunt, Hand-held optoacoustic imaging: a review. Photoacoustics **11**, 14–27 (2018)
11. L.V. Wang, Multiscale photoacoustic microscopy and computed tomography. Nat. Photonics **3**(9), 503 (2009)
12. X. Wang et al., Noninvasive laser-induced photoacoustic tomography for structural and functional in vivo imaging of the brain. Nat. Biotechnol. **21**(7), 803 (2003)
13. L.V. Wang, S. Hu, Photoacoustic tomography: in vivo imaging from organelles to organs. Science **335**(6075), 1458–1462 (2012)
14. D. Razansky, C. Vinegoni, V. Ntziachristos, Multispectral photoacoustic imaging of fluorochromes in small animals. Opt. Lett. **32**(19), 2891–2893 (2007)
15. H.F. Zhang et al., Functional photoacoustic microscopy for high-resolution and noninvasive in vivo imaging. Nat. Biotechnol. **24**(7), 848 (2006)
16. J. Jo et al., In vivo quantitative imaging of tumor pH by nanosonophore assisted multispectral photoacoustic imaging. Nat. Commun. **8**(1), 471 (2017)
17. C.H. Lee et al., Ion-selective nanosensor for photoacoustic and fluorescence imaging of potassium. Anal. Chem. **89**(15), 7943–7949 (2017)
18. T.J. Allen, P.C. Beard, High power visible light emitting diodes as pulsed excitation sources for biomedical photoacoustics. Biomed. Opt. Express **7**(4), 1260–1270 (2016)
19. A. Hariri et al., The characterization of an economic and portable LED-based photoacoustic imaging system to facilitate molecular imaging. Photoacoustics **9**, 10–20 (2018)
20. K. Sivasubramanian, M. Pramanik, High frame rate photoacoustic imaging at 7000 frames per second using clinical ultrasound system. Biomed. Opt. Express **7**(2), 312–323 (2016)
21. Y.-H. Wang, P.-C. Li, SNR-dependent coherence-based adaptive imaging for high-frame-rate ultrasonic and photoacoustic imaging. IEEE Trans. Ultrason. Ferroelectr. Freq. Control **61**(8), 1419–1432 (2014)
22. Y. Zhu et al., Light emitting diodes based photoacoustic imaging and potential clinical applications. Sci. Rep. **8**(1), 9885 (2018)
23. X. Wang et al., Noninvasive imaging of hemoglobin concentration and oxygenation in the rat brain using high-resolution photoacoustic tomography. J. Biomed. Opt. **11**(2), 024015 (2006)
24. S. Yang et al., Functional imaging of cerebrovascular activities in small animals using high-resolution photoacoustic tomography. Med. Phys. **34**(8), 3294–3301 (2007)
25. C. Kim, C. Favazza, L.V. Wang, In vivo photoacoustic tomography of chemicals: high-resolution functional and molecular optical imaging at new depths. Chem. Rev. **110**(5), 2756–2782 (2010)
26. J.E. Tooke, Peripheral microvascular disease in diabetes. Diabetes Res. Clin. Pract. **30**, S61–S65 (1996)
27. J.E. Tooke, Microvascular function in human diabetes: a physiological perspective. Diabetes **44**(7), 721–726 (1995)
28. E. Tateishi-Yuyama et al., Therapeutic angiogenesis for patients with limb ischaemia by autologous transplantation of bone-marrow cells: a pilot study and a randomised controlled trial. The Lancet **360**(9331), 427–435 (2002)
29. B.A. Lipsky et al., Diagnosis and treatment of diabetic foot infections. Clin. Infect. Dis., 885–910 (2004)

30. A.E. Jones et al., Lactate clearance vs central venous oxygen saturation as goals of early sepsis therapy: a randomized clinical trial. JAMA **303**(8), 739–746 (2010)
31. G. Hernandez-Cardoso et al., Terahertz imaging for early screening of diabetic foot syndrome: a proof of concept. Sci. Rep. **7**, 42124 (2017)
32. I.R. Mahy et al., Disturbance of peripheral microvascular function in congestive heart failure secondary to idiopathic dilated cardiomyopathy. Cardiovasc. Res. **30**(6), 939–944 (1995)
33. I.R. Mahy, J.E. Tooke, Peripheral microvascular function in human heart failure. Clin. Sci. **88**(5), 501–508 (1995)
34. M.J.U. Corrêa et al., Quantification of basal digital blood flow and after cold stimulus by laser doppler imaging in patients with systemic sclerosis. Revista brasileira de reumatologia **50**(2), 128–134 (2010)
35. A.L. Herrick, S. Clark, Quantifying digital vascular disease in patients with primary Raynaud's phenomenon and systemic sclerosis. Ann. Rheum. Dis. **57**(2), 70–78 (1998)
36. T. Kanetaka et al., Laser Doppler skin perfusion pressure in the assessment of Raynaud's phenomenon. Eur. J. Vasc. Endovasc. Surg. **27**(4), 414–416 (2004)
37. X. Wang et al., Imaging of joints with laser-based photoacoustic tomography: an animal study. Med. Phys. **33**(8), 2691–2697 (2006)
38. H.F. Zhang et al., Imaging of hemoglobin oxygen saturation variations in single vessels in vivo using photoacoustic microscopy. Appl. Phys. Lett. **90**(5), 053901 (2007)
39. Y. Sun, E.S. Sobel, H. Jiang, Quantitative three-dimensional photoacoustic tomography of the finger joints: an in vivo study. J. Biomed. Opt. **14**(6), 064002 (2009)
40. G. Xu et al., Photoacoustic and ultrasound dual-modality imaging of human peripheral joints. J. Biomed. Opt. **18**(1), 010502 (2012)
41. P. van Es et al., Initial results of finger imaging using photoacoustic computed tomography. J. Biomed. Opt. **19**(6), 060501 (2014)
42. Y. Sun, E.S. Sobel, H. Jiang, First assessment of three-dimensional quantitative photoacoustic tomography for in vivo detection of osteoarthritis in the finger joints. Med. Phys. **38**(7), 4009–4017 (2011)
43. J.R. Rajian, G. Girish, X. Wang, Photoacoustic tomography to identify inflammatory arthritis. J. Biomed. Opt. **17**(9), 096013 (2012)
44. J.R. Rajian et al., Characterization and treatment monitoring of inflammatory arthritis by photoacoustic imaging: a study on adjuvant-induced arthritis rat model. Biomed. Opt. Express **4**(6), 900–908 (2013)
45. N. Beziere et al., Optoacoustic imaging and staging of inflammation in a murine model of arthritis. Arthritis Rheumatol. **66**(8), 2071–2078 (2014)
46. P. Brenchley, Antagonising angiogenesis in rheumatoid arthritis. Ann. Rheum. Dis. **60**(suppl 3), iii71–iii74 (2001)
47. Z. Szekanecz, A.E. Koch, Mechanisms of disease: angiogenesis in inflammatory diseases. Nat. Rev. Rheumatol. **3**(11), 635 (2007)
48. M. Drouart et al., High serum vascular endothelial growth factor correlates with disease activity of spondylarthropathies. Clin. Exp. Immunol. **132**(1), 158–162 (2003)
49. G.S. Firestein, Starving the synovium: angiogenesis and inflammation in rheumatoid arthritis. J. Clin. Investig. **103**(1), 3–4 (1999)
50. O. FitzGerald et al., Morphometric analysis of blood vessels in synovial membranes obtained from clinically affected and unaffected knee joints of patients with rheumatoid arthritis. Ann. Rheum. Dis. **50**(11), 792–796 (1991)
51. S. Hirohata, J. Sakakibara, Angioneogenesis in rheumatoid arthritis. The Lancet **354**(9176), 423–424 (1999)
52. P. Jones, R. Makki, J. Weiss, Endothelial cell stimulating angiogenesis factor—a new biological marker for disease activity in ankylosing spondylitis? Rheumatology **33**(4), 332–335 (1994)
53. A.E. Koch, Angiogenesis: implications for rheumatoid arthritis. Arthritis Rheum.: Official J. Am. Coll. Rheumatol. **41**(6), 951–962 (1998)
54. J. Levick, Hypoxia and acidosis in chronic inflammatory arthritis; relation to vascular supply and dynamic effusion pressure. J. Rheumatol. **17**(5), 579–582 (1990)

55. K. Lund-Olesen, Oxygen tension in synovial fluids. Arthritis Rheum.: Official J. Am. Coll. Rheumatol **13**(6), 769–776 (1970)
56. E.M. Paleolog, J.M. Miotla, Angiogenesis in arthritis: role in disease pathogenesis and as a potential therapeutic target. Angiogenesis **2**(4), 295–307 (1998)
57. C. Stevens et al., Hypoxia and inflammatory synovitis: observations and speculation. Ann. Rheum. Dis. **50**(2), 124 (1991)
58. H. Taylor et al., Raised endothelial cell stimulating angiogenesis factor in ankylosing spondylitis. Clin. Exp. Rheumatol. **11**(5), 537–539 (1993)
59. P.S. Treuhaft, D.J. McCarty, Synovial fluid pH, lactate, oxygen and carbon dioxide partial pressure in various joint diseases. Arthritis Rheum.: Official J. Am. Coll. Rheumatol. **14**(4), 475–484 (1971)
60. D. Walsh, Angiogenesis and arthritis. Rheumatology (Oxford, England) **38**(2), 103–112 (1999)
61. J. Jo et al., Detecting joint inflammation by an LED-based photoacoustic imaging system: a feasibility study. J. Biomed. Opt. **23**(11), 110501 (2018)

Diagnosis and Treatment Monitoring of Port-Wine Stain Using LED-Based Photoacoustics: Theoretical Aspects and First In-Human Clinical Pilot Study

Qian Cheng, Menglu Qian, Xiuli Wang, Haonan Zhang, Peiru Wang, Long Wen, Jing Pan, Ya Gao, Shiying Wu, Mengjiao Zhang, Yingna Chen, Naoto Sato, and Xueding Wang

Abstract Port-wine stain (PWS) is categorized as a benign capillary vascular mal-formation, which is difficult to cure. In general, PWS appears on the face, but it can affect other areas of the body too. The affected skin surface may thicken slightly and develop an irregular, pebbled surface in adulthood. PWS's cosmetic appearance causes substantial mental stress for the patients. Currently, characterization and treatment evaluation of PWS are generally conducted using physical examination and using imaging tools like digital camera, ultrasound imaging, dermoscopy, and tristimulus colorimeters. All these commonly used imaging techniques do not offer enough imaging depth and contrast required for the accurate evaluation of PWS. In this clinical pilot study, we demonstrated for the first time that LED-based photoacoustics can be used as a point-of-care tool for clinical evaluation and PDT-treatment monitoring of PWS disease.

Haonan Zhang, Peiru Wang, Long Wen, Jing Pan, Ya Gao, Shiying Wu, Mengjiao Zhang, Yingna Chen are contributed equally to this work.

Q. Cheng (✉) · M. L. Qian · H. N. Zhang · J. Pan · Y. Gao · S. Y. Wu · M. J. Zhang · Y. N. Chen
Institute of Acoustics, School of Physics Science and Engineering, Tongji University, Shanghai, China
e-mail: q.cheng@tongji.edu.cn

X. L. Wang · P. R. Wang · L. Wen
Department of Dermatology and Venereology, Shanghai Skin Diseases Hospital, Shanghai, China

N. Sato
Research and Development Department, CYBERDYNE, INC., Tsukuba, Japan

X. D. Wang
Department of Biomedical Engineering, University of Michigan, Ann Arbor, MI, USA

© Springer Nature Singapore Pte Ltd. 2020
M. Kuniyil Ajith Singh (ed.), *LED-Based Photoacoustic Imaging*,
Progress in Optical Science and Photonics 7,
https://doi.org/10.1007/978-981-15-3984-8_15

1 Introduction

Skin diseases, such as port-wine stain (PWS) can be diagnosed and treated using techniques like photoacoustic imaging (PAI) [1–3], photodynamic therapy (PDT) and acoustic dynamic therapy (ADT) because the lesions are located at the superficial skin layer.

PWS is a discoloration of human skin caused by a vascular anomaly (i.e., capillary malformation in the skin). In the past years, several techniques have been developed for characterization and treatment evaluation of PWS. In current dermatology clinical practice of China, physicians diagnose and evaluate the status of PWS dominantly based on subjective observation. The primary assisting tools used in the evaluation process of PWS include digital camera (DC), high-frequency ultrasound, Dermoscopy, and Tristimulus colorimeters (VISIA-CR™ system). Each of them has some limitations. Optical methods working in the ballistic regime, such as dermoscopy and VISIA, do not have sufficient penetration to cover the entire scale of PWS. High-frequency ultrasound, although with better imaging depth, does not offer sufficient contrast to differentiate PWS and healthy skin tissue.

The emerging photoacoustic (PA) imaging technology is capable of mapping the optical absorption contrast in deep biological tissue with excellent ultrasonic resolution. Combining the advantages of US imaging and fluorescence imaging, PA imaging offers great potential to offer a new way for the evaluation of PWS, quantitatively.

2 Theory

The human skin is mainly composed of three layers: epidermis, dermis, and hypodermis (Fig. 1). The epidermis is divided into two sub-layers: The stratum corneum (~10 μm thickness) with high lipids and low water content and living epidermis (~80 μm thickness) containing melanosomes for light absorption and scattering. The dermis (~2 mm thickness) has two sub-layers too: The papillary dermis and the reticular dermis, which contains two vascular plexuses, i.e., upper and deep blood plexuses in the upper and lower reticular dermis. The thickness of hypodermis is around 3 mm [4, 5].

The thickness d, light absorption coefficient η, and optical depth (ηd) of each skin layer at 840 and 532 nm are shown in Table 1 [2, 4, 6]. The optical depth of the dermis layer is the largest and contributes most to the optical absorption of the whole skin.

Especially for 840 nm, the light absorption coefficients of upper and deep blood plexuses are much larger than those of other layers, meaning that the change of the blood plexuses thickness has more effect on the total optical absorption. While not only is the epidermis thin, it also absorbs very little light. On the other hand, although the hypodermis and muscle layers are relatively thick, their absorption is much less than that of the dermis.

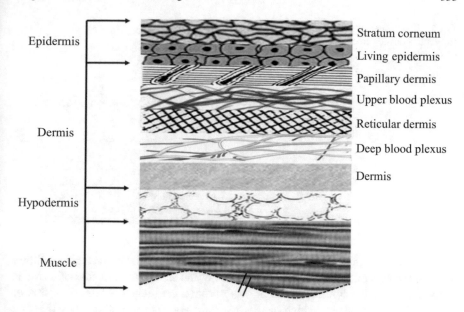

Fig. 1 Schematic of the multilayered skin

Table 1 Thickness, light absorption coefficients and optical depths of the normal skin [4, 6] and PWS [2] at 840 and 532 nm

Layers		d (mm)	$\lambda = 840$ nm		$\lambda = 532$ nm	
			η (mm^{-1})	$\eta \cdot d$	η (mm^{-1})	$\eta \cdot d$
Epidermis	Stratum corneum	0.01	0.00091	0.010	4.0	0.36
	Living epidermis	0.08	0.13		4.0	
Dermis	Papillary dermis	0.1	0.105	0.23	0.5	2.34
	Upper blood plexus	0.08	0.15875		2.45	
	Reticular dermis	1.50	0.105		0.5	
	Deep blood plexus	0.07	0.4443		18.1	
	Dermis	0.16	0.105		0.5	
Hypodermis		3.0	0.009	0.027	0.4778	1.43
Muscle tissues		3.0	0.029	0.087	0.1366	0.41
PWS (upper blood plexus)		0.1 ~ 1.5	0.15875	0.016 ~ 0.24	2.5	0.25 ~ 3.75

Fig. 2 Theoretical model
for photoacoustic detection
of skin tissue

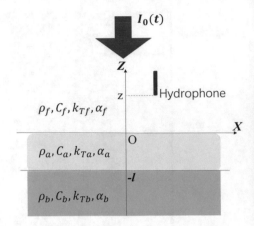

PWS is a benign neoplasia formed by hyperplasia and dilatation of postcapillary venules throughout the epidermis and dermis of the skin. The thickness of lesion is usually about 0.1–1.5 mm, and its light absorption coefficient is similar to that of upper capillary plexuses, which causes the light absorption of skin in visible and near-infrared bands to increase significantly.

The light penetration depths $\mu_\eta(1/\bar{\eta})$ of the lesion at 840 and 532 nm are estimated at around 6.3 mm and 0.41 mm, respectively. Comprehensively considering the optical attenuation and imaging depth, it is evident that 840 nm is more suitable for PWS detection than 532 nm.

Therefore, we approximately treated the light absorption layer (lesions) as composed of the epidermis and the dermis. The hypodermis was treated with weak light absorption and the muscle tissue were took as the backing tissue without light absorption. Hence, the three layers of transparent liquid (coupling layer)—light absorption tissue (lesions)—backing tissue formed the physical model for PA signal excitation and detection (Fig. 2).

2.1 Temperature Field

A pulsed plane-wave laser beam with light intensity $I_0(t)$, wavelength λ, and pulse width τ_L is incident perpendicularly on the skin surface after passing through a transparent coupling layer f (liquid) [7]. Skin tissue is composed of two layers. The upper layer a is lesion with the thickness l, the density ρ_a, the specific heat capacity C_{Ta}, the thermal conductivity κ_{Ta}, the thermal diffusivity $\alpha_{Ta} = \kappa_{Ta}/\rho_a C_{Ta}$, and light absorption coefficient $\eta(\lambda)$, which absorbs light energy and form a heat source distributing along the depth z. The thermal power density of the heat source is,

$$g(z,t) = \eta I(z,t) = \eta I_0(t)e^{\eta z}. \quad (-l < z < 0) \tag{1.1}$$

The second layer is a semi-infinite transparent medium b, whose density, specific heat capacity, thermal conductivity, and thermal diffusivity are ρ_b, C_{Tb}, κ_{Tb}, and α_{Tb}. The heat source in medium a heats the adjacent medium b and liquid layer f through heat conduction. The PA signal is formed by the thermoelastic signal of medium a, and the thermal expansion of liquid layer f. This PA signal can be received by the acoustic probe being placed at a distance 'Z' away from the interface of layer a and f in the liquid layer. When the detector scans along x-axis (or using an ultrasound array probe), the photoacoustic images of different components or structures in the tissue can be obtained together with the ultrasound images.

The temperature fields $T_a(z, t)$, $T_b(z, t)$, and $T_f(z, t)$ in media a, b and f respectively satisfy the heat conduction equation:

$$\frac{\partial^2}{\partial z^2} T_a(z, t) - \frac{1}{\alpha_{Ta}} \frac{\partial}{\partial t} T_a(z, t) = -\frac{g(z, t)}{k_{Ta}} = -\frac{\eta}{k_{Ta}} I_0(t) e^{\eta z}, \quad (0 \geq z > -l)$$

$$(1.2)$$

$$\frac{\partial^2}{\partial z^2} T_b(z, t) - \frac{1}{\alpha_{Tb}} \frac{\partial}{\partial t} T_b(z, t) = 0, \quad (-l \geq z > -\infty) \tag{1.3}$$

$$\frac{\partial^2}{\partial z^2} T_f(z, t) - \frac{1}{\alpha_{Tf}} \frac{\partial}{\partial t} T_f(z, t) = 0 \quad (z > 0) \tag{1.4}$$

Here, $\alpha_{Tj} = \kappa_{Tj}/\rho_j C_{Tj}$ and κ_{Tj} are the thermal diffusivity and conductivity of medium j ($=a, b, f$), respectively. ρ_j and C_{Tj} are the density, specific heat capacity accordingly. The Initial conditions at $t = 0$,

$$T_a(z, 0) = T_b(z, 0) = T_f(z, 0) = T_\infty = 0, \tag{1.5a}$$

$$\frac{\partial}{\partial t} T_a(z, 0) = \frac{\partial}{\partial t} T_b(z, 0) = \frac{\partial}{\partial t} T_f(z, 0) = 0, \tag{1.5b}$$

And the boundary conditions at $Z = -l$ and $Z = 0$,

$$T_b(-l, t) = T_a(-l, t), \tag{1.6a}$$

$$T_a(0, t) = T_f(0, t), \tag{1.6b}$$

$$k_{Tb} \frac{\partial}{\partial z} T_b(-l, t) = k_{Ta} \frac{\partial}{\partial z} T_a(-l, t), \tag{1.6c}$$

$$k_{Ta} \frac{\partial}{\partial z} T_a(0, t) = k_{Tf} \frac{\partial}{\partial z} T_f(0, t). \tag{1.6d}$$

The Laplace transform of heat conduction equations:

$$\frac{\partial^2}{\partial z^2} T_a(z, s) - \sigma_a^2 T_a(z, s) = -\frac{\eta}{k_{Ta}} I_0(s) e^{\eta z}, \quad (0 \geq z > -l) \tag{1.7a}$$

$$\frac{\partial^2}{\partial z^2} T_b(z, s) - \sigma_b^2 T_b(z, s) = 0, \quad (-l \geq z > -\infty) \tag{1.7b}$$

$$\frac{\partial^2}{\partial z^2} T_f(z, s) - \sigma_f^2 T_f(z, s) = 0, \quad (z > 0) \tag{1.7c}$$

Here, $\sigma_j^2 = s/\alpha_{Tj}$, $j = a, b, f$.

And the Laplace transform of boundary conditions at $Z = -l$ and $Z = 0$,

$$T_a(-l, s) = T_b(-l, s), \tag{1.8a}$$

$$T_a(0, s) = T_f(0, s), \tag{1.8b}$$

$$k_{Ta} \frac{\partial}{\partial z} T_a(-l, s) = k_{Tb} \frac{\partial}{\partial z} T_b(-l, s), \tag{1.8c}$$

$$k_{Ta} \frac{\partial}{\partial z} T_a(0, s) = k_{Tf} \frac{\partial}{\partial z} T_f(0, s), \tag{1.8d}$$

The general solution of Eq. (1.6a) is

$$T_a^h(z, s) = A(s)e^{\sigma_a z} + B(s)e^{-\sigma_a z}.$$

Assume the particular solution of Eq. (1.6a) is

$$T_a^*(z, s) = -M(s)I_0(s)e^{\eta z},$$

and substitute it to Eq. (1.6a) to obtain

$$-M(s)I_0(s)e^{\eta z}\left(\eta^2 - \sigma_a^2\right) = -\frac{\eta}{k_{Ta}} I_0(s)e^{\eta z}.$$

Then we can solve the particular solution coefficient

$$M(s) = \frac{\eta}{k_{Ta}\left(\eta^2 - \sigma_a^2\right)}, \tag{1.9}$$

So, the Laplace transform solution of temperature field in medium a is

$$T_a(z, s) = A(s)e^{\sigma_a z} + B(s)e^{-\sigma_a z} - M(s)I_0(s)e^{\eta z}. \quad (0 \geq z > -l) \tag{1.10a}$$

And the Laplace transform solutions of temperature field in medium b and f are

$$T_b(z, s) = D(s)e^{\sigma_b(z+l)}, \quad (-l \geq z > -\infty) \tag{1.10b}$$

$$T_f(z, s) = F(s)e^{-\sigma_f z}, \quad (z > 0) \tag{1.10c}$$

Substitute Eq. (1.10a)–(1.10c) into Eq. (1.8a)–(1.8d), the simultaneous equations of coefficients $A(s)$, $B(s)$, $C(s)$ and $D(s)$ can be obtained as follows,

$$
\begin{aligned}
A(s) + B(s) - F(s) &= M(s)I_0(s), \\
A(s) - B(s) + \xi F(s) &= \gamma M(s)I_0(s), \\
A(s)e^{-\sigma_a l} + B(s)e^{\sigma_a l} - D(s) &= M(s)I_0(s)e^{-\eta l}, \\
A(s)e^{-\sigma_a l} - B(s)e^{\sigma_a l} - \chi D(s) &= \gamma M(s)I_0(s)e^{-\eta l}.
\end{aligned}
\tag{1.11}
$$

Here,

$$\xi = \frac{k_{Tf}\sigma_f}{k_{Ta}\sigma_a}, \quad \chi = \frac{k_{Tb}\sigma_b}{k_{Ta}\sigma_a}, \quad \gamma = \frac{\eta}{\sigma_a}. \tag{1.11a}$$

From the coefficient determinant,

$$
\Delta = \begin{vmatrix}
1 & 1 & 0 & -1 \\
1 & -1 & 0 & \xi \\
e^{-\sigma_a l} & e^{\sigma_a l} & -1 & 0 \\
e^{-\sigma_a l} & -e^{\sigma_a l} & -\chi & 0
\end{vmatrix} = e^{-\sigma_a l}(1 - \chi)(1 - \xi) - e^{\sigma_a l}(1 + \chi)(1 + \xi),
$$

$$
\Delta_A = M(s)I_0(s) \begin{vmatrix}
1 & 1 & 0 & -1 \\
r & -1 & 0 & \xi \\
e^{-\eta l} & e^{\sigma_a l} & -1 & 0 \\
\gamma e^{-\eta l} & -e^{\sigma_a l} & -\chi & 0
\end{vmatrix},
$$

$$
\Delta_B = M(s)I_0(s) \begin{vmatrix}
1 & 1 & 0 & -1 \\
1 & \gamma & 0 & \xi \\
e^{-\sigma_a l} & e^{-\eta l} & -1 & 0 \\
e^{-\sigma_a l} & \gamma e^{-\eta l} & -\chi & 0
\end{vmatrix},
$$

$$
\Delta_D = M(s)I_0(s) \begin{vmatrix}
1 & 1 & 1 & -1 \\
1 & -1 & \gamma & \xi \\
e^{-\sigma_a l} & e^{\sigma_a l} & e^{-\eta l} & 0 \\
e^{-\sigma_a l} & -e^{\sigma_a l} & \gamma e^{-\eta l} & 0
\end{vmatrix},
$$

$$
\Delta_F = M(s)I_0(s) \begin{vmatrix}
1 & 1 & 0 & 1 \\
1 & -1 & 0 & \gamma \\
e^{-\sigma_a l} & e^{\sigma_a l} & -1 & e^{-\eta l} \\
e^{-\sigma_a l} & -e^{\sigma_a l} & -\chi & \gamma e^{-\eta l}
\end{vmatrix},
$$

$$\Delta_A = M(s)I_0(s)\left[(1 - \xi)(\gamma - \chi)e^{-\eta l} - (\gamma + \xi)(1 + \chi)e^{\sigma_a l}\right],$$

$$\Delta_B = M(s)I_0(s)\left[(1 + \xi)(\gamma - \chi)e^{-\eta l} - (\gamma + \xi)(1 - \chi)e^{-\sigma_a l}\right],$$

$$\Delta_D = M(s)I_0(s)\left\{\left[(1 + \gamma)(1 + \xi)e^{\sigma_a l} - (1 - \gamma)(1 - \xi)e^{-\sigma_a l}\right]e^{-\eta l} - 2(\gamma + \xi)\right\},$$

$$\Delta_F = M(s)I_0(s)\left[(1+\chi)(1-\gamma)e^{\sigma_a l} - (1+\gamma)(1-\chi)e^{-\sigma_a l} + 2(\gamma-\chi)e^{-\eta l}\right].$$

We get the solution of each coefficient,

$$A(s) = \frac{\Delta_A}{\Delta} = \frac{\left[(1-\xi)(\gamma-\chi)e^{-\eta l} - (\gamma+\xi)(1+\chi)e^{\sigma_a l}\right]}{(1-\xi)(1-\chi)e^{-\sigma_a l} - (1+\xi)(1+\chi)e^{\sigma_a l}}M(s)I_0(s), \quad (1.12a)$$

$$B(s) = \frac{\Delta_B}{\Delta} = \frac{\left[(1+\xi)(\gamma-\chi)e^{-\eta l} - (\gamma+\xi)(1-\chi)e^{-\sigma_a l}\right]}{(1-\xi)(1-\chi)e^{-\sigma_a l} - (1+\xi)(1+\chi)e^{\sigma_a l}}M(s)I_0(s), \quad (1.12b)$$

$$D(s) = \frac{\Delta_D}{\Delta} = \frac{\left\{\left[(1+\gamma)(1+\xi)e^{\sigma_a l} - (1-\gamma)(1-\xi)e^{-\sigma_a l}\right]e^{-\eta l} - 2(\gamma+\xi)\right\}}{(1-\xi)(1-\chi)e^{-\sigma_a l} - (1+\xi)(1+\chi)e^{\sigma_a l}}M(s)I_0(s), \quad (1.12c)$$

$$F(s) = \frac{\Delta_F}{\Delta} = \frac{\left[(1-\gamma)(1+\chi)e^{\sigma_a l} - (1+\gamma)(1-\chi)e^{-\sigma_a l} + 2(\gamma-\chi)e^{-\eta l}\right]}{(1-\xi)(1-\chi)e^{-\sigma_a l} - (1+\xi)(1+\chi)e^{\sigma_a l}}M(s)I_0(s). \quad (1.12d)$$

Equations (1.10) and (1.12) show that the temperature field in each layer is related to the incident laser pulse power, the absorbed light energy by medium a, and the thermophysical properties of each layer of the medium. By using the Laplace transform solution of the temperature field, the displacement field and stress field in the media a, b and f can be solved by the thermoelastic equations of the solid and liquid media, then the PA signal at position Z can be deduced.

2.2   Displacement Field in Media and Photoacoustic Signal in Liquid

One-dimensional thermoelastic equation and constitutive equation along the z direction in isotropic biological tissue a (Fig. 2) are:

$$(\lambda_a + 2\mu_a)\frac{\partial^2 u_a}{\partial z^2} - (3\lambda_a + 2\mu_a)\beta_{Ta}\frac{\partial T_a}{\partial z} = \rho_a\frac{\partial^2 u_a}{\partial t^2},$$

$$\tau_{zza} = (\lambda_a + 2\mu_a)\frac{\partial u_a}{\partial z} - (3\lambda_a + 2\mu_a)\beta_{Ta}T_a,$$

from which it can be derived that:

$$\frac{\partial^2 u_a}{\partial z^2} - \frac{1}{c_a^2}\frac{\partial^2 u_a}{\partial t^2} = G_a\frac{\partial T_a}{\partial z}, \quad (2.1a)$$

$$\tau_{zza} = \rho_a c_a^2\left[\frac{\partial u_a}{\partial z} - G_a T_a\right], \quad (2.1b)$$

$$c_a^2 = \frac{\lambda_a + 2\mu_a}{\rho_a}, \quad G_a = \beta_{Ta}\frac{(3\lambda_a + 2\mu_a)}{(\lambda_a + 2\mu_a)}. \quad (2.1c)$$

Similarly, one-dimensional thermoelastic equation and constitutive equation of biological tissue b (Fig. 2) are as follows:

$$\frac{\partial^2 u_b}{\partial z^2} - \frac{1}{c_b^2}\frac{\partial^2 u_b}{\partial t^2} = G_b \frac{\partial T_b}{\partial z}, \tag{2.2a}$$

$$\tau_{zzb} = \rho_b c_b^2 \left[\frac{\partial u_b}{\partial z} - G_b T_b\right], \tag{2.2b}$$

$$c_b^2 = \frac{\lambda_b + 2\mu_b}{\rho_b}, \quad G_b = \beta_{Tb}\frac{(3\lambda_b + 2\mu_b)}{(\lambda_b + 2\mu_b)}. \tag{2.2c}$$

Here, c_j is speed of longitudinal wave of medium j, G_j is thermoelastic coefficient of medium j, β_{Tj} is coefficient of linear expansion of medium j, λ_j and μ_j are Lamé elastic constants of medium j ($j = a, b$).

Acoustic wave equation in liquid f,

$$\rho_f \frac{\partial^2 u_f}{\partial t^2} = -\frac{\partial}{\partial z}p(z, t). \tag{2.3}$$

One-dimensional state equation of liquid,

$$-p(z, t) = B_f \frac{\partial u_f(z, t)}{\partial z} - B_f \beta T_f(z, t) \tag{2.3a}$$

From the derivative of Eq. (2.3a) with respect to z and substitute into Eq. (2.3), the thermoelastic equation describing liquid particle displacement can be obtained,

$$\rho_f \frac{\partial^2 u_f}{\partial t^2} = B_f \frac{\partial^2 u_f(z, t)}{\partial z^2} - B_f \beta \frac{\partial}{\partial z}T_f(z, t),$$

Or

$$\frac{\partial^2 u_f(z, t)}{\partial z^2} - \frac{1}{c_f^2}\frac{\partial^2 u_f(z, t)}{\partial t^2} = \beta\frac{\partial}{\partial z}T_f(z, t), \tag{2.4}$$

Here, β is the volume thermal expansion coefficient of liquid, ρ_f is the density of liquid, B_f is the volume elasticity coefficient of liquid, and c_f is the speed of sound of liquid,

$$c_f^2 = \frac{B_f}{\rho_f}. \tag{2.4a}$$

The initial conditions for particles motion in the three-layer media:

$$u_a(z, 0) = u_b(z, 0) = u_f(z, 0) = 0, \tag{2.5a}$$

$$\frac{\partial u_a(z,0)}{\partial t} = \frac{\partial u_b(z,0)}{\partial t} = \frac{\partial u_f(z,0)}{\partial t} = 0, \qquad (2.5b)$$

The boundary condition are continuous displacement and continuous normal stress at the boundary $z = 0$ and $z = -l$,

$$u_a(0,t) = u_f(0,t), \quad \tau_{zza}(0,t) = -p(0,t), \qquad (2.6a)$$

$$u_a(-l,t) = u_b(-l,t), \quad \tau_{zza}(-l,t) = \tau_{zzb}(-l,t). \qquad (2.6b)$$

The Laplace transform of thermoelastic displacement Eqs. (2.1), (2.2), (2.4) and constitutive Eqs. (2.1a), (2.2a), (2.3a),

$$\frac{\partial^2 u_a(s,z)}{\partial z^2} - \frac{s^2}{c_a^2} u_a(s,z) = G_a \frac{\partial T_a(s,z)}{\partial z}, \quad (-l < z < 0) \qquad (2.7a)$$

$$\frac{\partial^2 u_b(s,z)}{\partial z^2} - \frac{s^2}{c_b^2} u_b(s,z) = G_b \frac{\partial T_b(s,z)}{\partial z}, \quad (z < -l) \qquad (2.7b)$$

$$\frac{\partial^2 u_f(s,z)}{\partial z^2} - \frac{s^2}{c_f^2} u_f(s,z) = \beta \frac{\partial T_f(s,z)}{\partial z}, \quad (z > 0) \qquad (2.7c)$$

$$\tau_{zza}(s,z) = \rho_a c_a^2 \left[\frac{\partial u_a(s,z)}{\partial z} - G_a T_a(s,z) \right], \qquad (2.8a)$$

$$\tau_{zzb}(s,z) = \rho_b c_b^2 \left[\frac{\partial u_b(s,z)}{\partial z} - G_b T_b(s,z) \right], \qquad (2.8b)$$

$$p(s,z) = -\rho_f c_f^2 \left[\frac{\partial u_f(s,z)}{\partial z} - \beta T_f(s,z) \right]. \qquad (2.8c)$$

Substitute Laplace transform solution (1.10) of temperature field into Eq. (2.7), and get:

$$\frac{\partial^2 u_a(s,z)}{\partial z^2} - \frac{s^2}{c_a^2} u_a(s,z) = G_a \left[A(s)\sigma_a e^{\sigma_a z} - B(s)\sigma_a e^{-\sigma_a z} - M(s)I_0(s)\eta e^{\eta z} \right],$$
$$(2.9a)$$

$$\frac{\partial^2 u_b(s,z)}{\partial z^2} - \frac{s^2}{c_b^2} u_b(s,z) = G_b D(s)\sigma_b e^{\sigma_b(z+l)}, \qquad (2.9b)$$

$$\frac{\partial^2 u_f(s,z)}{\partial z^2} - \frac{s^2}{c_f^2} \frac{\partial^2 u_f(s,z)}{\partial t^2} = -\sigma_f \beta F(s) e^{-\sigma_f z}. \qquad (2.9c)$$

To solve the displacement field Eq. (2.9) in the three-layer media, the particular solutions of Eq. (2.9) can be set as:

$$u_a^*(s, z) = G_a \left[a(s)\sigma_a e^{\sigma_a z} - b(s)\sigma_a e^{-\sigma_a z} - m(s)I_0(s)\eta e^{\eta z} \right],$$
$$u_b^*(s, z) = G_b d(s)\sigma_b e^{\sigma_b (z+l)},$$
$$u_f^*(s, z) = -\sigma_f \beta f(s)e^{-\sigma_f z}. \tag{2.10}$$

Substitute Eq. (2.10) to Eq. (2.9) and determine coefficients $a(s), b(s), m(s), d(s)$, and $f(s)$ as

$$a(s)\sigma_a e^{\sigma_a z}\left(\sigma_a^2 - \frac{s^2}{c_a^2}\right) = A(s)\sigma_a e^{\sigma_a z}, \quad a(s) = \frac{A(s)}{\left(\sigma_a^2 - \frac{s^2}{c_a^2}\right)},$$

$$b(s)\sigma_a e^{-\sigma_a z}\left(\sigma_a^2 - \frac{s^2}{c_a^2}\right) = B(s)\sigma_a e^{-\sigma_a z}, \quad b(s) = \frac{B(s)}{\left(\sigma_a^2 - \frac{s^2}{c_a^2}\right)},$$

$$m(s)I_0(s)\eta e^{\eta z}\left(\eta^2 - \frac{s^2}{c_a^2}\right) = M(s)I_0(s)\eta e^{\eta z}, \quad m(s) = \frac{M(s)}{\left(\eta^2 - \frac{s^2}{c_a^2}\right)},$$

$$\left(\sigma_b^2 - \frac{s^2}{c_b^2}\right)G_b d(s)\sigma_b e^{\sigma_b(z+l)} = G_b D(s)\sigma_b e^{\sigma_b(z+l)}, \quad d(s) = \frac{D(s)}{\left(\sigma_b^2 - \frac{s^2}{c_b^2}\right)},$$

$$- \sigma_f \beta f(s)e^{-\sigma_f z}\left(\sigma_f^2 - \frac{s^2}{c_f^2}\right) = -\sigma_f \beta F(s)e^{-\sigma_f z}, \quad f(s) = \frac{F(s)}{\left(\sigma_f^2 - \frac{s^2}{c_f^2}\right)}. \tag{2.11}$$

Suppose the homogeneous general solution of Eq. (2.9) as

$$u_a^h(s, z) = a^h(s)e^{(s/c_a)z} + b^h(s)e^{-(s/c_a)z},$$
$$u_b^h(s, z) = d^h(s)e^{(s/c_b)(z+l)},$$
$$p^h(s, z) = f^h(s)e^{-(s/c_f)z}. \tag{2.12}$$

Therefore, the displacement transformation solutions in media a and b and the sound pressure transformation solution in the liquid are:

$$u_a(s, z) = a^h(s)e^{\left(\frac{s}{c_a}\right)z} + b^h(s)e^{-\left(\frac{s}{c_a}\right)z}$$

$$+ G_a \left[\frac{A(s)\sigma_a}{\left(\sigma_a^2 - \frac{s^2}{c_a^2}\right)}e^{\sigma_a z} - \frac{B(s)\sigma_a}{\left(\sigma_a^2 - \frac{s^2}{c_a^2}\right)}e^{-\sigma_a z} - \frac{\eta I_0(s)M(s)}{\left(\eta^2 - \frac{s^2}{c_a^2}\right)}e^{\eta z} \right], \tag{2.13a}$$

$$u_b(s, z) = d^h(s)e^{\left(\frac{s}{c_b}\right)(z+l)} + G_b \frac{D(s)\sigma_b}{\left(\sigma_b^2 - \frac{s^2}{c_b^2}\right)}e^{\sigma_b(z+l)}, \tag{2.13b}$$

$$u_f(s, z) = f^h(s)e^{-(s/c_f)z} - \beta\frac{F(s)\sigma_f}{\left(\sigma_f^2 - \frac{s^2}{c_f^2}\right)}e^{-\sigma_f z}. \tag{2.13c}$$

Substitute the displacement solution (2.13) into the Laplace transform of the boundary condition:

$$u_a(0, s) = u_f(0, s), \quad \tau_{zza}(0, s) = -p(0, s),$$
$$u_a(-l, s) = u_b(-l, s), \quad \tau_{zza}(-l, s) = \tau_{zzb}(-l, s),$$

and get the coefficients $a^h(s), b^h(s), d^h(s)$ and $f^h(s)$. Normal stress in biological tissue is:

$$\tau_{zzj}(s, z) = \rho_j c_j^2\left[\frac{\partial u_j(s, z)}{\partial z} - G_j T_j(s, z)\right], \quad j = a, b, \tag{2.13d}$$

And acoustic pressure in liquid is

$$p(s, z) = -\rho_f c_f^2\left[\frac{\partial u_f(s, z)}{\partial z} - \beta T_f(s, z)\right],$$

$$p(s, z) = \left[sz_f f^h(s)e^{-(s/c_f)z} - \rho_f s^2\frac{\beta F(s)}{\left(\sigma_f^2 - \frac{s^2}{c_f^2}\right)}e^{-\sigma_f z}\right], \tag{2.13e}$$

Here, $z_f = \rho_f c_f$ is acoustic impedance of the liquid.

From the boundary conditions, the linear equations of the unknown coefficients are

$$a^h(s) + b^h(s) - f^h(s) = -G_a\left[\frac{A(s)\sigma_a}{\left(\sigma_a^2 - \frac{s^2}{c_a^2}\right)} - \frac{B(s)\sigma_a}{\left(\sigma_a^2 - \frac{s^2}{c_a^2}\right)} - \frac{\eta I_0(s)M(s)}{\left(\eta^2 - \frac{s^2}{c_a^2}\right)}\right] - \beta\frac{F(s)\sigma_f}{\left(\sigma_f^2 - \frac{s^2}{c_f^2}\right)}, \tag{2.14a}$$

$$a^h(s) - b^h(s) + Z_{af}f^h(s) = Z_{af}\left(\frac{s}{c_f}\right)\frac{\beta F(s)}{\left(\sigma_f^2 - \frac{s^2}{c_f^2}\right)} - \left(\frac{s}{c_a}\right)G_a\left[\frac{A(s) + B(s)}{\left(\sigma_a^2 - \frac{s^2}{c_a^2}\right)} - \frac{I_0(s)M(s)}{\left(\eta^2 - \frac{s^2}{c_a^2}\right)}\right]. \tag{2.14b}$$

$$a^h(s)e^{-\left(\frac{s}{c_a}\right)l} - b^h(s)e^{\left(\frac{s}{c_a}\right)l} - Z_{ab}d^h(s) = G_b Z_{ab}\left(\frac{s}{c_b}\right)\frac{D(s)}{\left(\sigma_b^2 - \frac{s^2}{c_b^2}\right)}$$

$$- G_a\left(\frac{s}{c_a}\right)\left[\frac{A(s)}{(\sigma_a^2 - \frac{s^2}{c_a^2})}e^{-\sigma_a l} + \frac{B(s)}{(\sigma_a^2 - \frac{s^2}{c_a^2})}e^{\sigma_a l} - \frac{I_0(s)M(s)}{\left(\eta^2 - \frac{s^2}{c_a^2}\right)}e^{-\eta l}\right], \tag{2.14c}$$

$$a^h(s)e^{-\left(\frac{s}{c_a}\right)l} + b^h(s)e^{\left(\frac{s}{c_a}\right)l} - d^h(s) = G_b\left[\frac{D(s)\sigma_b}{\left(\sigma_b^2 - \frac{s^2}{c_b^2}\right)}\right]$$

$$- G_a \left[\frac{A(s)\sigma_a}{\left(\sigma_a^2 - \frac{s^2}{c_a^2}\right)} e^{-\sigma_a l} - \frac{B(s)\sigma_a}{\left(\sigma_a^2 - \frac{s^2}{c_a^2}\right)} e^{\sigma_a l} - \frac{\eta I_0(s) M(s)}{\left(\eta^2 - \frac{s^2}{c_a^2}\right)} e^{-\eta l} \right], \qquad (2.14d)$$

or

$$a^h(s) + b^h(s) - f^h(s) = N_1(s), \qquad (2.14a)$$

$$a^h(s) - b^h(s) + Z_{af} f^h(s) = N_2(s), \qquad (2.12b)$$

$$a^h(s) e^{-\left(\frac{s}{c_a}\right) l} - b^h(s) e^{\left(\frac{s}{c_a}\right) l} - Z_{ab} d^h(s) = N_3(s), \qquad (2.14c)$$

$$a^h(s) e^{-\left(\frac{s}{c_a}\right) l} + b^h(s) e^{\left(\frac{s}{c_a}\right) l} - d^h(s) = N_4(s), \qquad (2.14d)$$

Here,

$$Z_{af} = \frac{z_f}{z_a} = \left(\frac{\rho_f c_f}{\rho_a c_a} \right), \quad Z_{ab} = \left(\frac{\rho_b c_b}{\rho_a c_a} \right), \qquad (2.14e)$$

where, $z_j = \rho_j c_j$ is the acoustic impedance of medium j ($j = a, b$ and f). And

$$N_1(s) = -G_a \left[\frac{A(s)\sigma_a}{(\sigma_a^2 - \frac{s^2}{c_a^2})} - \frac{B(s)\sigma_a}{(\sigma_a^2 - \frac{s^2}{c_a^2})} - \frac{\eta I_0(s) M(s)}{\left(\eta^2 - \frac{s^2}{c_a^2}\right)} \right] - \beta \frac{F(s)\sigma_f}{\left(\sigma_f^2 - \frac{s^2}{c_f^2}\right)}, \qquad (2.15a)$$

$$N_2(s) = Z_{af} \left(\frac{s}{c_f} \right) \frac{\beta F(s)}{\left(\sigma_f^2 - \frac{s^2}{c_f^2}\right)} - \left(\frac{s}{c_a} \right) G_a \left[\frac{A(s) + B(s)}{(\sigma_a^2 - \frac{s^2}{c_a^2})} - \frac{I_0(s) M(s)}{\left(\eta^2 - \frac{s^2}{c_a^2}\right)} \right], \qquad (2.15b)$$

$$N_3(s) = Z_{ab} \left(\frac{s}{c_b} \right) G_b \frac{D(s)}{\left(\sigma_b^2 - \frac{s^2}{c_b^2}\right)}$$

$$- \left(\frac{s}{c_a} \right) G_a \left[\frac{A(s)}{(\sigma_a^2 - \frac{s^2}{c_a^2})} e^{-\sigma_a l} + \frac{B(s)}{(\sigma_a^2 - \frac{s^2}{c_a^2})} e^{\sigma_a l} - \frac{I_0(s) M(s)}{\left(\eta^2 - \frac{s^2}{c_a^2}\right)} e^{-\eta l} \right], \qquad (2.15c)$$

$$N_4(s) = G_b \left[\frac{D(s)\sigma_b}{\left(\sigma_b^2 - \frac{s^2}{c_b^2}\right)} \right]$$

$$- G_a \left[\frac{A(s)\sigma_a}{\left(\sigma_a^2 - \frac{s^2}{c_a^2}\right)} e^{-\sigma_a l} - \frac{B(s)\sigma_a}{\left(\sigma_a^2 - \frac{s^2}{c_a^2}\right)} e^{\sigma_a l} - \frac{\eta I_0(s) M(s)}{\left(\eta^2 - \frac{s^2}{c_a^2}\right)} e^{-\eta l} \right]. \quad (2.15d)$$

The coefficient determinant Δ and Δ_f of Eqs. (2.14a–2.14d)

$$\Delta = \begin{vmatrix} 1 & 1 & 0 & -1 \\ 1 & -1 & 0 & Z_{af} \\ e^{-\left(\frac{s}{c_a}\right)l} & -e^{\left(\frac{s}{c_a}\right)l} & -Z_{ab} & 0 \\ e^{-\left(\frac{s}{c_a}\right)l} & e^{\left(\frac{s}{c_a}\right)l} & -1 & 0 \end{vmatrix}, \quad \Delta_f = \begin{vmatrix} 1 & 1 & 0 & N_1(s) \\ 1 & -1 & 0 & N_2(s) \\ e^{-\left(\frac{s}{c_a}\right)l} & -e^{\left(\frac{s}{c_a}\right)l} & -Z_{ab} & N_3(s) \\ e^{-\left(\frac{s}{c_a}\right)l} & e^{\left(\frac{s}{c_a}\right)l} & -1 & N_4(s) \end{vmatrix},$$

So,

$$\Delta = \left[(1 + Z_{af})(1 + Z_{ab}) e^{\left(\frac{s}{c_a}\right)l} - (1 - Z_{af})(1 - Z_{ab}) e^{-\left(\frac{s}{c_a}\right)l} \right], \quad (2.16a)$$

$$\Delta_f = [N_2(s) - N_1(s)](1 + Z_{ab}) e^{\left(\frac{s}{c_a}\right)l}$$

$$+ [N_2(s) + N_1(s)](1 - Z_{ab}) e^{-\left(\frac{s}{c_a}\right)l} - 2N_3(s) + 2Z_{ab}N_4(s). \quad (2.16b)$$

The coefficient of the general solution of displacement field in liquid is

$$f^h(s) = \frac{\Delta_f}{\Delta} = \left\{ \frac{[N_2(s) - N_1(s)](1 + Z_{ab}) e^{\left(\frac{s}{c_a}\right)l} + [N_2(s) + N_1(s)](1 - Z_{ab}) e^{-\left(\frac{s}{c_a}\right)l} - 2N_3(s) + 2Z_{ab}N_4(s)}{\left[(1 + Z_{af})(1 + Z_{ab}) e^{\left(\frac{s}{c_a}\right)l} - (1 - Z_{af})(1 - Z_{ab}) e^{-\left(\frac{s}{c_a}\right)l} \right]} \right\}.$$

$$(2.17a)$$

Thus, the Laplace transform solution of the photoacoustic signal in the liquid can be obtained from Eq. (2.13e):

$$p(s, z) = \left[s z_f f^h(s) e^{-(s/c_f)z} - s^2 \frac{\rho_f \beta F(s)}{\left(\sigma_f^2 - \frac{s^2}{c_f^2}\right)} e^{-\sigma_f z} \right]. \quad (2.17b)$$

Let $s = i\omega$ in the above equation, photoacoustic signal spectrum can be expressed as,

$$p(\omega, z) = i z_f \omega f^h(\omega) e^{-ik_f z} + \rho_f \omega^2 \frac{\beta}{\left(\sigma_f^2 + k_f^2\right)} F(\omega) e^{-\sigma_f(\omega)z}], \quad k_f^2 = \frac{\omega^2}{c_f^2}. \quad (2.18a)$$

where, the acoustic displacement amplitude of liquid particle is:

$$f^h(\omega) = \left\{ \frac{[N_2(\omega) - N_1(\omega)](1 + Z_{ab})e^{ik_al} + [N_2(\omega) + N_1(\omega)](1 - Z_{ab})e^{-ik_al} - 2N_3(\omega) + 2Z_{ab}N_4(\omega)}{(1 + Z_{af})(1 + Z_{ab})e^{ik_al} - (1 - Z_{af})(1 - Z_{ab})e^{-ik_al}} \right\}.$$

(2.18b)

While

$$F(\omega) = \frac{[(1 - \gamma)(1 + \chi)e^{\sigma_a l} - (1 + \gamma)(1 - \chi)e^{-\sigma_a l} + 2(\gamma - \chi)e^{-\eta l}]}{(1 - \xi)(1 - \chi)e^{-\sigma_a l} - (1 + \xi)(1 + \chi)e^{\sigma_a l}} M(\omega)I_0(\omega).$$

(2.18c)

Here,

$$M(\omega) = \frac{\eta}{k_{Ta}(\eta^2 - \sigma_a^2)}, \quad \sigma_j(\omega) = (1 + i)\sqrt{\frac{\omega}{2\alpha_j}} = \frac{(1 + i)}{\mu_j(\omega)}, \quad (j = a, b, f)$$

Here, $\sigma_j(\omega)$ is he heat wave vector in medium j, and $\mu_j(\omega)$ is its heat diffusion length. And

$$N_1(\omega) = -\beta \frac{F(\omega)\sigma_f}{(\sigma_f^2 + k_f^2)} - G_a\left[\frac{A(\omega)\sigma_a}{(\sigma_a^2 + k_a^2)} - \frac{B(\omega)\sigma_a}{(\sigma_a^2 + k_a^2)} - \frac{\eta I_0(\omega)M(\omega)}{(\eta^2 + k_a^2)}\right],$$

(2.19a)

$$N_2(\omega) = iZ_{af}\frac{k_f\beta F(\omega)}{(\sigma_f^2 + k_f^2)} - ik_aG_a\left[\frac{A(\omega)}{(\sigma_a^2 + k_a^2)} + \frac{B(\omega)}{(\sigma_a^2 + k_a^2)} - \frac{I_0(\omega)M(\omega)}{(\eta^2 + k_a^2)}\right],$$

(2.19b)

$$N_3(\omega) = iG_bZ_{ab}\frac{k_bD(\omega)}{(\sigma_b^2 + k_b^2)}$$
$$- ik_aG_a\left[\frac{A(\omega)}{(\sigma_a^2 + k_a^2)}e^{-\sigma_a l} + \frac{B(\omega)}{(\sigma_a^2 + k_a^2)}e^{\sigma_a l} - \frac{I_0(\omega)M(\omega)}{(\eta^2 + k_a^2)}e^{-\eta l}\right],$$

(2.19c)

$$N_4(\omega) = G_b\left[\frac{D(\omega)\sigma_b}{(\sigma_b^2 + k_b^2)}\right] - G_a\left[\frac{A(\omega)\sigma_a}{(\sigma_a^2 + k_a^2)}e^{-\sigma_a l} - \frac{B(\omega)\sigma_a}{(\sigma_a^2 + k_a^2)}e^{\sigma_a l} - \frac{\eta I_0(\omega)M(\omega)}{(\eta^2 + k_a^2)}e^{-\eta l}\right], \quad (2.19d)$$

$$A(\omega) = \frac{[(1 - \xi)(\gamma - \chi)e^{-\eta l} - (\gamma + \xi)(1 + \chi)e^{\sigma_a l}]}{(1 - \xi)(1 - \chi)e^{-\sigma_a l} - (1 + \xi)(1 + \chi)e^{\sigma_a l}} M(\omega)I_0(\omega), \quad (2.20a)$$

$$B(\omega) = \frac{[(1 + \xi)(\gamma - \chi)e^{-\eta l} - (\gamma + \xi)(1 - \chi)e^{-\sigma_a l}]}{(1 - \xi)(1 - \chi)e^{-\sigma_a l} - (1 + \xi)(1 + \chi)e^{\sigma_a l}} M(\omega)I_0(\omega), \quad (2.20b)$$

$$D(\omega) = \frac{\{[(1 + \gamma)(1 + f)e^{\sigma_a l} - (1 - \gamma)(1 - \xi)e^{-\sigma_a l}]e^{-\eta l} - 2(\gamma + \xi)\}}{(1 - \xi)(1 - \chi)e^{-\sigma_a l} - (1 + \xi)(1 + \chi)e^{\sigma_a l}} M(\omega)I_0(\omega), \quad (2.20c)$$

Equation (2.18) is the theoretical solution of PA signal generated by a two-layer medium (light absorbing medium/semi-infinite non-absorbing substrate) in a semi-infinite fluid. Equations (2.18)–(2.20) show that the light energy absorbed by the medium can produce a PA signal consisting of two parts, i.e. acoustic and thermal wave. It is relevant to the incident light intensity, light absorption coefficient, the thickness of medium a, thermal diffusion lengths, acoustic impedance of three kinds of medium, acoustic frequency etc. Through the measurement of PA spectrum, the mechanical and thermal properties of medium can be nondestructively tested and evaluated.

2.3   Photoacoustic Signal of PWS

PWS is a congenital abnormal proliferation of capillaries that does not resolve on its own. The lesion thickens and darkens with age. Clinically, the lesion thickness (l) will be around 0.1–3 mm. Under 840 nm wavelength, due to the difference of the capillary network density, the average light absorption coefficients of isolated lesions are different. The light penetration depth μ_η at 840 nm of capillaries is about 5–10 mm, according to the data from Table 1 $\left(\mu_\eta = 1/\bar{\eta}\right)$.

In the range of frequency from 1 to 10 MHz, the wavelength of sound wave in water is 1.5–0.15 mm, and the thermal diffusion length μ_a of skin and vasculature are around 1.04–0.33 μm and 0.286–0.090 μm when their thermal diffusivity α_{Ta} are about 6.75 mm^2 s^{-1} and 0.513 mm^2 s^{-1}, respectively.

Therefore, for PWS, the light penetration depth μ_η is much higher than the acoustic wavelength λ_{ac} and lesions thickness l, which are higher than the thermal diffusion length μ_a. So PWS is a kind of weak light absorption and thermally "thick" sample for the wavelength 840 nm. That is:

$$\mu_\eta \gg (\lambda_{ac}, l) \gg (\mu_a, \mu_f), \text{i.e.} \eta \ll (k_a, k_f) \ll (\sigma_a, \sigma_{f.}). \tag{3.1}$$

But the low-frequency approximation of $k_a l \ll 1$ is not satisfied. So, using the formula

$$e^{\pm i k_a l} = cos(k_a l) \pm i sin(k_a l),$$

Equation (2.18b) can be rewritten as

$$f^h(\omega) = \frac{\Delta_f}{\Delta} = \frac{[N_2(\omega) - Z_{ab}N_1(\omega)]cos(k_a l) - i[N_1(\omega) - Z_{ab}N_2(\omega)]sin(k_a l) - N_3(\omega) + Z_{ab}N_4(\omega)}{(Z_{af} + Z_{ab})cos(k_a l) + i(1 + Z_{ab}Z_{af})sin(k_a l)}. \tag{3.2}$$

Here,

$$\Delta_f = [N_2(\omega) - Z_{ab}N_1(\omega)]cos(k_a l) - i[N_1(\omega) - Z_{ab}N_2(\omega)]sin(k_a l) - N_3(\omega) + Z_{ab}N_4(\omega) \tag{3.2a}$$

$$\Delta = (Z_{af} + Z_{ab})\cos(k_a l) + i(1 + Z_{ab}Z_{af})\sin(k_a l). \qquad (3.2b)$$

Photoacoustic spectrum generated by light absorption of PWS in the fluid can be written as

$$p(\omega, z) \approx i z_f \omega f^h(\omega) e^{-ik_f z}, \quad k_f^2 = \frac{\omega^2}{c_f^2}. \qquad (3.3)$$

Using Eq. (3.1) and

$$e^{-\eta l} \approx (1 - \eta l), \quad e^{-\sigma_a l} \approx 0,$$

Equations (2.18)–(2.20) can be written as

$$f = \frac{k_{Tf}\sigma_f}{k_{Ta}\sigma_a}, \quad b = \frac{k_{Tb}\sigma_b}{k_{Ta}\sigma_a}, \quad \sigma_j(\omega) = (1 + i)\sqrt{\frac{\omega}{2\alpha_j}} = \frac{(1 - i)}{\mu_j}, \quad (j = a, b, f)$$

$$Z_{af} = \frac{\rho_f c_f}{\rho_a c_a}, \quad Z_{ab} = \frac{\rho_b c_b}{\rho_a c_a}, \quad G_a = \beta_{Ta}\frac{(3\lambda_a + 2\mu_a)}{(\lambda_a + 2\mu_a)}, \quad G_b = \beta_{Tb}\frac{(3\lambda_b + 2\mu_b)}{(\lambda_b + 2\mu_b)},$$

$$r = \frac{\eta}{\sigma_a} \ll 1, \quad M(\omega) \approx -\frac{\eta}{k_{Ta}\sigma_a^2}, \qquad (3.4a)$$

$$A(\omega) \approx \frac{(r + f)}{(1 + f)}M(\omega)I_0(\omega) \approx \frac{f}{(1 + f)}M(\omega)I_0(\omega), \qquad (3.4b)$$

$$B(\omega) \approx -\frac{(r - b)(1 - \eta l)}{(1 + b)e^{\sigma_a l}}M(\omega)I_0(\omega) \approx \frac{b(1 - \eta l)}{(1 + b)e^{\sigma_a l}}M(\omega)I_0(\omega), \qquad (3.4c)$$

$$D(\omega) \approx -\frac{(1 + r)(1 - \eta l)}{(1 + b)}M(\omega)I_0(\omega) \approx -\frac{(1 - \eta l)}{(1 + b)}M(\omega)I_0(\omega), \qquad (3.4d)$$

$$F(\omega) \approx -\frac{(1 - r)}{(1 + f)}M(\omega)I_0(\omega) \approx -\frac{1}{(1 + f)}M(\omega)I_0(\omega), \qquad (3.4e)$$

and

$$N_1(\omega) \approx -\beta\frac{F(\omega)}{\sigma_f} - G_a\left[\frac{A(\omega)}{\sigma_a} - \frac{B(\omega)}{\sigma_a} - \frac{\eta I_0(\omega)M(\omega)}{k_a^2}\right]$$

$$\approx -\beta\frac{F(\omega)}{\sigma_f} - G_a\left[\frac{A(\omega)}{\sigma_a} - \frac{\eta I_0(\omega)M(\omega)}{k_a^2}\right]$$

$$\approx \left[\frac{\beta}{\sigma_f(1 + f)} + \frac{\eta G_a}{k_a^2}\right]I_0(\omega)M(\omega) \approx \frac{\eta G_a}{k_a^2}I_0(\omega)M(\omega), \qquad (3.5a)$$

$$N_2(\omega) \approx iZ_{af}\frac{k_f\beta F(\omega)}{\sigma_f^2} - ik_a G_a\left[\frac{A(\omega)}{\sigma_a^2} + \frac{B(\omega)}{\sigma_a^2} - \frac{I_0(\omega)M(\omega)}{k_a^2}\right]$$

$$\approx iZ_{af}\frac{k_f\beta F(\omega)}{\sigma_f^2} - ik_aG_a\left[\frac{A(\omega)}{\sigma_a^2} - \frac{I_0(\omega)M(\omega)}{k_a^2}\right] \ll iN_1(\omega), \quad (3.5\text{b})$$

$$N_3(\omega) \approx iG_bZ_{ab}\frac{k_bD(\omega)}{\sigma_b^2} - ik_aG_a\left[\frac{A(\omega)}{\sigma_a^2}e^{-\sigma_a l} + \frac{B(\omega)}{\sigma_a^2}e^{\sigma_a l} - \frac{I_0(\omega)M(\omega)}{k_a^2}(1-\eta l)\right]$$

$$\approx iG_bZ_{ab}\frac{k_bD(\omega)}{\sigma_b^2} - ik_aG_a\left[\frac{B(\omega)}{\sigma_a^2}e^{\sigma_a l} - \frac{I_0(\omega)M(\omega)}{k_a^2}(1-\eta l)\right] \ll iN_4(\omega), \quad (3.5\text{c})$$

$$N_4(\omega) \approx G_b\left[\frac{D(\omega)}{\sigma_b}\right] - G_a\left[\frac{A(\omega)}{\sigma_a}e^{-\sigma_a l} - \frac{B(\omega)}{\sigma_a}e^{\sigma_a l} - \frac{\eta I_0(\omega)M(\omega)}{k_a^2}(1-\eta l)\right]$$

$$\approx G_b\left[\frac{D(\omega)}{\sigma_b}\right] + G_a\left[\frac{B(\omega)}{\sigma_a}e^{\sigma_a l} + \frac{\eta I_0(\omega)M(\omega)}{k_a^2}(1-\eta l)\right].$$

$$\approx (1-\eta l)\left[\frac{\eta G_a}{k_a^2} - \frac{G_b}{\sigma_b(1+b)}\right]I_0(\omega)M(\omega) \approx \frac{\eta(1-\eta l)G_a}{k_a^2}I_0(\omega)M(\omega).$$

$$(3.5\text{d})$$

So,

$$\Delta_f \approx [N_2(\omega) - Z_{ab}N_1(\omega)]\cos(k_al)$$
$$- i[N_1(\omega) - Z_{ab}N_2(\omega)]\sin(k_al) - N_3(\omega) + Z_{ab}N_4(\omega)$$
$$\approx -[Z_{ab}\cos(k_al) + i\sin(k_al)]N_1(\omega) + Z_{ab}N_4(\omega)$$
$$\approx \{Z_{ab}[1 - \eta l - \cos(k_al)] - i\sin(k_al)\}\frac{\eta G_a}{k_a^2}I_0(\omega)M(\omega)$$

Using

$$M(\omega) \approx -\frac{\eta}{k_{Ta}\sigma_a^2}, \quad \sigma_a^2(\omega) = -\frac{i\omega}{\alpha_{Ta}}, \quad \alpha_{Ta} = \frac{k_{Ta}}{\rho_aC_{Ta}},$$

to substitute into the above equation and get:

$$\Delta_f \approx \frac{\eta^2 G_a}{i\omega k_a^2}\frac{1}{\rho_aC_{Ta}}\{Z_{ab}[1 - \eta l - \cos(k_al)] - i\sin(k_al)\}I_0(\omega). \quad (3.6)$$

Therefore, the displacement amplitude of the fluid particle is:

$$f^h(\omega) = \frac{\Delta_f}{\Delta} \approx \frac{\eta^2 G_a}{i\omega k_a^2}\frac{1}{\rho_aC_{Ta}}\left\{\frac{Z_{ab}[1 - \eta l - \cos(k_al)] - i\sin(k_al)}{(Z_{af} + Z_{ab})\cos(k_al) + i(1 + Z_{ab}Z_{af})\sin(k_al)}\right\}I_0(\omega). \quad (3.7)$$

The corresponding normalized PA signal spectrum is:

$$\bar{p}(\omega, z) = i\omega z_f f^h(\omega)e^{-ik_f z}/I_0(\omega)$$

$$\bar{p}(\omega, z) \approx G_a\frac{\eta^2 z_f}{k_a^2}\frac{1}{\rho_aC_{Ta}}\left\{\frac{Z_{ab}[1 - \eta l - \cos(k_al)] - i\sin(k_al)}{(Z_{af} + Z_{ab})\cos(k_al) + i(1 + Z_{ab}Z_{af})\sin(k_al)}\right\}e^{-ik_f z}, \quad k_a^2 = \frac{\omega^2}{c_a^2}. \quad (3.8)$$

The results show that the PA signal of PWS is proportional to the light absorption coefficient η^2 and inversely proportional to the frequency ω^2, which is associated with the mechanical and thermal properties of lesions. Therefore, PWS can be diagnosed quantitatively with PA imaging technique.

For most patients with mild lesions, whose thickness satisfies $(k_a l) \ll 1$, so

$$\cos(k_a l) \approx 1, \quad \sin(k_a l) \approx k_a l \ll 1, \quad \eta l \ll 1.$$

Equation (3.7) can be further simplified to:

$$\bar{p}(\omega, z) \approx -G_a \frac{\eta^2 z_f}{k_a^2} \frac{1}{\rho_a C_{Ta}} \left\{ \frac{\eta l Z_{ab} + i(k_a l)}{(Z_{af} + Z_{ab}) + i(1 + Z_{ab} Z_{af})(k_a l)} \right\} e^{-i k_f z},$$

or

$$\bar{p}(\omega, z) \approx -G_a \frac{z_f z_a}{(z_f + z_b)} \frac{\eta^2 l}{\rho_a C_{Ta}} \left(\eta \frac{z_b}{z_a} \frac{c_a^2}{\omega^2} + i \frac{c_a}{\omega} \right) e^{-i k_f z}. \tag{3.9}$$

This suggests that the PA signal amplitude is proportional to the lesion thickness l, increases rapidly with the increase of light absorption coefficient η and reduces rapidly with the increase of frequency ω. So, it's very sensitive to the changes of capillary density in the lesion. Therefore, PA diagnosis technique is a highly sensitive detection method for PWS.

3 Clinical Pilot Study

3.1 Material & Methods

The US and PA images were acquired by an LED-based PA and US imaging system (AcousticX, CYBERDYNE, INC., Tsukuba, Japan), which has been introduced in a previous publication [8]. The LED array has 144 elements, each with a size of 1 mm × 1 mm. The four lines of 36 elements distribute on an area of 6.88 mm × 50.4 mm, providing 200 µJ pulse energy at 850-nm wavelength. Working with a pulse repetition rate of 4 kHz, extensive signal averaging can be conducted to enhance the signal-to-noise ratio (SNR). The LED-produced PA image and US image from a sample can be acquired simultaneously by this dual-modality system. A 128-element linear probe working at 9 MHz central frequency was used to acquire the PA and US images. The light from the LED array illuminated the skin with a power density of 2.6 kW/m², which is below the safety limit of 5.98 kW/m², according to the international electrotechnical commission (IEC) 6247117 [9].

A coupling water bag (Fig. 3a) is sterilized by alcohol before each imaging procedure. Figure 3b shows a typical data acquisition scene with a performer (left) and

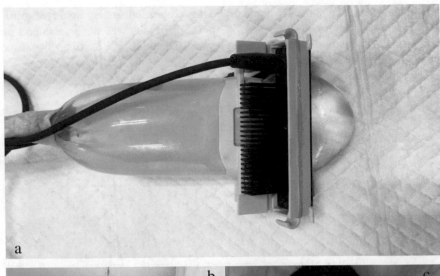

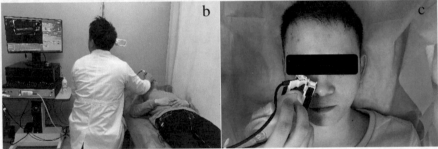

Fig. 3 **a** Photograph of coupling water bag and imaging probe, **b** clinical imaging scene, and **c** patient positioning

a patient (right). During the imaging period, the patient is required to lie flatly on a testing bed to fully expose the PWS region of interest (ROI), as is shown in Fig. 3c. For each ROI, we choose one healthy region (HR) with the most similar anatomical condition to it (we pick the symmetric position of ROI if available) as well for one additional imaging process to perform offline data processing. Before each imaging procedure, the performer will mark the ROI, as shown in Fig. 4, with the red arrow, and the corresponding HR, as indicated by the blue arrow. The arrows' orientations indicate the right side of resulting images.

During the imaging process, the performer scans the ROI with the imaging probe, and then record all the images involved. The result with the most appropriate contrast will then be picked for data processing. The same procedure is performed for each matched HR of every ROI.

In every selected image, firstly, an empty area away from ROI is circled out for the measurement of averaged PA signal amplitude (E_{pa}), which indicates the noise level of present image that is required for signal calibration. Afterwards, the performer

Fig. 4 Photograph of a PWS
patient with imaging
assisting marks

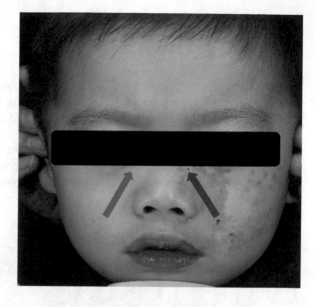

observes the results of ROI and matched HR to circle out areas suspended for the
incidence of PWS in the ROI and the symmetric healthy region in matched HR. The
E_{pa} values of ROI and HR are then divided by the noise level, yielding to V_{tar} and
V_{nor}. We define a parameter—PWS level by V_{tar}/V_{nor}, to quantitatively describe the
status of each ROI. Figure 5 gives the typical comparison of ROI and HR in an adult
patient.

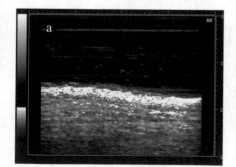
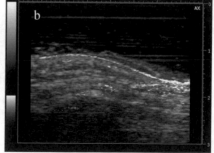

Fig. 5 Typical photoacoustic/ultrasound overlay image of **a** PWS region (ROI) and **b** control region
(HR)

3.2   Results and Discussion

3.2.1   PA Evaluation for Various Age Groups

As a typical capillary disorder, continuous development with age is an important feature of PWS. Figure 6 shows the photograph of 3 patients, age of 4, 13 and 33 years, respectively. It is clear that the darkness and the diseased skin thickness both grow with the age of patients.

In this study, a total of 22 patients were included. The clinical evaluation for each patient given by DC, dermoscopy, and VISIA were collected, as well as the newly developed PWS level parameter acquired by the PAI system. The typical data set for one patient is shown in Fig. 7.

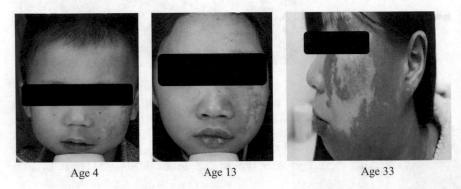

Age 4   Age 13   Age 33

Fig. 6 Photograph of patients with various ages

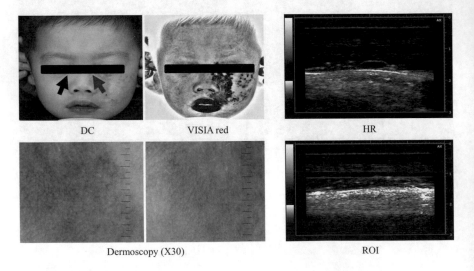

DC   VISIA red   HR

Dermoscopy (X30)   ROI

Fig. 7 Typical data set from a patient

Fig. 8 PWS level
comparison of the 2 groups

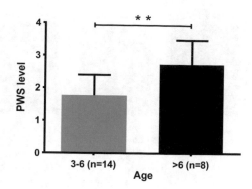

The enrolled 22 patients were divided into two groups, according to their ages, as 3–6 years old and above 6 years old. The PWS level comparison of the two groups is shown in Fig. 8. The mean PWS levels of 3–6 years group and >6 years group are 1.77 ± 0.63 and 2.73 ± 0.75, respectively.

The significant difference of the two different age groups correspond well with the given knowledge of PWS disease. Based on this result, the new parameter PWS level holds good potential in evaluation of PWS. To further demonstrate this point, study of dynamic monitoring of PWS before and after PDT treatment was conducted, which is detailed in next section.

3.2.2 PA Evaluation for PDT Treatment Efficacy

In current clinical practices in China, hematoporphyrin monomethyl ether photodynamic therapy (HMME-PDT) is proved to be an effective method for treating PWS [10, 11].

Two out of the 22 patients volunteered this study for dynamic PWS level monitoring, as shown in Figs. 9 and 10. The immediate reduction of PWS level corresponds to the edema instantly after the PDT treatment. At 3-day point, the PWS level of each patient shows a recovering status, or even getting worse (Fig. 9), which corresponds to the damage-repairing period that starts at that time point. The repairing period often comes with a hyperemia status, which is the reason why the PWS level gets back here. Nonetheless, after that period, the PWS level goes down steadily, and eventually gets to the 2-month endpoint.

Five out of the 22 patients volunteered for PWS level monitoring before and 2-month after PDT treatment, the result of which is shown in Fig. 11. A mean PWS level reduction of $41.08\% \pm 11.30\%$ was observed after one PDT treatment. From our results, it is clear that approximately 3–4 PDT treatments are required to completely cure the patient.

The dynamic monitoring results show the good correspondence of PWS level to the actual patient's status, and the 2-month PWS level monitor shows the efficacy of PDT treatment. Based on these results, it is clear that the PWS level parameter

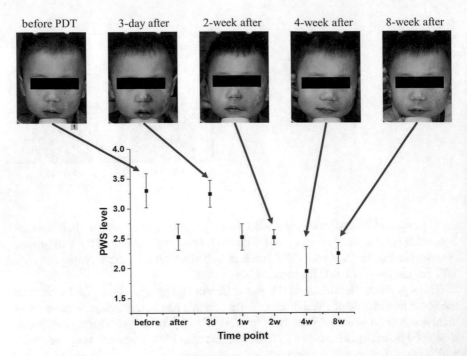

Fig. 9 Dynamic PWS level monitoring #Patient 1

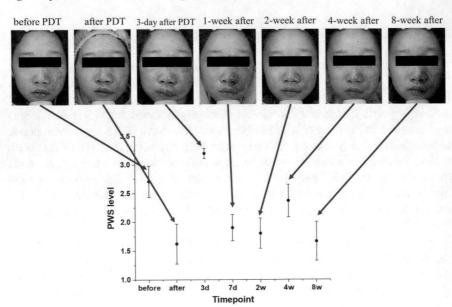

Fig. 10 Dynamic PWS level monitoring #Patient 2

Fig. 11 2-months PWS level monitor

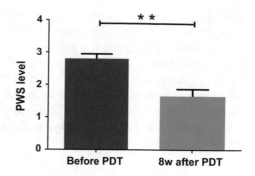

holds good potential to be used as a guiding tool for dose control in the treatment process of PDT. Additionally, as a quantitative tool, the PWS level of patients can be collected to serve as the source for big-data analysis as well, which will lead to a better understanding of this disease, or even other capillary disorders.

Other than mapping the hemoglobin, by combing different wavelengths that have optical absorption contrast on oxyhemoglobin and deoxyhemoglobin, the PA imaging is also capable of mapping the oxygen saturation [12, 13]. By scanning the imaging probe, it is technically feasible to perform volumetric imaging as well. In addition, for the practice of dermatology, the proposed protocol, which is already in a clinical-procedure-equivalent manner, is easily translatable to clinics.

4   Summary and Outlook

PAI has inherent advantages for vascular recognition. Firstly, compared with other biological macromolecules, hemoglobin has a very high absorption for the near-infrared light which leads to the high specificity for hemoglobin with PAI using these wavelengths. Secondly, due to the process of light-in and sound-out, the sensitivity and resolution of PAI is much higher than that of ultrasonic imaging, which makes it possible to image capillaries. But, because of the tiny size of the capillaries, traditional PA theory is no longer applicable for their quantitative diagnosis. Instead, the optical, thermal, mechanical and acoustic properties of different biological tissues need to be analyzed in the micro-scale. Meanwhile, the effects of some different parameters, such as light penetration depth, thermal diffusion length, wavelength of sound wave, and vascular size, on photoacoustic signals are very important and these are covered in the theoretical section of this chapter. This theoretical analysis can also be applied to other similar layered tissue with strong light absorption.

Current clinical results have already demonstrated the safety, functional contrast and clinic-friendly protocol of this LED-based PAI strategy. This ongoing research has already shown its great translational potential. Moreover, other than the discussed application of PWS evaluation, this imaging strategy is also potentially useful to assist doctors in assessing many other skin diseases, vascular tumor or skin carcinomas that come with microvessels malformation or angiotelectasis, like actinic keratosis (AK), Bowen's disease (BD), superficial basal cell carcinoma (BCC), squamous cell carcinoma (SCC) and extramammary Paget's disease (EMPD). Besides, some inflammatory skin diseases, like psoriasis, are accompanied with vascular abnormity as well, which could also be a potential application. Furthermore, the application could be extended to fundus oculi lesion diagnosis, angiogenesis monitoring, and so on. We believe that PAI, as an emerging imaging strategy, would promisingly play a greater role in the early stage evaluation of a lot of diseases located in the light-accessible depth range.

5 Conclusions

PWS levels acquired using LED-based PAI holds strong potential to be a quantitative parameter for the evaluation of PWS status. We demonstrated for the first time that LED-based photoacoustics can be used as a point-of-care tool for clinical evaluation of PWS disease. Our results also give a direct indication that LED-based PAI is useful for guiding PDT-treatments.

Acknowledgements Authors gratefully acknowledge the patients who participated in this clinical study.

Ethical Approval All procedures for human subjects in this study were approved by the Ethics Committee of Shanghai Skin Disease Hospital (2019-06).

References

1. R.G.M. Kolkman, M.J. Mulder, C.P. Glade et al., Photoacoustic imaging of port-wine stains. Lasers Surg. Med. **40**, 178–182 (2008)
2. J.A. Viator, B. Choi, M. Ambrose et al., In vivo port-wine stain depth determination with a photoacoustic probe. Appl. Opt. **42**, 3215 (2003)
3. H. Zhang, J. Pan, L. Wen et al., A novel light emitting diodes based photoacoustic evaluation method for port wine stain and its clinical trial (Conference Presentation), in *Photons Plus Ultrasound: Imaging and Sensing 2019*. International Society for Optics and Photonics (2019), p. 1087807
4. J.A. Iglesias-Guitian, C. Aliaga, A. Jarabo et al., A biophysically-based model of the optical properties of skin aging. Computer Graphics Forum **34**, 45–55 (2015)
5. A. Bhowmik, R. Repaka, S.C. Mishra et al., Analysis of radiative signals from normal and malignant human skins subjected to a short-pulse laser. Int. J. Heat Mass Transf. **68**, 278–294 (2014)
6. A.N. Bashkatov, E.A. Genina, V.V. Tuchin, Optical properties of skin, subcutaneous, and muscle tissues: a review. J Innov Opt Health Sci **04**, 9–38 (2011)
7. D.L. Balageas, J.C. Krapez, P. Cielo, Pulsed photothermal modeling of layered materials. J. Appl. Phys. **59**, 348–357 (1986)

8. Y. Zhu, G. Xu, J. Yuan et al., Light Emitting Diodes based Photoacoustic Imaging and Potential Clinical Applications. *Sci Rep*; 8. Epub ahead of print December 2018. https://doi.org/10.1038/s41598-018-28131-4

9. Commission, I. E. IEC 62471: 2006 Photobiological safety of lamps and lamp systems. International Standard (2006). http://tbt.testrust.com/image/zt/123/100123_2.pdf. Accessed 26 Nov 2018

10. Y. Zhao, P. Tu, G. Zhou et al., Hemoporfin photodynamic therapy for port-wine stain: a randomized controlled trial. PLoS ONE **11**, e0156219 (2016)

11. L. Wen, Y. Zhang, L. Zhang et al., Application of different noninvasive diagnostic techniques used in HMME-PDT in the treatment of port wine stains. Photodiagn. Photodyn. Ther. **25**, 369–375 (2019)

12. X. Wang, G. Ku, X. Xie et al., Noninvasive functional photoacoustic tomography of blood-oxygen saturation in the brain, in ed. A.A. Oraevsky, L.V. Wang (San Jose, CA), p. 69

13. X. Wang, X. Xie, G. Ku et al., Noninvasive imaging of hemoglobin concentration and oxygenation in the rat brain using high-resolution photoacoustic tomography. J. Biomed. Opt. **11**, 024015 (2006)

Clinical Translation of Photoacoustic Imaging—Opportunities and Challenges from an Industry Perspective

Mithun Kuniyil Ajith Singh, Naoto Sato, Fumiyuki Ichihashi, and Yoshiyuki Sankai

Abstract Photoacoustic imaging, the fastest growing biomedical imaging modality of the decade holds strong potential in creating a significant impact in the field of medicine. This non-invasive technique with optical spectroscopic contrast and ultrasonic spatial resolution can be a potential tool for diagnosis and treatment monitoring of several devastating diseases like cancer. Even though the growth of this imaging modality in a research setting is exemplary, clinical translation is not happening at an expected pace. In this chapter, after briefly discussing about the technology and its market, we will discuss about important components in a photoacoustic imaging system (light source, ultrasound probes, DAQ etc.) and conclude about the key strategies in this direction and the improvements that may potentially help for fast clinical translation. In the next part, we elaborate about the major steps in the clinical translation process—focusing on key opportunities and challenges to be solved.

1 Introduction

Photoacoustic (PA) imaging [1, 2] is increasingly becoming popular and slowly translating from benchtop to bedside after demonstrating strong potential in clinical medicine (diagnosis, treatment monitoring, interventional guidance) [3]. However,

M. Kuniyil Ajith Singh (✉)
Research & Business Development Division, Cambridge Innovation Center, CYBERDYNE INC., Stationsplein 45, A4.004, 3013 AK Rotterdam, The Netherlands
e-mail: mithun_ajith@cyberdyne.jp

N. Sato · F. Ichihashi · Y. Sankai
Research & Development Division, CYBERDYNE INC., 2-2-1, Gakuen-Minami, Tsukuba 305-0818, Ibaraki, Japan
e-mail: sato_naoto@cyberdyne.jp

F. Ichihashi
e-mail: ichihashi@cyberdyne.jp

Y. Sankai
e-mail: sankai@cyberdyne.jp

© Springer Nature Singapore Pte Ltd. 2020
M. Kuniyil Ajith Singh (ed.), *LED-Based Photoacoustic Imaging*,
Progress in Optical Science and Photonics 7,
https://doi.org/10.1007/978-981-15-3984-8_16

number of companies involved in PA commercialization is less and it is also worth mentioning that there is no system with an FDA pre-market approval yet. From these points, it is clear that the translation process is not straightforward and fast. At this point of time, it is of paramount importance to debate about the challenges and opportunities in clinical translation of PA imaging and the focus of this chapter is the same.

Clinical market demand for a high resolution real-time deep-tissue functional imaging modality is quite high and thus PA imaging is expected to revolutionize the field of medical diagnostics in coming years. This technique with optical spectroscopic contrast and acoustic resolution offers several advantages over other medical imaging modalities [3]:

Non-invasive and safe: Compared to gold standard modalities like X-ray, PA imaging can be repeatedly used on tissue, thus suitable for treatment monitoring.
Scalable resolution: Spatial resolution is dependent on the ultrasound detection and can be scaled for imaging cells to organs.
Endogenous contrast: Optical spectroscopic contrast offered by hemoglobin, fat, lipid etc. No need of contrast agents for imaging vasculature.
High imaging depth: Attenuation of ultrasound in tissue is orders of magnitude lower than that of light, thus PA imaging can easily image deep (for example: whole breast 3D imaging).

In general, there are 4 different implementations (Fig. 1) of PA imaging, each with different resolution, imaging depth, and field of view. PA microscopy or PAM is suitable for imaging superficial tissue structures (1–5 mm) with very high resolution [4]. Two implementations are common in a microscopy set up namely: optical resolution PAM (OR-PAM) and also acoustic resolution PAM (AR-PAM), with differences in scanning and image generation methods [4]. In an endoscopic method, pulsed light is delivered inside the body through an endoscope and functional PA-based tissue characterization can be performed. Minimally invasive PA imaging in

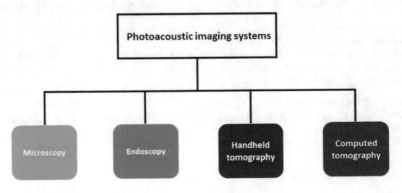

Fig. 1 Four different implementations of photoacoustic imaging

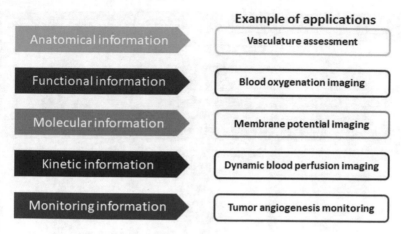

Fig. 2 Different applications of biomedical photoacoustic imaging

which light is delivered through a percutaneous needle for guiding surgical procedures are also common in a research-setting [5, 6]. Because of the synergy with pulse echo ultrasound imaging, most common implementation is a handheld reflection mode arrangement in which light delivery is attached to the ultrasound probe, which is also called photoacoustic tomography (PAT) with imaging depth in cm scale [7]. When pulsed light is illuminated on the tissue, PA signal generated from the absorber will propagate in all directions. If there are detectors around the tissue, it is feasible to generate a full view 3D image. In PA computed tomography, image slices from all locations are collected and stitched together to generate images. This is the most popular configuration for breast-imaging, one of the promising applications of photoacoustics [8]. Figure 2 shows the range of information that can be obtained from PA images and some of applications associated with it.

2 Commercially Available Photoacoustic Imaging Systems

As discussed before, number of companies involved in commercialization of PA imaging is not high (less than 20 entities are active or visible). Figure 3 shows a list of commercially available research-based PA imaging systems and their approximate list price [9]. Most of these systems are currently used for pre-clinical applications and clinical pilot studies in a research setting. Vevo LAZR X (Fujifilm Visualsonics, Canada), MSOT InVision 128 (iThera Medical Gmbh, Germany), and Nexus 128+ (Endra Life Sciences, USA) can provide real-time volumetric vascular/functional information and are specifically designed for pre-clinical applications.

LOUISA 3D (TomoWave, Inc., USA) is designed for breast cancer imaging research. All these commercial systems mentioned above use solid-state lasers for tissue illumination, which may be one of the key factors for high cost [9]. AcousticX

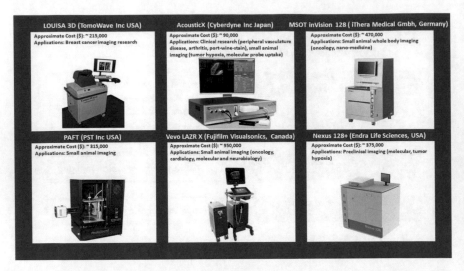

Fig. 3 Commercially available photoacoustic imaging systems—applications and list price [9]

(Cyberdyne Inc., Japan) utilizes LED arrays for tissue illumination and is portable, affordable, and potentially eye/skin safe because of this reason [10–13]. However, this system also is currently used only for pre-clinical applications [10] and clinical pilot studies [14] in a research setting as it is yet to obtain medical device safety approvals.

In recent years, leading companies in the field are working hard for translating PA imaging technology to clinic. Imagio® breast imaging system (Seno Medical Instruments Inc., USA) has received CE Mark (April 2014) and is undergoing a post-market surveillance and clinical follow-up study in Europe [15]. Figure 4 shows live use of this system which is also expecting a pre-market approval from US FDA very soon. Only other system which received European CE marking is MSOT Acuity (iThera Medical Gmbh, Germany) [16] which has already shown strong potential in detection of Crohn's disease, breast cancer, melanoma, scleroderma and vascular diseases (Fig. 5).

3 Photoacoustic Imaging Technology: Market Trend

In 2016, the total biomedical PA imaging market was worth $35 M, mainly due to pre-clinical and analytical imaging segments (Photoacoustic Imaging: Technology, Market and Trends, Report by Tematys with Laser & Medical Devices Consulting, 2017). It is forecasted to reach around $240 M in 2022. Release of clinical products (pending FDA approval) in coming years will have a significant impact in market growth. Considering the successful early clinical pilot studies, within few years from now, it is expected that clinical PA imaging market will be ahead of preclinical

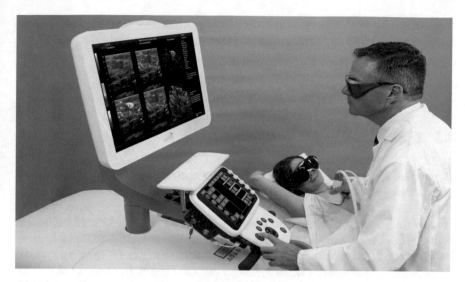

Fig. 4 Imagio® breast imaging system being used by a clinician (image provided by Seno Medical Instruments, Inc.)

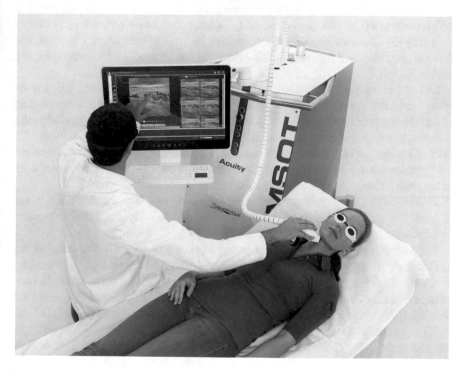

Fig. 5 MSOT Acuity system being used by a clinician (image provided by iThera Medical Gmbh)

segment owing to the foreseen release of many clinical PA systems suitable for multiple applications in oncology, rheumatology, cardiology, and dermatology.

4 Key Components in a Photoacoustic Imaging System

4.1 Optical Source for Tissue Illumination

To generate PA signals effectively, the thermal expansion caused by optical absorption needs to be time variant. This requirement can be achieved by using either a pulsed laser or a continuous-wave (CW) laser with intensity modulation at a constant or variable frequency. Pulsed excitations are the most commonly used because of high SNR offered by them (when compared with CW excitation). Typically, a Q-switched Nd:YAG pulsed laser is used as the optical source in PA imaging systems. Costs of these light sources are in the range of $45–$100 K USD depending on the level of energy/pulse and the pulse width [9]. Some of the commercially available laser sources provide built-in fixed wavelength options, typically at 532 and 1064 nm. Continuous wavelength tuning (650–950 nm) is usually provided by an optical parametric oscillator (OPO). Size and cost of these laser sources are one of the key factors hindering the clinical translation of PA imaging and it is of paramount importance to develop portable and affordable light sources that can be used for tissue illumination in PA imaging [3].

Use of pulsed laser diodes for PA imaging has been well explored recently by several research groups [7, 17]. In a European project FULLPHASE, a consortium of academic and industrial partners succeeded in developing a fully integrated multi-wavelength PA imaging (up to 4 wavelengths) probe that can acquire and display 2D US and PA images in an interleaved manner at high frame rates. Figure 6 shows the US (gray scale) and PA image (hot scale) of human proximal interphalangeal joint of a human volunteer acquired using this probe [7].

Even though this project succeeded in developing an integrated US/PA probe with smaller footprint and demonstrated its potential in early clinical pilot studies, translation to the clinic never happened.

Another category of light sources being explored heavily for PA imaging are LED's. Even though LED's are similar to laser diodes in terms of light generation, they do not create stimulated emission and thus are not considered as laser sources (broader bandwidth and less coherent). Use of high-power LED arrays for superficial PA imaging has shown good potential [10–13] and we (CYBERDYNE, INC.) are commercializing multiwavelength LED-based PA and US imaging system (AcousticX) for research use. From the initial clinical pilot studies, we believe that LED elements holds strong potential as light sources in PA imaging, especially in a hospital-setting where laser safe-rooms and goggles are not appropriate. Figure 7a shows the AcousticX PA/US probe with 850 nm LED arrays and a 7 MHz linear array probe. Figure 7c shows the 3D MIP PA image of area of a human foot

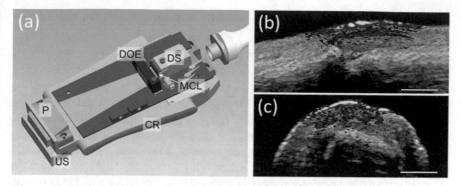

Fig. 6 **a** A schematic of the handheld PACT probe. US ultrasound array transducer, P deflecting prism, DOE diffractive optical elements, DS diode stack, MCL micro-cylindrical lenses, CR aluminum cooling rim. Photoacoustic/ultrasound images of a human proximal interphalangeal joint in **b** sagittal and **c** transverse planes. Adapted with permission from Ref. [7]

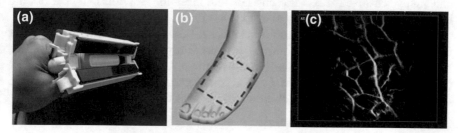

Fig. 7 **a** Integrated US/PA probe with LED arrays fixed on both sides of a linear array probe, **b** area of a human foot where handheld scanning was performed, **c** 3D MIP PA image of the marked imaging location—acquisition time: 6 s, image reconstruction time: 10 s

marked in Fig. 7b. In a completely handheld operation, the probe was linearly scanned through the marked region and 3D image was rendered using the in-built GPU-based reconstruction algorithm of AcousticX.

With wide range of available wavelengths (470–980 nm), possibility to tune pulse widths based on US probe frequencies, compactness, affordability, and energy efficiency, LED-based PA imaging holds promise in functional and molecular clinical imaging. We expect that addition of LED-based PA imaging to conventional pulse-echo US imaging in a clinical scanner will have profound impact in point-of-care diagnostic imaging and also accelerate the clinical translation of PA imaging.

Xenon flash lamps, solid-state diode pumped lasers (instead of bulky water-cooled lasers) and even intensity modulated continuous wave lasers are explored as PA-illumination source by different research labs around the globe [9]. It is expected that fast advancements in solid-state device technology will further improve these light sources and also help acceleration of clinical translation of PA imaging.

4.2   Ultrasound Detection

US probe is the key component of a PA imaging system as its characteristics (sensitivity, bandwidth etc.) defines the image quality to a large extent. Even though several configurations are used (curved, ring, hemispherical arrays etc.!) in a research setting, piezo electric linear array US probes are the most commonly used PA detectors because of its affordability and portability (easy to develop a handheld US/PA probe in this configuration) [9]. Clinicians are also used to this configuration and this may help to introduce PA imaging as an add on modality to conventional pulse echo US imaging easily [3]. However, clinical linear array probes are designed to perform US imaging which deals with pulse echoes in the pressure range of Mega pascals and PA signals are often in the range of pascals to Kilo pascals. Thus, the sensitivity will be a key issue if one picks a clinical US probe and use it for PA imaging [3]. Also, the bandwidth of PA signals is often very high for the conventional US probes. It is of paramount importance to develop affordable detectors which are broadband and highly sensitive (in both transmission and reception) for efficiently performing interleaved US and PA imaging of deep tissue [18].

Recently, other types of acoustic sensors have been also explored as a detector for the PA imaging. Some of the promising ones are Fabry–Perot interferometers (FPIs), micro ring resonators (MRRs), and capacitive micromachined ultrasound transducers (CMUTs) [19]. Out of these CMUT's are considered to be the future because of the ability to offer real-time and high-resolution images of larger field of views [19].

4.3   Data Acquisition System

Since both PA and US imaging involves acoustic detection, electronic DAQ can be shared for both modalities (by switching between two modalities) to obtain naturally overlaid PA and US images with structural and functional contrast. In this regard, the most economical way will be to integrate a light source to a clinical US scanner (light source can be controlled by a trigger signal from the scanner) and perform imaging (when light is on, US must be off and vice versa). In terms of image reconstruction, one can ideally reuse a planewave US reconstruction algorithm for PA imaging (by reducing the time of flight by 2—as PA imaging only involves one side travel of acoustic signals from tissue to probe) [3].

As discussed before, PA signals are quite weak in nature and it is important to have low noise preamplification and multiple steps of signal enhancement for extracting real PA signals from noise. Conventional US scanner may not be handy if the detection sensitivity (front end electronics) is low. Custom-made electronics for parallel PA data acquisition and US transmission is currently quite expensive. It is expected that fast growing field of electronics will help circumvent this problem and help accelerate the translation of PA imaging from benchtop to bedside [3].

5 Clinical Translation of Photoacoustic Imaging: Steps, Opportunities and Challenges

In a general sense, fate of any new medical imaging modality depends upon how effectively it can solve an unmet clinical need and also its usefulness in a hospital setting (rich information, minimal disruption etc.). PA imaging naturally possess all advantages of optical imaging techniques (the famous optical contrast) combined with depth of penetration and spatial resolution of US. It is evident that combined PA/US imaging holds strong potential in detection of vasculature and hypoxia, two key biomarkers for different disease conditions (inflammation, cancer etc.). However, there is a need to optimize the system based on application. For example, 1 MHz spherical array US probe and 1064 nm laser light source with 10 mJ/pulse energy would be ideal for imaging a whole breast [20]. On the other hand, a linear array 7 MHz US probe and 850 nm LED light with 0.5 mJ/pulse optical energy is sufficient for real-time imaging of rheumatoid arthritis in a human finger [14]. This is different from established radiological imaging modalities (x-ray, CT, MRI, PET) which are usually centralized systems that are used for different clinical applications. Considering this, PA imaging faces different challenges and opportunities for clinical translation and commercialization.

As discussed before, no PA imaging device has got US FDA approval for use as a medical device. Imagio® system from Seno Medical Inc. and MSOT Acuity system from iThera medical Gmbh recently obtained CE marking and clinical studies are progressing in Europe. Even though some companies are focusing on one application (for example Imagio® system targets breast cancer), academia and industry are working together to explore other killer applications that can give rich medical information that cannot be obtained using gold-standard techniques.

Real-time imaging of deep tissue with structural, functional, and molecular contrast is not easy to obtain with any other conventional imaging modalities and this is the key reason for huge amount of technical, R&D in the field of PA imaging in recent years. Over the past 20 years, PA-based medical imaging has progressed significantly in terms of imaging speed and sensitivity, thanks to the advances in component hardware and software. With all these developments, PA imaging can now image a whole organ (for example breast) in 3D with micrometer resolution and imaging depth exceeding 4 cm. Figure 8 shows such an image from the group of Lihong Wang, one of the pioneers in the field [20]. This shows the potential of the technology in breast cancer screening and also other clinical applications in which angiogenesis is an early biomarker.

Commercialization or in other words "evolution of an idea to a product which is marketable" is the key factor for any new technology to reach patients and impact healthcare. When compared to nonclinical products, commercialization of a medical device is complex and expensive. Apart from all important commercialization steps for a nonclinical device, additional tasks including clinical trials, regulatory approvals (including vast amount of documentation), negotiations with insurance companies, and post use safety monitoring are required when any company is trying

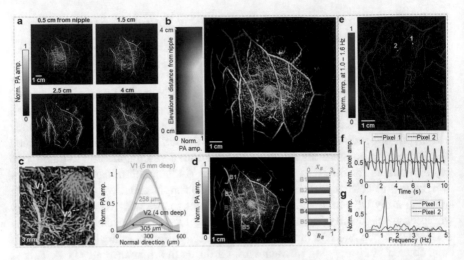

Fig. 8 Breast imaging results from Caltech Single-Breath-Hold ring-shaped imager **a** Vasculature in the right breast of a 27-year-old healthy female volunteer. Images at four depths are shown in increasing depth order from the nipple to the chest wall, **b** the same breast image with color-encoded depths, **c** a close-up view of the region outlined by the magenta dashed box in (**b**), with selected thin vessels and their line spread plots, **d** a selected vessel tree with five vessel bifurcations, labeled from B1 to B5, **e** heartbeat-encoded arterial network mapping of a breast cross-sectional image (red = artery, blue = vein), **f** amplitude fluctuation in the time domain of the two pixels highlighted by yellow and green dots in (**e**), and **g** Fourier domain of the pixel value fluctuations in (**f**). The oscillation of the arterial pixel value shows the heartbeat frequency at ~1.2 Hz. Reproduced with permission from [20]

to commercialize a PA system, which is a medical imaging device. It is clear from these factors that huge investment in terms of time and finance are required upfront for any companies to translate an optical technology like PA imaging into clinic. Since PA imaging holds potential in several clinical applications, it is also difficult for companies to identify the right application to target, conduct clinical trials and obtain safety approvals. At this point, it is still important for both researchers and academia to explore all possibilities and identify right clinical applications of PA imaging, which can consequently accelerate the translation of the technology from a research-setting to clinic.

Figure 9 shows generalized steps towards clinical translation of any optical clinical technologies [21]. Some of the steps in this can be restructured or carried out in parallel for decreasing cost and speed up the whole process. Even though PA imaging is well developed in a research setting, speed of clinical translation is rather slow, and most companies involved have reached only the stage of clinical trials (step 5 in Fig. 9). Couple of exceptions are systems from iThera Medical Gmbh and Seno Medical Instruments Inc. which got CE marking and are validated by clinical studies in large scale. Some of the key steps in clinical translation [21] and its associated challenges are briefly described in following sections.

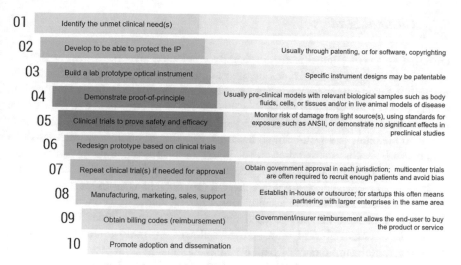

Fig. 9 Steps toward clinical translation of optical technologies. Reproduced with permission from [21]

5.1 Validation Using Preclinical Models

After building a prototype, preclinical models (ex vivo body fluids, tissue or animal models of disease) are usually used for validating the technology. Results of these studies are generally very important for supporting the valorizations. For PA imaging, companies involved must invest carefully on this step as each application may need separate preclinical models. Also, it is important to test different system configurations (US probes, light wavelengths etc.) with relevant animal models focused on targeted contrast, spatial resolution and imaging depth.

The basic requirement of a preclinical model is to accurately replicate the intended human use and also the associated clinical utility claims. In PA imaging, lot of papers are published in which this aspect is not taken care of. For example, transplanted tumors in rodents are commonly used to demonstrate high sensitivity and specificity of the technology and this is often not useful since the tumors are far different when compared to human tumors in terms of bio-chemical or structural characteristics [21]. At least in some situations, it may be worth investigating the possibility of using human tissue/tumors ex vivo (for example, tumor embedded in tissue mimicking phantoms) to validate the technology. However, this may require safety and ethical approval for the system well in advance.

5.2 Clinical Studies

Different stages of clinical trials are usually conducted after preclinical validation of the technology. Phase I studies involve validation of initial safety and technical feasibility. Since pulsed-light sources are used for tissue illumination in PA imaging, it may be very important to prove that the system is safe for skin and eyes of the users and patients (ANSI standards are well defined in this regard) [22]. Laser-safe rooms in hospitals and well-defined laser goggles (based on light wavelengths) for patients and users are a must when pulsed lasers are used as illumination sources [22], consequently resulting in additional hurdles for clinical translation process. In this regard, use of LED's which are non-coherent (not defined as lasers) is a wise option at least for superficial imaging applications. Phase II studies are conducted to prove that the system can efficiently be used for the intended application. This is usually conducted by small clinical pilot studies followed by multi-center clinical trials involving significant number of patients to avoid bias. Last but not the least, phase III studies are used for comparing the performance of the technology with existing approved alternatives (gold standard techniques). It is important that technology is at least equivalent to a gold standard in terms of performance.

For companies, these steps are cumbersome, however, the motivation is the anticipation of significant revenues and high clinical impact once the technology is translated to clinics. Results of this phase may result in redesign of system or additional preclinical studies. In some cases, these studies may conclude that the technology being validated is not worth commercializing. However, PA imaging clinical studies until now are quite promising and results show that when compared to gold standard modalities, PA imaging can offer higher specificity and sensitivity in several applications (breast cancer, arthritis, crohn's disease, skin cancer etc.).

One key challenge foreseen is the burden of companies in acquiring regulatory approvals separately for different applications that can be tackled using the same system. For example, a linear array 7 MHz US probe and two LED arrays of 850 nm (AcousticX, Cyberdyne Inc.) are sufficient for imaging rheumatoid arthritis in a human finger and also detection/staging of port wine stain. The interesting question or challenge is that whether the company should go through all regulatory process separately for accessing these two markets. In any case, using clinical trials, it is important to identify whether the market is large enough for the product to be commercially viable.

5.3 Standardization of the Technology

Even though not a routine step of clinical translation, standardizing the technology is of profound importance considering the diverse system configurations and associated clinical applications. When compared to gold standard medical imaging

Fig. 10 Different focus areas of IPASC

modalities (Nuclear imaging, MRI, CT etc.), optical imaging technology in general lacks established methods for standardization and calibration. However, a new consortium (IPASC: International PhotoAcoustic Standardization Consortium) was recently formed specifically for accelerating clinical translation of PA imaging [23]. The overall goal of IPASC is to reach consensus on PA imaging standardization to improve the quality of preclinical studies and also to speed up the efforts in clinical translation. Furthermore, by establishing standards, IPASC hopes to facilitate open access, use, and exchange of data between different research groups and companies. This consortium is currently represented globally in over 15 countries with more than 90 academic and industrial members. It is expected that the standardization efforts of this consortium will immensely help accelerating the clinical translation of this powerful biomedical imaging modality.

IPASC is focused on

1. Define the recipe for tissue mimicking phantoms that can be used for testing any preclinical or clinical PA imaging systems
2. Use these phantoms to quantitatively compare PA imaging data acquired using different commercially available and lab-based systems
3. Provide open access reference data sets for validating and comparing different reconstruction and spectral processing algorithms
4. Developing standardized test methods for testing and validating new PA imaging systems.

To achieve the above goals, activities of IPASC are focused on three different areas as shown in Fig. 10. Considering the participation of US FDA, renowned research labs, and companies involved in clinical translation of PA imaging, this excellent cooperation is expected to have a significant impact in accelerating the process of bringing this technology to clinic.

6 Conclusions

PA imaging is one of the fastest growing biomedical imaging modalities of the decade with excellent potential in wide range of preclinical and clinical applications. With its rich optical absorption contrast and high ultrasonic scalability, PA imaging provides a complete toolbox for the life sciences, complementing other imaging methods in its contrast mechanism, spatial-temporal resolution, and penetration depth. Compared to the tremendous scientific developments, clinical translation of PA imaging is still in a premature stage (no product with a US FDA approval). Use of expensive and bulky pulsed lasers that demands the use of laser-safe rooms and eye-safety goggles is one of the key factors hindering the translation process.

This chapter details about the main components of a PA imaging system and the developments/strategies required for accelerating the technology from benchtop to bedside. Cost and applications of some of the commercially available PA imaging systems are briefly discussed. Also, vital steps in the clinical translation of the technology are detailed with specific focus on challenges and opportunities.

US and PA imaging involves acoustic detection, dual-mode PA/US imaging with structural and functional contrast can be seamlessly implemented in a clinical US scanner. US imaging is a well-accepted modality worldwide and integration of illumination unit to this will be the easiest way for PA imaging to hit the clinic. Naturally overlaid PA and US images will offer complementary contrast which then may be the key for early detection of several diseases including cancer. Impact and importance of a point-of-care imaging device that can provide deep-tissue structural, functional, and molecular contrast with high spatial and temporal resolution is very high. Even though there are diverse hurdles and challenges to be solved in the clinical translation process, it is expected that this promising technique will revolutionize the field of medical imaging and touch lives in coming years.

Acknowledgements Authors gratefully acknowledge iThera Medical Gmbh, Germany and Seno Medical Instruments Inc., USA for providing latest images of MSOT Acuity and Imagio® respectively.

References

1. P. Beard, Biomedical photoacoustic imaging. Interface Focus **1**, 602–631 (2011)
2. X. Wang, Y. Pang, G. Ku, X. Xie, G. Stoica, L.V. Wang, Noninvasive laser-induced photoacoustic tomography for structural and functional in vivo imaging of the brain. Nat. Biotechnol. **21**, 803 (2003)
3. M. Kuniyil Ajith Singh, W. Steenbergen, S. Manohar, Handheld probe-based dual mode ultrasound/photoacoustics for biomedical imaging, in *Frontiers in Biophotonics for Translational Medicine* (Springer, Singapore, 2016), pp. 209–247
4. S. Jeon, J. Kim, D. Lee, J.W. Baik, C. Kim, Review on practical photoacoustic microscopy. Photoacoustics **15**, 100141 (2019)
5. M. Kuniyil Ajith Singh, V. Parameshwarappa, E. Hendriksen, W. Steenbergen, S. Manohar, Photoacoustic-guided focused ultrasound for accurate visualization of brachytherapy seeds with the photoacoustic needle. J. Biomed. Opt. **21**(12), 120501 (2016)
6. T. Zhao, A. Desjardins, S. Ourselin, T. Vercauteren, W. Xia, Minimally invasive photoacoustic imaging: current status and future perspectives. Photoacoustics **16**, 100146 (2019)
7. K. Daoudi, P. Van Den Berg, O. Rabot, A. Kohl, S. Tisserand, P. Brands, W. Steenbergen, Handheld probe integrating laser diode and ultrasound transducer array for ultrasound/photoacoustic dual modality imaging. Opt. Express **22**, 26365–26374 (2014)

8. S. Manohar, M. Dantuma, Current and future trends in photoacoustic breast imaging. Photoacoustics **16**, 100134 (2019)
9. A. Fatima, K. Kratkiewicz, R. Manwar, M. Zafar, R. Zhang, B. Huang, N. Dadashzadeh, J. Xia, K. Avanaki, Review of cost reduction methods in photoacoustic computed tomography. Photoacoustics **15**, 100137 (2019)
10. A. Hariri, J. Lemaster, J. Wang, A.S. Jeevarathinam, D.L. Chao, J.V. Jokerst, The characterization of an economic and portable LED-based photoacoustic imaging system to facilitate molecular imaging. Photoacoustics **9**, 10–20 (2018)
11. W. Xia, M. Kuniyil Ajith Singh, E. Maneas, N. Sato, Y. Shigeta, T. Agano, S. Ourselin, S. J. West, A. E. Desjardins, Handheld real-time LED-based photoacoustic and ultrasound imaging system for accurate visualization of clinical metal needles and superficial vasculature to guide minimally invasive procedures. Sensors **18**, 1394 (2018)
12. S. Agrawal, C. Fadden, A. Dangi, X. Yang, H. Albahrani, N. Frings, S. Heidari Zadi, S.R. Kothapalli, Light-emitting-diode-based multispectral photoacoustic computed tomography system. Sensors **19**(22), 4861 (2019)
13. Y. Zhu, G. Xu, J. Yuan, J. Jo, G. Gandikota, H. Demirci, T. Agano, N. Sato, Y. Shigeta, X. Wang, Light emitting diodes based photoacoustic imaging and potential clinical applications. Sci. Rep. **8**, 9885 (2018)
14. J. Jo, G. Xu, Y. Zhu, M. Burton, J. Sarazin, E. Schiopu, G. Gandikota, X. Wang, Detecting joint inflammation by an LED-based photoacoustic imaging system: a feasibility study. J. Biomed. Opt. **23**, 110501 (2018)
15. G.L.G. Menezes, R.M. Pijnappel, C. Meeuwis, R. Bisschops, J. Veltman, P.T. Lavin, M.J. van de Vijver, R.M. Mann, Downgrading of breast masses suspicious for cancer by using optoacoustic breast imaging. Radiology **288**(2), 355–365 (2018)
16. G. Diot, S. Metz, A. Noske, E. Liapis, B. Schroeder, S.V. Ovsepian, R. Meier, E. Rummeny, V. Ntziachristos, Multispectral optoacoustic tomography (MSOT) of human breast cancer. Clin. Cancer Res. **23**(22), 6912–6922 (2017)
17. P.K. Upputuri, M. Pramanik, Performance characterization of low-cost, high-speed, portable pulsed laser diode photoacoustic tomography (PLD-PAT) system. Biomed. Opt. Express **10**, 4118–4129 (2015)
18. W. Xia, D. Piras, J.C.G. van Hespen, S. van Veldhoven, C. Prins, T.G. van Leeuwen, W. Steenbergen, S. Manohar, An optimized ultrasound detector for photoacoustic breast tomography. Med. Phys. **40**, 032901 (2013)
19. J. Chan, Z. Zheng, K. Bell, M. Le, P.H. Reza, J.T.W. Yeow, Photoacoustic imaging with capacitive micromachined ultrasound transducers: principles and developments. Sensors **19**, 3617 (2019)
20. L. Lin, P. Hu, J. Shi, C.M. Appleton, K. Maslov, L. Li, R. Zhang, L.V. Wang, Single-breath-hold photoacoustic computed tomography of the breast. Nat. Commun. **9**, 2352 (2018)
21. B.C. Wilson, M. Jermyn, F. Leblond, Challenges and opportunities in clinical translation of biomedical optical spectroscopy and imaging. J. Biomed. Opt. **23**(3), 030901 (2018)
22. R. Sheikh, M. Cinthio, U. Dahlstrand, T. Erlöv, M. Naumovska, B. Hammar, S. Zackrisson, T. Jansson, N. Reistad, M. Malmsjö, Clinical translation of a novel photoacoustic imaging system for examining the temporal artery. IEEE Trans. Ultrason. Ferroelectr. Freq. Control **66**(3), 472–480 (2019)
23. S. Bohndiek, Addressing photoacoustics standards. Nat. Photonics **13**, 298 (2019)

Printed in the United States
by Baker & Taylor Publisher Services